AF411420

Quality Evaluation of Plant-Derived Foods

Quality Evaluation of Plant-Derived Foods

Editor

Ivo Vaz de Oliveira

MDPI • Basel • Beijing • Wuhan • Barcelona • Belgrade • Manchester • Tokyo • Cluj • Tianjin

Editor
Ivo Vaz de Oliveira
University of Trás-os-Montes
and Alto Douro (UTAD)
Vila Real
Portugal

Editorial Office
MDPI
St. Alban-Anlage 66
4052 Basel, Switzerland

This is a reprint of articles from the Special Issue published online in the open access journal *Plants* (ISSN 2223-7747) (available at: www.mdpi.com/journal/plants/special_issues/quality_foods).

For citation purposes, cite each article independently as indicated on the article page online and as indicated below:

LastName, A.A.; LastName, B.B.; LastName, C.C. Article Title. *Journal Name* **Year**, *Volume Number*, Page Range.

ISBN 978-3-0365-1830-5 (Hbk)
ISBN 978-3-0365-1829-9 (PDF)

Contents

Preface to "Quality Evaluation of Plant-Derived Foods"

It is well established that preference (but not exclusivity) for plant-derived foods can result in both health and environmental benefits. However, it must be acknowledged that not all plant-derived foods present the same quality to consumers. Hence, traditional and novel tools to assure high quality standards have to be applied to these types of foods. At the same time, the definition of quality may be different from product to product, and must be studied accordingly. The composition in terms of bioactive compounds content, fat content or fatty acid profile, vitamins, carbohydrates, and volatile compounds, as well as microbial safety and sensorial characteristics, are some of the parameters that can provide insight into the quality of plant-derived foods. Of course, these types of foods are usually subjected to some kind of post-harvest processing or storage that can alter their properties. This has also led to the need to study how these procedures change the characteristics of the original food.

This new MDPI book publishes important manuscripts dealing with "Quality Evaluation of Plant-Derived Foods". This includes novel approaches to this line of research, as well as the use of established methodologies to analyze novel plant foods, understudied species, or to obtain new data on known plant foods.

Ivo Vaz de Oliveira
Editor

Article

Apple (*Malus domestica* Borkh.) Cultivar 'Majda', a Naturally Non-Browning Cultivar: An Assessment of Its Qualities

Anka Cebulj [1,*], Andreja Vanzo [1], Joze Hladnik [1], Damijana Kastelec [2] and Urska Vrhovsek [3]

1   Agricultural Institute of Slovenia, Department of Fruit Growing, Viticulture and Oenology, Hacquetova ulica 17, 1000 Ljubljana, Slovenia; andreja.vanzo@kis.si (A.V.); joze.hladnik@kis.si (J.H.)
2   Biotechnical Faculty, University of Ljubljana, Jamnikarjeva 101, 1000 Ljubljana, Slovenia; damijana.kastelec@bf.uni-lj.si
3   Research and Innovation Centre, Fondazione Edmund Mach, Via E. Mach 1, 38010 San Michele all'Adige, Italy; urska.vrhovsek@fmach.it
*   Correspondence: anka.cebulj@kis.si

Abstract: Browning of apple and apple products has been a topic of numerous research and there is a great number of methods available for browning prevention. However, one of the most efficient ways, and the one most acceptable for the consumers, is the selection of a non-browning cultivar. Cultivar 'Majda' is a Slovenian cultivar, a cross between 'Jonatan' and 'Golden Noble'. In this study, it was thoroughly examined and compared to the well-known cultivar 'Golden Delicious' with the aim to decipher the reason for non-browning. We have determined the content of sugars, organic acids, vitamin C, glutathione and phenolics in apple flesh, with the addition of phenolic content in apple peel and leaves. The change in color in halves and pomace was also measured and the activity of peroxidase (POX) and polyphenol oxidase (PPO) were determined. Additionally, the analyses of flesh were repeated post-storage. The most prominent results were high acidity (malic acid), low phenol content, especially hydroxycinnamic acid and flavan-3-ol content of cultivar 'Majda' in comparison to 'Golden Delicious', and no difference in PPO activity between cultivars. After the overview of the results, we believe that both low phenol content and high reduced glutathione content impact the non-browning characteristics of cultivar 'Majda'.

Keywords: non-browning; polyphenol oxidase; phenolics; vitamin C; glutathione

Citation: Cebulj, A.; Vanzo, A.; Hladnik, J.; Kastelec, D.; Vrhovsek, U. Apple (*Malus domestica* Borkh.) Cultivar 'Majda', a Naturally Non-Browning Cultivar: An Assessment of Its Qualities. *Plants* **2021**, *10*, 1402. https://doi.org/10.3390/plants10071402

Academic Editor: Ivo Vaz de Oliveira

Received: 22 June 2021
Accepted: 6 July 2021
Published: 9 July 2021

Publisher's Note: MDPI stays neutral with regard to jurisdictional claims in published maps and institutional affiliations.

1. Introduction

Browning is associated with deterioration. Its prevention, either by using additives or choosing resistant cultivars, has taken a great part in horticultural and food research. The conventional method to inhibit browning in fruit has been to utilize sulfites [1]. However, due to health concerns, alternative means of controlling enzymatic browning are required [2]. In addition to numerous food processing techniques for prevention of browning, the initial decision on cultivar selection is crucial for all further steps. We now have several known cultivars with a smaller rate or lack of browning, such as arctic apples, 'Ambrosia', 'Eden', 'Aori27', etc. [3–5]. The mechanism behind browning is oxidation of polyphenols. Oxidation occurs when tissues are damaged, either by improper handling causing bruising or by processing, cutting, peeling or grinding. Due to damaged cells, the phenolics come into contact with polyphenol oxidase (PPO). In intact cells, PPO seems to have little activity towards phenolics [6]. In addition to the phenolic content, the activity of PPO is the reason for the development of browning [7]. PPO interacts with phenolic substrates and molecular oxygen, since it is a bi-metalloenzyme with two copper-binding domains [8]. The primary reaction is initiated by PPO accumulated in plastids, and not by *de novo* formed enzyme, although high activation occurs in time after a cell-damaging event. Furthermore, phenolic concentration increases in time after wounding [9]. What are the main mechanisms behind non-browning cultivars? Arctic apple cultivars were genetically engineered with a transgene that produces specific RNAs to silence PPO genes [5]. Cultivar 'Eden' has a low phenolic content [3]. In addition to the low phenolic content, a low PPO activity is also thought to be behind a lack of browning in cultivar 'Aori 27' [4]. The lack of browning for cultivar 'Ambrosia' is explained by the lower activity of PPO enzyme [10].

Another important quality of a cultivar and its part of defense metabolism are two major low-molecular-weight antioxidants, ascorbic acid (Vitamin C) and glutathione (GSH) [11,12]. Ascorbic acid has an ability to reduce quinones back to phenolic compounds prior to their subsequent reaction to form pigments. While GSH is directly linked to cellular ascorbic acid metabolism through the ascorbate-glutathione cycle, GSH is used as a source of reducing power for the enzymatic regeneration of oxidized ascorbic acid [13]. Furthermore, glutathionyl-chlorogenic acid conjugate was reported in apple juice [14]. Glutathionyl conjugates of hydroxycinnamic acids are known for limiting the browning of grape juice, where GSH interferes by trapping the caftaric acid quinones produced by oxidation in the form of 2-s-glutathionylcaftaric acid [15]. In addition, GSH has a role in biosynthetic pathways, detoxification, antioxidant biochemistry, and redox homeostasis [16].

All the above-mentioned metabolites have different preserving abilities during storage that depend firstly on cultivar and pre- and post-harvest parameters. According to Awad and Jager [17], total phenolics are relatively stable during storage. A good stability of the main antioxidants (including GSH and ascorbic acid) was also reported [18]. Davey and Keulemans [19] reported on the increased GSH content after 3 months of cold storage of several apple cultivars, as well as increased vitamin C content. The increase, and in some cases the decrease, of GSH and vitamin C mainly depended on the cultivar. However, a weak correlation to the harvest time was implicated, as well.

Cultivar 'Majda' was confirmed as a variety in 1986 and was made from the cross of 'Jonatan' and 'Golden Noble' [20]. The apple has a dark green basic color with a dark red top color (Figure 1). Even though its non-browning characteristics were described when it was introduced, it is not a well-known or a widely used cultivar. Only a few growers have cv. 'Majda' planted in orchards. One of the reasons is probably the color of fruit, which is not as appealing as the color of modern apple cultivars. It has high acidity and is therefore mainly known for its use in processing. Cultivar 'Golden Delicious' is a well-known cultivar and in numerous countries, the time of harvest and basic characteristic of cultivars are compared to this cultivar. With respect to this, cultivar 'Golden Delicious' was chosen as a comparison to cultivar 'Majda'. The first phenolic analysis of cultivar 'Majda' were made by Persic et al. [21], where they compared several cultivars in terms of phenolic content and browning. They have correlated a stronger oxidation to the high total phenolic content in apples. Their results urged us to focus on this cultivar, to further explore the phenomenon of non-browning of cultivar 'Majda'.

Figure 1. 'Majda' apples.

The aim of this study was to establish the main parameters of inner quality of cultivar 'Majda' in comparison to 'Golden Delicious'. We have determined the content of sugars, organic acids, vitamin C, glutathione (GSH and GSSG), its precursors cysteine, methionine, and phenolics in apple flesh, with the addition of phenolic content in apple peel. Additionally, to establish if the low phenolic content also reflects in leaves, the content of phenolics in leaves was determined. Furthermore, we wanted to decipher the main reasons for its low susceptibility to browning. The change in color in halves and pomace was also measured and the activity of peroxidase and polyphenol oxidase were determined. Moreover, we have repeated the analysis on flesh post-storage to determine if the trait persists after storage.

2. Results

2.1. Sugars

The results of total sugars and organic acids are presented in Figure 2 and their statistical analysis in Table 1. All three factors (cultivar, time, and location) have a statistically significant influence on sugar content. Moreover, the interactions between the cultivar and time as well as the cultivar and location are significant. The contrast analysis showed that there are statistically significant differences in the mean total sugar content between 'Golden Delicious' and 'Majda' at L2, whereas there were no statistically significant differences between cultivars at L1. However, when also looking at confidence interval, one can see that it is also close to a statistically significant difference at location L1 (Table S1). Following storage, there are statistically significant differences between L1 and L2 with the 'Golden Delicious' cultivar. Time had a statistically significant influence on the content of total sugars in 'Majda' at both locations. The composition of individual sugars also differs between cultivars (Figure S1), with 'Majda' having a higher content of sucrose and sorbitol and a lower content of glucose and fructose.

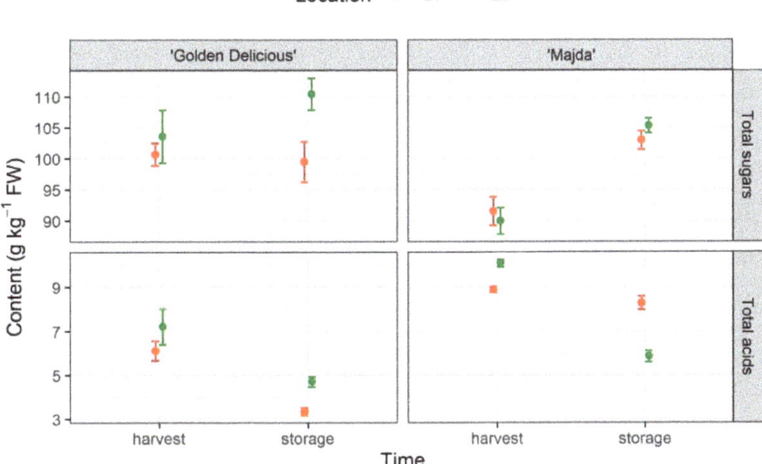

Figure 2. Content of total sugars (g kg^{-1} FW; mean $\pm$ SE) and total organic acids (g kg^{-1} FW; mean $\pm$ SE) of cultivars 'Golden Delicious' and 'Majda' at two locations (L1 and L2) at harvest and following storage.

Table 1. Three-factor ANOVA for cultivar (Cul: 'Golden Delicious' and 'Majda'), location (Loc: L1 and L2), and time (T: harvest and storage), as well as their interactions ($p < 0.05$).

	Cul	Loc	T	Cul:Loc	T:Cul	T:Loc	T:Cul:Loc
Total sugars	**	.	***	NS	**	.	NS
Total acids	***	NS	***	**	NS	**	***
pH	***	NS	***	NS	**	NS	NS
Vitamin C	***	**	***	**	***	*	NS
Methionine	***	***	***	NS	***	***	*
Cysteine	NS	NS	***	NS	***	NS	NS
GSH	***	*	***	NS	.	**	*
GSSG	**	NS	NS	NS	NS	NS	NS
Hydroxycinnamic acids	***	**	***	NS	NS	NS	NS
Dihydrochalcones	***	NS	***	NS	**	NS	NS
Flavonols	***	**	.	*	NS	NS	NS
Flavan-3-ols	***	NS	NS	NS	NS	NS	NS
PPO	*	**	**	NS	NS	**	NS
POX	NS	NS	***	**	NS	NS	NS

	Cul	T	Trt	Cul:T	Trt:Cul	Trt:T	Tr:Cul:T
$\Delta E\ \Delta t^{-1}$	***	NS	***	.	NS	***	NS

., Statistically significant differences at $p < 0.1$; *, statistically significant differences at $p < 0.05$; **, statistically significant differences at $p < 0.01$; ***, statistically significant differences at $p < 0.001$; NS: Not significant.

2.2. Organic Acids

Malic and citric acid were quantified among organic acids (Table S2). Fumaric acid was also determined, but its amount was exceptionally low and thereby not quantified. Malic acid is the prevalent acid in apples, which is why its content has a decisive influence on the total acid content in apple. When analyzing all the factors, we can see that there is a statistically significant interaction between all of them. After the contrast analysis was made, statistically significant differences were confirmed between cultivars within locations at harvest and at location L1 following storage. Significant differences in the total acid content following storage in 'Golden Delicious' at both locations were noted and in cv. 'Majda' at location L2.

2.3. pH

The results for apple juice mean that the pH values ($\pm$SE) are presented in Figure 3 and statistical analysis in Table 1. The cultivar and time have a statistically significant influence on pH as well as their interaction. Apple juice pH is not influenced by location. The analysis of contrasts (Table S3) confirmed statistically significant differences between cultivars.

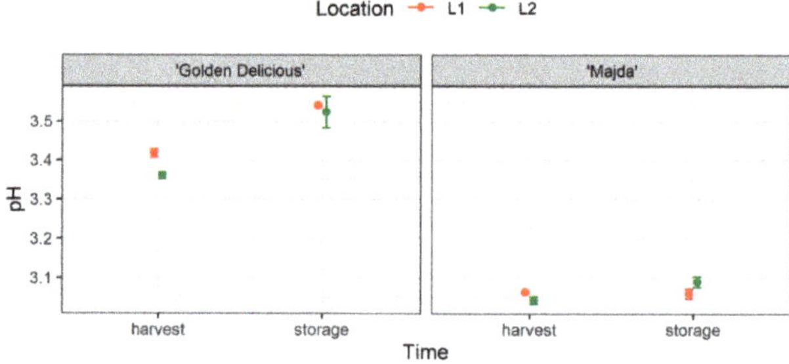

Figure 3. Content of apple juice pH (mean $\pm$ SE) of cultivars 'Golden Delicious' and 'Majda' at two locations (L1 and L2) at harvest and following storage.

2.4. Color Change

The change in color is notably different between cultivars (Figure 4 and Table 1). Halves and pomace of cv. 'Majda' changed color far less in comparison to 'Golden Delicious'. The difference was more pronounced after harvest, but even following storage, cv. 'Majda' kept its potential of a lower rate of color change. The contrast analysis confirmed statistical differences between cultivars (Table S5). Interestingly, following storage, there were no statistically significant differences between halves and pomace of cv. 'Majda'.

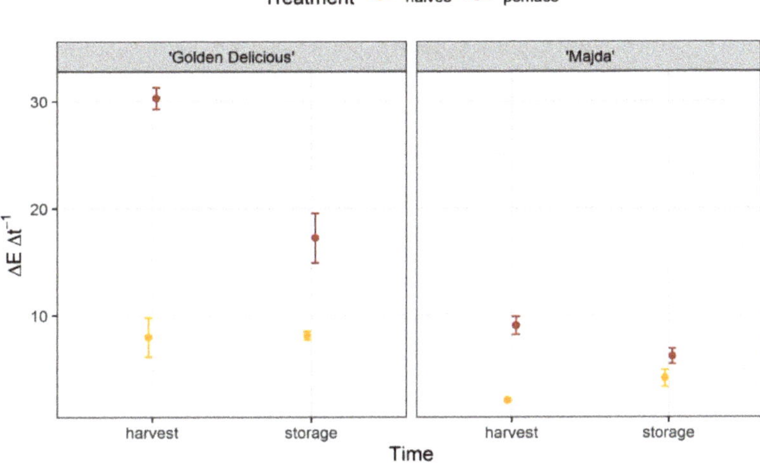

Figure 4. Change in color ($\Delta E\ \Delta t^{-1}$; mean $\pm$ SE) of apple halves after 1 h and pomace after 10 min for cultivars 'Golden Delicious' and 'Majda' at harvest and following storage.

2.5. Vitamin C, Methionine, L-cysteine, GSH and GSSG

The content of two major antioxidants in plants, vitamin C and glutathione (GSH and GSSG) as well as GSH precursors L-cysteine and methionine, presented in Figure 5 and their statistics in Table 1. All the three factors (cultivar, time, and location) have a statistically significant influence on the vitamin C content as well as on the methionine and GSH content. The most interesting statistical difference for our study is the vitamin C content in apple flesh following storage, where cv. 'Majda' flesh contains a higher content of vitamin C in comparison to cv. 'Golden Delicious' (Table S5). In addition, interestingly, there were no statistically significant changes in the vitamin C content in cv. 'Majda' between harvest and storage. The same pattern is also visible with methionine and GSH. With the latter content differed already at harvest (between cultivars within locations). Cysteine was highly influenced by the time and interaction between time and cultivar, its content being statistically lower in cv. 'Golden Delicious' after storage in comparison to harvest at both locations. The GSSG content depended on the cultivar, the content of GSSG in cv. 'Golden Delicious' flesh representing 69% of the content of GSSG in cv. 'Majda' flesh. Differences are also represented with the heatmap (Figure S2).

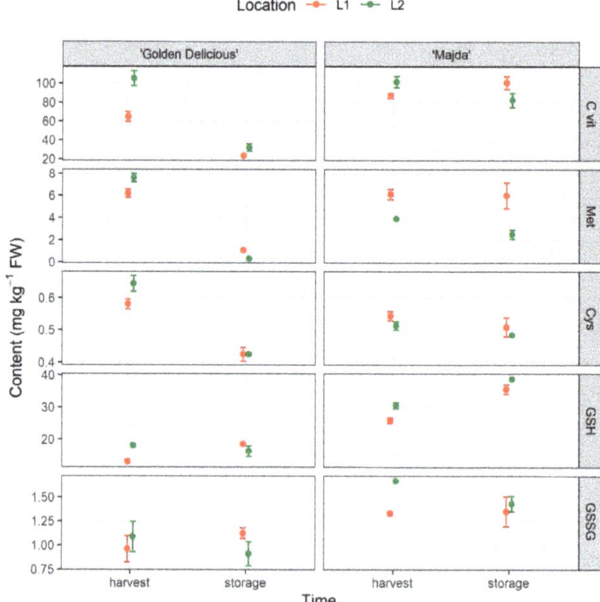

Figure 5. Vitamin C (C vit), methionine (Met), cysteine (Cys), reduced glutathione (GSH), and oxidised glutathione (GSSG) content in apple flesh (mg kg^{-1} FW; mean ± SE) for cultivars 'Golden Delicious' and 'Majda' at two different locations (L1 and L2) at harvest and following storage.

2.6. Phenolic Content

Phenolic content was determined in flesh (Figure 6), peel (Figure 7), and apple leaves (Figure 8). Determined phenolics were arranged into four groups (for leaves, five) in which they are presented in Figures and mean contents of individual phenols are presented in Table S6 and Figure S3. Hydroxycinnamic acids: Cryptochlorogenic acid, chlorogenic acid, p-coumaric acid, and neochlorogenic acid; dihydrochalcones: Phloridzin, phloretin 2'-O-xylosyl-glucoside, 3-hydroxyphloridzin, and 3-hydroxyphloretin; flavonols: Quercetin-3-rhamnoside, quercetin-3-rutinoside, and quercetin-3-glycoside + quercetin-3-galactoside; and flavan-3-ols: Catechin, epicatechin, procyanidin B1, procyanidin B2 + B4. In the peel of cv. 'Majda', cyanidin-3-galactoside was also determined, but was not included in the results. The 3-hydroxyphloretin was not determined in peel and the phenols neochlorogenic acid, 3-hydroxyphloridzin, 3-hydroxyphloretin, and quercetin-3-rutinoside were not determined in apple flesh. There is a pronounced difference in the phenolic content between cultivars (Table 1). The strongest difference is visible in hydroxycinnamic acid and flavan-3-ol content. The hydroxycinnamic group of phenolics is influenced by all three factors, but there are no interactions between them. Cultivar 'Golden Delicious' has on average 11 times higher hydroxycinnamic acid content in comparison to 'Majda' (Table S8). At L1, the average content of hydroxycinnamic acids represents 75% content of hydroxycinnamic acids at location L2. On the contrary, the flavan-3-ol content was influenced only by factor cultivar, but with high significance. The dihydrochalcone content differed between cultivars only following storage but not at harvest. However, it was also statistically significantly different within both cultivars comparing apple flesh at harvest and following storage (Table S7). Flavonols, on the other hand, were not influenced by time, but by location in addition to the cultivar. At L1, the contrast analysis confirmed the statistical difference between cultivars.

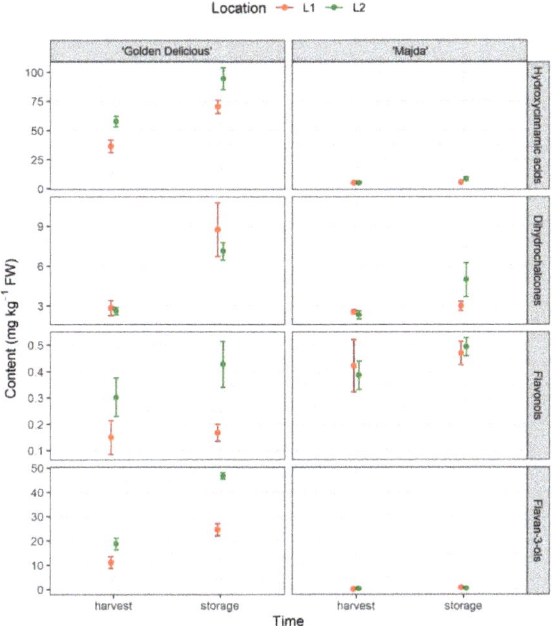

Figure 6. Phenolic content in apple flesh presented in four groups: Hydroxycinnamic acids, dihydrochalcones, flavonols, and flavan-3-ols (mg kg^{-1} FW; mean $\pm$ SE) for cultivars 'Golden Delicious' and 'Majda' at two different locations (L1 and L2) at harvest and following storage.

The pattern is quite similar in apple peel, but the differences are not that extreme (Figure 7 and Table 2). Here, the highest difference is in the content of hydroxycinnamic acids. Compared to cv. 'Golden Delicious', cv. 'Majda' has about 13 times lower content of hydroxycinnamic acids (Table S8). Location 1 represents 67% content of hydroxycinnamic acids at L2. There were no statistical differences between cultivars in dihydrochalcone and flavonol content, but the latter did vary between locations. Again, flavan-3-ol content differed in cv. 'Golden Delicious' (four times higher content) in comparison to cv. 'Majda'.

Table 2. Two-factor ANOVA for cultivar (Cul: 'Golden Delicious' and 'Majda') and location (Loc: L1 and L2) and their interactions ($p < 0.05$).

	Cul	Loc	Cul:Loc
Peel			
Hydroxycinnamic acids	***	***	NS
Dihydrochalcones	NS	NS	NS
Flavonols	NS	*	NS
Flavan-3-ols	***	***	NS
Leaves			
Arbutin	*	**	NS
Hydroxycinnamic acids	*	.	NS
Dihydrochalcones	***	*	NS
Flavonols	***	.	NS
Flavan-3-ols	***	***	NS

., statistically significant differences at $p < 0.1$; *, statistically significant differences at $p < 0.05$; **, statistically significant differences at $p < 0.01$; ***, statistically significant differences at $p < 0.001$; NS: Not significant.

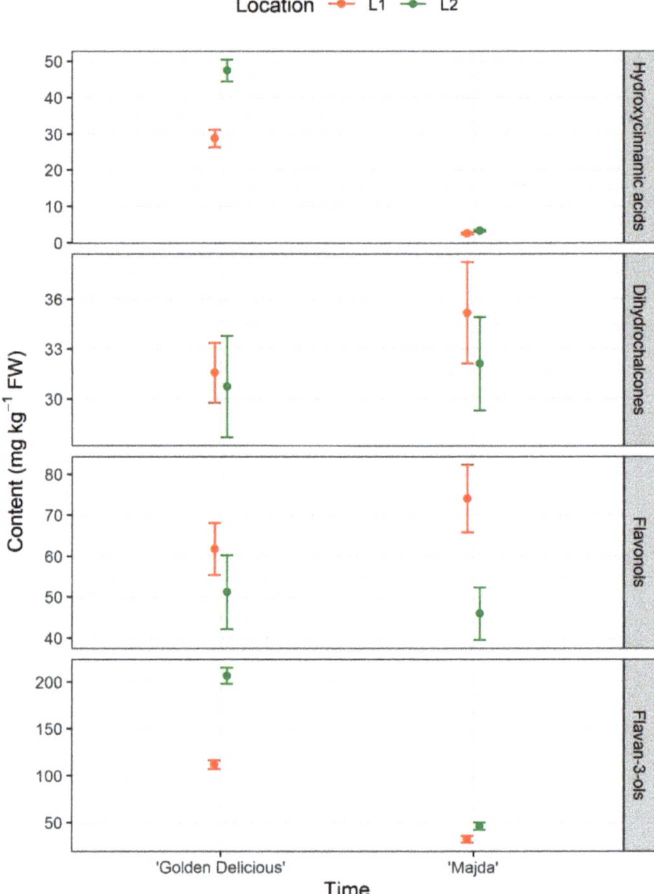

Figure 7. Phenolic content in apple peel presented in four groups: Hydroxycinnamic acids, dihydrochalcones, flavonols, and flavan-3-ols (mg kg^{-1} FW; mean ±SE) for cultivars 'Golden Delicious' and 'Majda' at two different locations (L1 and L2).

In apple leaves, the differences are not as drastic as in fruits but are still present (Figure 8 and Table 2). All the phenolics were influenced by both cultivar and location, with hydroxycinnamic acids and flavonols having weak, statistically significant differences regarding location. However, there were no interactions between factors cultivar and location. Cultivar 'Golden Delicious' had twice as much arbutin, 1.4 times higher flavonol content and 2.9 times higher flavan-3-ol content as cv. 'Majda' (Table S8). On the other hand, cv. 'Golden Delicious' had 70% and 92% content of the cv. 'Majda' content of hydroxycinnamic acids and dihydrochalcones, respectively.

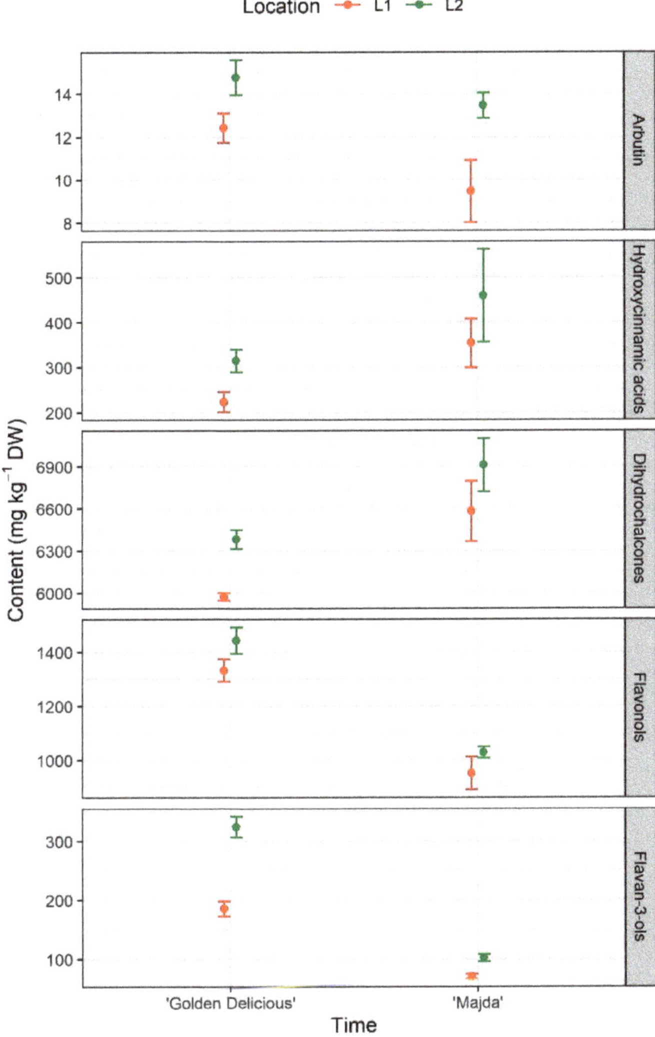

Figure 8. Phenolic content in apple leaves presented as arbutin and four groups: Hydroxycinnamic acids, dihydrochalcones, flavonols, and flavan-3-ols (mg kg^{-1} DW) for cultivars 'Golden Delicious' and 'Majda' at two different locations (L1 and L2).

2.7. PPO and POX

The PPO and POX activities are presented in Figure 9 and statistics in Table 1. There were statistically significant differences in the PPO activity in all three factors. However, the contrast analysis showed no statistically significant differences between cultivars (Table S9). There was a difference in the enzyme activity between apple flesh at harvest and following storage. With POX, time was the only statistically different factor and cultivar:location was the statistically significant interaction.

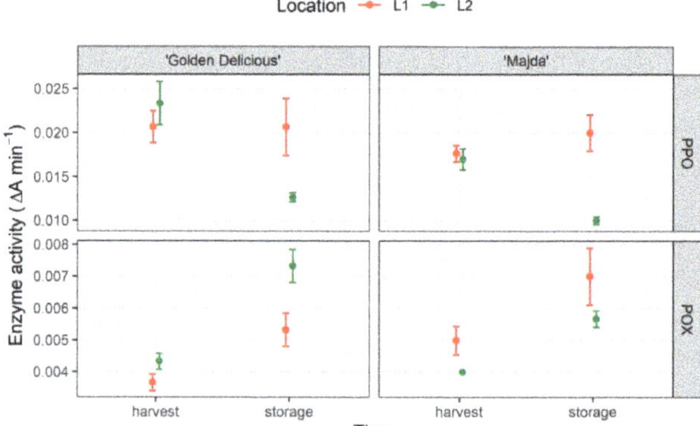

Figure 9. Activity of PPO and POX ($\Delta A\ min^{-1}$) in apple flesh of cultivars 'Golden Delicious' and 'Majda' at harvest and following storage.

3. Discussion

The results show some promising differences between cultivars. Cultivar 'Majda' contains a lower sugar content than cv. 'Golden Delicious'. Cultivars also differ in individual sugar composition. Even though according to Aprea et al. [22] sorbitol correlates to sweetness the most, its low share in total sugar composition and lower fructose and glucose content in cv. 'Majda' do not contribute so much to the sweetness. In addition, the study by Rymenants et al. [23] reported that the perceived sweetness was greatly influenced by the acidity and vice versa. Apples with low acidity are perceived to be sweeter, whereas apples with a strong acidic taste are perceived as less sweet. This is most likely behind the cv. 'Majda' taste perception, further supported by Aprea et al. [22], who assessed a negative correlation between malic acid and perceived sweetness (r = −0.449). The higher malic acid content in cv. 'Majda' could also be connected to non-browning. In the study by Morimoto et al. [24], they reported on the connection between high acidity alleles and the alleles for non-browning. The major locus controlling apple fruit acidity has been designated as Ma (malic acid), where Ma corresponds to the dominant high acidity or the low pH allele and ma is the low acidity or the high pH allele [25]. For cv. 'Majda', it was also believed that the high acidity and low pH are the main factors for its lack of browning. However, in accordance with the previously mentioned data, it might just be the consequence of heritability rather than the key reason for non-browning.

The difference in color change between cultivars is evident. Similar results were also reported by Joshi et al. [26] for cultivar 'Eden'. In our research, the halves and pomace of cv. 'Majda' both had minimal change in color in comparison to cv. 'Golden Delicious'. Next, we will discuss the main possible causes for this trait.

Both vitamin C and GSH are major antioxidants and this characteristic is also recognized and used in the food industry [2]. However, according to Nicolas [27] the endogenous vitamin C has a marginal role in the prevention of browning. On the contrary, Joshi et al. [26] did find the correlation between the vitamin C content and the whiteness index. In our study, the cultivars did not differ in vitamin C content at harvest, but they did differ in retaining of vitamin C following storage. Following storage, the content of vitamin C remained higher in the cv. 'Majda' fruit. Davey and Keulemans [19] reported that cultivars differ substantially in their ability to maintain vitamin C during storage. In their study, the 'Golden Delicious' vitamin C decreased following storage, whereas in a few other cultivars it increased. Furthermore, they report on the correlation between fruit vitamin C contents and the harvest date, such that cultivars with the highest vitamin C contents were harvested latest in the season and the lowest contents were found among the early cultivars. It is suggested that this is linked to the cultivar's capacity for longer storage. However, at this point, we must emphasize on the better storage ability of cultivar 'Golden

Delicious' in comparison to that of cultivar 'Majda' in CA storage (non-published data and experience from growers). The contrast analysis also showed the difference in vitamin C content in the flesh of cv. 'Golden Delicious' between locations, whereas the content of vitamin C of cv. 'Majda' apples did not vary between locations. It is known that the content of vitamin C is highly influenced by temperature and radiation [28], so this might add an explanation to the higher vitamin C content at L2, but then again, since cv. 'Majda' did not show the similar response, this is just a speculation that needs to be addressed in future studies. What we can see, though, is also a higher content of the reduced form of GSH in cv. 'Golden Delicious' at harvest at L2. Therefore, location did influence these two antioxidants in cv. 'Golden Delicious'. However, what stands out with GSH is the statistically significant higher content of GSH in cv. 'Majda' at both locations and at both times. What is more, if we look at the phenolic content in flesh, we can significantly see a lower content of hydroxycinnamic acids. In a review on the role of glutathione in winemaking, Kritzinger et al. [29] reported on an interesting observation, the ratio of hydroxycinnamic acid and GSH (HCA/GSH) represents a good indication of the grape must susceptibility to oxidation. A ratio of 0.9–2.2 characterizes a lightly colored must, while the medium and dark must are characterized by 1.1–3.6 and 3.8–5.9 HCA/GSH ratio, respectively. If we compare the HCA/GSH ratio in our samples we can see a great difference, 2.7–3.2 and 0.2–0.3, for cv. 'Golden Delicious' and cv. 'Majda', respectively. Therefore, we have two aspects we need to take into account when discussing the lack of oxidation in cv. 'Majda': Low hydroxycinnamic acid content and high GSH content.

In addition to the low hydroxycinnamic acid, the flavan-3-ols content shows the significant differences in the content of main phenols responsible for browning, namely chlorogenic acid, catechin, and epicatechin [30,31]. Furthermore, at harvest, none of the procyanidins were determined nor were the cryptochlorogenic acid or p-coumaroylquinic acid from flavanols and hydroxycinnamic acids groups, respectively. As Khanizadeh et al. [3] described, the lack of substrate for PPO enzyme may be the cause of non-browning. This statement may be further supported by the results of PPO activity, where no statistically significant differences between our tested cultivars were determined. Following storage, the activity of PPO decreased in both cultivars. The low PPO activity is the main mechanism behind cv. 'Ambrosia' non-browning, but based on these results, it is rather the lack of substrate than the lower PPO activity that is the cause, therefore it is more similar to the 'Eden' cultivar's background. The content of dihydrochalcones differed between cultivars only following storage and the flavonol content varied between cultivars merely at L1. Both groups are known for their antioxidant role in apple [32,33], but the content of both groups in apple flesh is quite low. The POX activity was mostly influenced by time, and only the interaction between cultivar and location is statistically significant. With its various activities in plants, it is hard to make any assumptions without any further investigation. Nonetheless, the main differences between cultivars are well summarized by the heatmaps (Figures S1–S3).

Since the high phenolic content is preferred due to its beneficial role in human health [34], we also wanted to evaluate the apple peel from both cultivars. As done in flesh, the low sum of hydroxycinnamic acids and flavan-3-ols also stands out in the peel in cv. 'Majda'. There were no statistical differences in the contents of dihydrochalcones and flavonols between cultivars. Therefore, the peel has similar properties to flesh. This is not the best in the context of high antioxidant content. However, cv. 'Majda' contains anthocyanin cyanidin-3-O-galactoside. Anthocyanins are known for their beneficial role in human health [35].

With interesting results from flesh and peel, we wanted to broaden the knowledge on the phenolic content in cv. 'Majda', to see if this trait is present through the whole tree or if it is present just in the fruit. The group of flavonols and flavan-3-ols shows a similar pattern as flesh and peel, while the hydroxycinnamic acids and dihydrochalcones are even higher in cv. 'Majda' leaves in comparison to leaves of cv. 'Golden Delicious'. Both hydroxycinnamic acids and dihydrochalcones are known for their role as antioxidants and the role against various pathogens [32,36]. This might explain the overall lower sensitivity to pathogens of cv. 'Majda', which is an important factor for growers. This is encouraging, since in addition to the great fruit attributes, the sensitivity to diseases is a key concern for growers. Location also had an influence on the phenolic content, which was expected due to the different weather conditions.

4. Materials and Methods

4.1. Plant Material

Leaves and apples were harvested in 2019, at two different locations, in the experimental orchard Brdo pri Lukovici (L1: Continental Slovenia; 46°10'04.8" N, 14°40'55.2" E, altitude 368 m) and in the north-west part of Slovenia, in Sadovnjak Resje (L2: Latitude 46°20" N, longitude 14°12" E, altitude 500 m). Soils on location L1 are Dystric cambisol on fine sediments (clay and silt) with pH = 4.8 (0.01 M CaCl$_2$) and organic carbon content 2.1% (Corg) in top 30 cm. On location L2, soils are deep (Eutric cambisol) with a similar organic carbon content (Corg = 2.4%) but not so acidic (pH= 5.8) top soil. Weather conditions on both locations during the vegetation period in 2019 are presented in Table S10 and Figure S4. The amount of precipitation was very similar on both locations (731 mm on L1 and 753 mm on L2). The average daily air temperatures were 16.3 °C on L1 and 17.1 °C on L2, the average daily minimum air temperatures were 10.8 and 11.5 °C and the average daily maximum air temperatures were 21.8 and 23.5 °C, respectively. Rootstock M9 and the slender spindle training system were used in both orchards. In addition to cv. 'Majda', cv. 'Golden Delicious' was collected as a comparison due to its position in the European market. The leaves were incorporated in the study, to include another angle in comparison. They were sampled on 22 July 2019 at both locations. The leaves were ground in liquid nitrogen and lyophilized. The apples were harvested in their technological ripeness, as determined by firmness, the total soluble solids and starch test of cv. 'Golden Delicious' and cv. 'Majda' were harvested on 19 and 29 of September 2019, respectively. Both locations were harvested at the same time. Only apples of correct size (1st quality class) were collected and chosen for further handling and analysis. Approximately 30 kg of apples for each location and each cultivar were collected, half of which was immediately placed in cold storage and the rest were peeled, cut directly into liquid nitrogen and stored for the analysis of sugar, organic acid, phenolic content, and enzyme activity (POX and PPO). Apples were stored for 2 months in a cold storage with the temperature of 4 °C.

4.2. Color Analysis

Apples were either cut in half or ground with a commercial juicer (Sana juicer by Omega EUJ—707), and the change in color (Spectrophotometer CM-700d; Konica Minolta) was monitored over time, 1 h for the halves and 10 min for the pomace. Parameters a*, b*, and L* were recorded and parameter ΔE was calculated according to the following formula:

$$\Delta E = \sqrt{\Delta a^2 + \Delta b^2 + \Delta L^2}$$

4.3. Sugar and Organic Acid Extraction and Analysis

Frozen flesh was briefly chopped, and 5 g were placed in a 100 mL beaker and then homogenized with 25 mL of twice distilled water using Ultra-Turrax T-25 (Ika-Labortechnik, Staufen im Breisgau, Germany). After 30 min of shaking and 10 min of centrifuge (8400 rpm), the samples were filtered through a 45 μm cellulose filter (Chromafil®) into vials until further analysis.

Sugars and organic acids were measured using the Agilent 1100 Series. The Rezex RCM-monosaccharide column (300 × 7.8 mm; Phenomenex, Torrance, CA, USA) for sugars and sorbitol and Rezex ROA organic acid column (300 × 7.8 mm; Phenomenex, Torrance, CA, USA) for organic acids were used for analysis. To determine the sugar content, a method described by Mikulic-Petkovsek et al. [37] was used, with an adjustment of the column temperature to 80 °C. For organic acids, method OIV-MA-AS313-04: R2009 from the Compendium of international methods of analysis was used. The concentrations of sugars and organic acids were calculated with the help of corresponding external standards and were expressed in g kg^{-1} of fresh weight (FW).

4.4. Vitamin C Extraction and Analysis

Apples were processed on the day of the harvest to determine their vitamin C content. The method used for the determination is compliant with standard SIST EN 14130:2003 (determination of vitamin C using HPLC (Agilent Technologies, Santa Clara, CA, USA). Apples were quickly peeled, chopped, and immersed in liquid nitrogen prior to being ground in a mortar, using ceramic knife and mortar. Twenty grams of flesh powder were extracted with the metaphosphoric acid (Merck) and placed on the shaker for 30 min. The further procedure was as described in the standard. HPLC (Agilent 1200) with the UV-VIS detector at 265 nm was used for detection. The Gemini C18 110A (250 $\times$ 4.6 mm, 5 μm) column with the flow rate 0.7 mL min^{-1} was used. The results are in mg per kg of fresh sample.

4.5. Phenolic Content Extraction and Analysis

The phenolic content was determined from the apple powder and leaf powder gained from grounding in the mortar with the help of liquid nitrogen. The extraction and determination of phenolics was made following the protocol of Vrhovsek et al. [38]. Two grams of apple powder were extracted using aqueous 80% methanol (Honeywell, LC-MS Chromasolv). For leaves, a slight modification was made, 1 g of lyophilized leaf powder was used and extracted with 4.8 mL of MeOH:H$_2$O (2:1) and 3.2 mL of chloroform (Honeywell, LC-MS Chromasolv). An ultraperformance liquid chromatography-tandem mass spectrometry (UPLC-MS/MS) was used for the analysis. Two μL of samples were injected and processed through 100 $\times$ 2.1 mm, 1.8 μm column (Acquitiy HSS T3, Waters), maintained at 40 $^\circ$C, with the flow rate of 0.4 mL min^{-1}. The gradient profile of mobile phases was as described in Vrhovsek et al. [38]. The mass spectrometry detection was performed on Waters Xevo TQMS (Milford, MA, USA), equipped with an electrospray (ESI) source in positive and negative modes [38]. Results are presented in mg kg^{-1} of fresh weight (FW) for flesh and mg kg^{-1} of dry weight for leaves.

4.6. Glutathione (GSH, GSSH), Cystein and Methionine Extraction, and Analysis

The frozen apple flesh (-80 $^\circ$C storage) was ground under cryogenic conditions (-196 $^\circ$C) using AK11 mill (IKA), rapidly weighted into ice cold MeOH with sample weight to solvent volume (w/v) ratios of 1:4, shaken at room temperature for 15 min, and centrifuged (10,000$\times$ g, 4 $^\circ$C) for 10 min. The volume was adjusted, samples were filtered through a 0.2 μm PVDF filter (Agilent technologies) into dark vials and directly injected onto UHPLC-MS/MS (Agilent technologies), and analyzed as described [39].

4.7. POX and PPO Activity Determination

For the enzyme activity, the frozen material from -80 $^\circ$C of storage was used. We were following the protocol described in Zupan et al. [40] for the preparation of samples for the POX activity and Cebulj et al. [41] for the preparation of samples for the PPO activity. Measurements were made according to the Worthington manual (Worthington Biochemical Corporation, 1972). The enzyme activity was calculated in ΔA min^{-1}.

4.8. Statistics

A three-factor analysis of variance (ANOVA) with factors cultivar, location, and time was used for the flesh samples. All the factors had two levels: Cultivar ('Majda' and 'Golden Delicious'), location (L1 and L2), and time (harvest and storage). For peel and leaves, the two-factor ANOVA was used (factors cultivar and location). For the determination of sugars, organic acids, and phenolic compounds, the samples were prepared in five replicates. For leaves, four replicates were used. For the enzyme activity, three replicates were performed per treatment (each with three technical repetitions). Some variables were log-transformed before the statistical analysis to meet the assumption of constant variance between treatments. When the ANOVA showed a statistical significance, a contrast analysis was performed with user-defined contrasts. When the interactions between the factors did not show a statistically significant difference, Tukey's HSD test was used for comparison. The contrasts with statistical significances, heatmap of individual sugars, as well as the means with SE of individual phenolics are presented in supplementary materials (Tables S1–S9 and Figures S1–S3). If the *p*-value for differences between the means was less than 0.05, it was considered statistically significant.

5. Conclusions

In conclusion, cv. 'Majda' is one of the rare cultivars from traditional cross-breeding with a non-oxidation trait. We have screened some of the main actors of oxidation in apple flesh glutathione, vitamin C, phenolic content, and PPO activity. Based on the results, we conclude that the low phenolic content and high reduced glutathione content are the major reasons for the cv. 'Majda' lack of oxidation. We have upgraded the knowledge on cv. 'Majda' by also analyzing glutathione, and this feature will have to be further researched. Its high acidity is feasibly a consequence of heredity, rather than the main reason for non-oxidation, but one must not disregard that the high acidity with a low pH can contribute to a slower oxidation by reducing PPO efficiency. This will be further researched by the gene expression analysis. This work also touches upon another topic: Fruit processing. Today, the consumers seek products for a quick consumption, such as prepared snacks, but with as little as possible additives to preserve them. Cultivar 'Majda' is suitable for this role and has a great potential for use as a minimally processed product, as well as a regular apple product with no additives for the prevention of oxidation. This will also be addressed in our future research. Furthermore, cultivar 'Majda' is not an overly sensitive cultivar and is thereby undemanding for fruit growers.

Supplementary Materials: The following are available online at https://www.mdpi.com/article/10.3390/plants10071402/s1. Figure S1: Heatmap for standardized variables: Individual sugars: Sucrose, Suc; glucose, Glu; fructose, Fru; sorbitol, Sorb; organic acids: Citric acid, Cit_ac; malic acid, Mal_ac; and pH for cultivars 'Golden Delicious' (GD) and 'Majda' (MA) at two different locations (L1 and L2) at harvest and following storage, dendrogram based on Ward's clustering squared Euclidian distance. Figure S2: Heatmap for standardized variables: Vitamin C, C vit; methionine, Met; cysteine, Cys; reduced glutathione, GSH; oxidised glutathione, GSSG; polyphenol oxidase, PPO; peroxidase, POX for cultivars 'Golden Delicious' (GD) and 'Majda' (MA) at two different locations (L1 and L2) at harvest and following storage, dendrogram based on Ward's clustering squared Euclidian distance. Figure S3: Heatmap for standardized variables: Hydroxycinnamic acids: Cryptochlorogenic acid, X4-CQA; chlorogenic acid, X3-CQA; neochlorogenic acid, X5-CQA; and p-coumaric acid, p_coum_ac; dihydrochalcones: Phloridzin, Phl; phloretin 2'-O- xylosyl-glucoside, Phl_xy_phl; 3-hydroxyphloridzin, X3_Hy_phl; and 3-hydroxyphloretin, X3_Hyd_phl; flavonols: Quercetin-3-rhamnoside, Q-Rha; quercetin-3-rutinoside, Q-Rut; and quercetin-3-glycoside + quercetin-3-galactoside, Q-Glc_Gal; flavan-3-ols: Catechin, Cat; epicatechin, Epicat; procyanidin B1, P_B1; procyanidin B2 + B4, P_B2_B4 for cultivars 'Golden Delicious' (GD) and 'Majda' (MA) at two different locations (L1 and L2) at harvest and following storage, dendrogram based on Ward's clustering squared Euclidian distance. Figure S4: Daily precipitations, daily minimum and maximum air temperature for the locations L1 and L2. Table S1: The contrast analysis for total sugars with estimated differences between averages and corresponding 95% confidence intervals (lwr, lower boundary; upr, upper boundary), for time (harvest, storage), location (L1 and L2), and cultivar (GD, 'Golden Delicious'; MA, 'Majda'). Table S2: The contrast analysis for total organic acids with estimated differences between averages and corresponding 95% confidence intervals (lwr, lower boundary; upr, upper boundary), for time (harvest, storage), location (L1 and L2), and cultivar (GD, 'Golden Delicious'; MA, 'Majda'). Table S3: The contrast analysis for pH with estimated differences between averages and corresponding 95% confidence intervals (lwr, lower boundary; upr, upper boundary), for time (harvest, storage), location (L1 and L2), and cultivar (GD, 'Golden Delicious'; MA, 'Majda'). Table S4: The contrast analysis for change in color ($\Delta E \, \Delta t^{-1}$) with estimated ratios of averages and corresponding 95% confidence intervals (lwr, lower boundary; upr, upper boundary), for time (harvest, storage), location (L1 and L2), and cultivar (GD, 'Golden Delicious'; MA, 'Majda'). Table S5: The contrast analysis for vitamin C, methionine, cysteine, and glutathione (GSH and GSSG) with estimated ratios of averages and corresponding 95% confidence intervals (lwr, lower boundary; upr, upper boundary), for time (harvest, storage), location (L1 and L2), and cultivar (GD, 'Golden Delicious'; MA, 'Majda'). Table S6: Individual phenols (mean $\pm$ SE) in flesh (mg kg^{-1} FW), peels (mg kg^{-1} FW), and leaves (mg kg^{-1} DW). Hydroxycinnamic acids: Cryptochlorogenic acid, 4-CQA; chlorogenic acid, 3-CQA; neochlorogenic acid, 5-CQA; and p-coumaric acid, p_coum_ac; dihydrochalcones: Phloridzin, Phl; phloretin 2'-O- xylosyl-glucoside, Phl_xy_phl; 3-hydroxyphloridzin, 3_Hy_phl; and 3-hydroxyphloretin, 3_Hyd_phl; flavonols: Quercetin-3-rhamnoside, Q-Rha; quercetin-3-rutinoside, Q-Rut; and quercetin-3-glycoside + quercetin-3-galactoside, Q-Glc_Gal; flavan-3-ols: Catechin, Cat; epicatechin, Epicat; procyanidin B1, P_B1; and procyanidin B2 + B4, P_B2_B4. Table S7: The contrast analysis for phenolic groups dihydrochalcones and flavonols in apple flesh with estimated ratios of averages and corresponding 95% confidence intervals (lwr, lower boundary; upr, upper boundary), for time (harvest, storage), location (L1 and L2), and cultivar (GD, 'Golden Delicious'; MA, 'Majda'). Table S8: HSD test for phenolic groups, hydroxycinnamic acids, dihydrochalcones, flavonols and flavan-3-ols in apple flesh, peel and leaves with estimated ratios of averages and corresponding 95% confidence intervals (lwr, lower boundary; upr, upper boundary), for time (harvest, storage), location (L1 and L2), and cultivar (GD, 'Golden Delicious'; MA, 'Majda'). Table S9: The contrast analysis for PPO and POX with estimated ratios of averages and corresponding 95% confidence intervals (lwr, lower boundary; upr, upper boundary), for time (harvest, storage), location (L1 and L2), and cultivar (GD, 'Golden Delicious'; MA, 'Majda'). Table S10: Meteorological conditions during the vegetation period (1. 4. 2019–30. 9. 2021) on locations L1 and L2.

Author Contributions: Conceptualization, A.C.; methodology, A.C., A.V., and U.V.; software, D.K.; validation, U.V. and A.V., formal analysis, A.C., A.V., and U.V.; investigation, A.C. and J.H.; resources, A.C. and U.V.; data curation, A.C. and D.K.; writing—original draft preparation, A.C.; writing—review and editing, U.V.; visualization, D.K.; supervision, U.V.; project administration, A.C.; funding acquisition, A.C. and U.V. All authors have read and agreed to the published version of the manuscript.

Funding: This work was funded by the Slovenian Research Agency (grants P4-0133 Sustainable agriculture) and the Republic of Slovenia (Ministry of Education, Science and Sport) and the European Regional Development Fund (project "Raziskovalci-2.1-KIS-952049", A.Č.).

Data Availability Statement: No additional data is available.

Acknowledgments: The authorsare grateful to Veronika Kmecl for the help with the vitamin C analysis, Martina Peršić for the valuable comments on the manuscript, and the FEM staff which were involved in our research.

Conflicts of Interest: The authors declare no conflict of interest.

References

1. Le Tien, C.; Vachon, C.; Mateescu, M.-A.; Lacroix, M. Milk protein coatings prevent oxidative browning of apples and potatoes. *J. Food Sci.* **2001**, *66*, 512–516. [CrossRef]
2. Gacche, R.N.; Warangkar, S.C.; Ghole, V.S. Glutathione and cinnamic acid: Natural dietary components used in preventing the process of browning by inhibition of polyphenol oxidase in apple juice. *J. Enzym. Inhib. Med. Chem.* **2004**, *19*, 175–179. [CrossRef]
3. Khanizadeh, S.; Groleau, Y.; Levasseur, A.; Charles, M.T.; Tsao, R.; Yang, R.; DeEll, J.; Hampson, C.; Toivonen, P.T. SJCA38R6A74 (Eden). *HortScience* **2006**, *41*, 1513–1515. [CrossRef]
4. Tazawa, J.; Oshino, H.; Kon, T.; Kasai, S.; Kudo, T.; Hatsuyama, Y. Genetic characterization of flesh browning trait in apple using the non-browning cultivar 'Aori 27'. *Tree Genet. Genomes* **2019**, *15*, 49. [CrossRef]
5. Waltz, E. Nonbrowning GM apple cleared for market. *Nat. Biotechnol.* **2015**, *33*, 326–327. [CrossRef] [PubMed]
6. Mesquita, V.L.V.; Queiroz, C. Enzymatic browning. In *Biochemistry of Foods*; Eskin, N.A.M., Shahidi, F., Eds.; Academic Press: London, UK, 2013; Volume 3, pp. 387–418. [CrossRef]
7. Amiot, M.J.; Tacchini, M.; Aubert, S.; Nicolas, J. Phenolic composition and browning susceptibility of various apple cultivars at maturity. *J. Food Sci.* **1992**, *57*, 958–962. [CrossRef]
8. Demeke, T.; Morris, C.F. Molecular characterization of wheat polyphenol oxidase (PPO). *Theor. Appl. Genet.* **2002**, *104*, 813–818. [CrossRef] [PubMed]

9. Di Guardo, M.; Tadiello, A.; Farneti, B.; Lorenz, G.; Masuero, D.; Vrhovsek, U.; Costa, G.; Velasco, R.; Costa, F. A multidisciplinary approach providing new insight into fruit flesh browning physiology in apple (Malus × domestica Borkh). *PLoS ONE* **2013**, *8*, e78004. [CrossRef] [PubMed]

10. Ambrosia, T.M. Why Do Apples Turn Brown? Available online: https://ambrosiaapples.ca/apples-turn-brown/ (accessed on 23 March 2021).

11. Davey, M.W.; Montagu, M.V.; Inzé, D.; Sanmartin, M.; Kanellis, A.; Smirnoff, N.; Benzie, I.J.; Strain, J.J.; Favell, D.; Fletcher, J. Plant L-ascorbic acid: Chemistry, function, metabolism, bioavailability and effects of processing. *J. Sci. Food Agric.* **2000**, *80*, 825–860. [CrossRef]

12. May, M.J.; Vernoux, T.; Leaver, C.; Van Montagu, M.; Inzé, D. Glutathione homeostasis in plants: Implications for environmental sensing and plant development. *J. Experiment. Bot.* **1998**, *49*, 649–667. [CrossRef]

13. Noctor, G.; Foyer, C.H. Ascorbate and glutathione: Keeping active oxygen under control. *Annu. Rev. Plant Biol.* **1998**, *49*, 249–279. [CrossRef] [PubMed]

14. McDougall, G.J.; Foito, A.; Dobson, G.; Austin, C.; Sungurtas, J.; Su, S.; Wang, L.; Feng, C.; Li, S.; Wang, L.; et al. Glutathionyl-S-chlorogenic acid is present in fruit of Vaccinium species, potato tubers and apple juice. *Food Chem.* **2020**, *330*, 127227. [CrossRef]

15. Singleton, V.L.; Salgues, M.; Zaya, J.; Trousdale, E. Caftaric acid disappearance and conversion to products of enzymic oxidation in grape must and wine. *Am. J. Enol. Vitic.* **1985**, *36*, 50–56.

16. Noctor, G.; Mhamdi, A.; Chaouch, S.; Han, Y.; Neukermans, J.; Marquez-Garcia, B.; Qeval, G.; Foyer, C.H. Glutathione in plants: An integrated overview. *Plant Cell Environ.* **2012**, *35*, 454–484. [CrossRef]

17. Awad, M.A.; Jager, A. Influences of air and controlled atmosphere storage on the concentration of potentially helathful phenolics in apples and other fruits. *Postharvest Biol. Technol.* **2003**, *27*, 53–58. [CrossRef]

18. Li, L.; Li, X.; Ban, Z.; Jiang, Y. Variation in antioxidant metabolites and enzymes of 'Red Fuji' apple pulp and peel during cold storage. *Int. J. Food Prop.* **2014**, *17*, 1067–1080. [CrossRef]

19. Davey, M.W.; Keulemans, J. Determining the potential to breeed for enhanced antioxidant status in *Malus*:mean inter- and intravarietal fruit vitamin C and glutathiones contents at harvest and their evolution during storage. *J. Agric. Food Chem.* **2004**, *52*, 8031–8038. [CrossRef]

20. Crnko, J.; Marn, M. 'Majda' nova Jugoslovanska jablanova sorta za dvojno uporabo. In *Zbornik Biotehniske Fakultete Univerze Edvarda Kardelja v Ljubljani: Kmetijstvo*; Biotechnical Faculty: Ljubljana, Slovenia, 1986; Volume 47, pp. 49–62.

21. Persic, M.; Mikulic-Petkovsek, M.; Slatnar, A.; Veberic, R. Chemical composition of apple fruit, juice and pomace and the correlation between phenolic content, enzymatic activity and browning. *LWT Food Sci. Technol.* **2017**, *82*, 23–31. [CrossRef]

22. Aprea, E.; Charles, M.; Endrizzi, I.; Laura Corollaro, M.; Betta, E.; Biasioli, F.; Gasperi, F. Sweet taste in apple: The role of sorbitol, individual sugars, organic acids and volatile compounds. *Sci. Rep.* **2017**, *7*, 44950. [CrossRef]

23. Rymenants, M.; van de Weg, E.; Auwerkerken, A.; De Wit, I.; Czech, A.; Nijland, B.; Heuven, H.; De Storme, N.; Keulemans, W. Detection of QTL for apple fruit acidity and sweetness using sensorial evaluation in multiple pedigreed full-sib families. *Tree Genet. Genomes* **2020**, *16*, 71. [CrossRef]

24. Morimoto, T.; Yonemushi, K.; Ohnishi, H.; Banno, K. Genetic and physical mapping of QTLs for fruit juice browning and fruit acidity on linkage group 16 in apple. *Tree Genet. Mol. Breed.* **2014**, *4*, 1–10. [CrossRef]

25. Maliepaard, C.; Alston, F.H.; van Arkel, G.; Brown, L.M.; Chevreau, E.; Dunemann, F.; Evans, K.M.; Gardiner, S.; Guilford, P.; van Heusden, A.W.; et al. Aligning male and female linkage maps of apple (Malus pumila Mill.) using multi-allelic markers. *Theor. Appl. Genet.* **1998**, *97*, 60–73. [CrossRef]

26. Joshi, A.P.K.; Rupasinghe, H.P.V.; Pitts, N.L.; Khanizadeh, S. Biochemical characterization of enzymatic browning in selected apple genotypes. *Can. J. Plant Sci.* **2007**, *87*, 1067–1074. [CrossRef]

27. Nicolas, J.J.; Richard-Forget, F.C.; Goupy, P.M.; Amiot, M.J.; Aubert, S.Y. Enzymatic browning reactions in apple and apple products. *Crit. Rev. Food Sci. Nutr.* **1994**, *2*, 109–157. [CrossRef] [PubMed]

28. Fenech, M.; Amaya, I.; Valpuesta, V.; Botella, M.A. Vitamin C content in fruits: Biosynthesis and regulation. *Front. Plant Sci.* **2018**, *9*, 2006. [CrossRef] [PubMed]

29. Kritziger, E.C.; Bauer, F.F.; du Toit, W.J. Role of glutathione in winemaking: A review. *J. Agric. Food Chem.* **2013**, *2*, 269–277.

30. Boss, P.K.; Gardner, R.C.; Janssen, B.-J.; Ross, G.S. An apple polypgenol oxidase cDNA is up-regulated in wounded tissues. *Plant Mol. Biol.* **1995**, *27*, 429–433. [CrossRef] [PubMed]

31. Janovitz-Klapp, A.H.; Richard, F.C.; Goupy, M.; Nicolas, J.J. Kinetic studies on apple polyphenol oxidase. *J. Agric. Food Chem.* **1990**, *38*, 1437–1441. [CrossRef]

32. Dugé de Bernonville, T.; Guyot, S.; Paulin, J.-P.; Gaucher, M.; Loufrani, L.; Henrion, D.; Derbré, S.; Guilet, D.; Richomme, P.; Dat, J.F.; et al. Dihydrochalcones: Implication in resistance to oxidative stress and bioactivities against advanced glycation end-products and vasoconstriction. *Phytochemistry* **2010**, *71*, 443–452. [CrossRef]

33. Yuri, A.J.; Neira, A.; Quilodran, A.; Razmilic, I.; Motomura, Y.; Torres, C.; Palomo, I. Sunburn on apples is associated with increases in phenolic compounds and antioxidant activity as a function of the cultivar and areas of the fruit. *J. Food Agric. Environ.* **2010**, *8*, 920–925.

34. Łata, B.; Tomala, K. Apple Peel as a Contributor to whole fruit quantity of potentially healthful bioactive compounds. cultivar and year Implication. *J Agric. Food Chem.* **2007**, *55*, 10795–10802. [CrossRef] [PubMed]

35. Li, D.; Wang, P.; Luo, Y.; Zhao, M.; Chen, F. Health benefits of anthocyanins and molecular mechanisms: Update from recent decade. *Crit. Rev. Food Sci. Nutr.* **2017**, *57*, 1729–1741. [CrossRef]
36. Gosch, C.; Halbwirth, H.; Stich, K. Phloridzin: Biosynthesis, distribution and physiological relevance in plants. *Phytochemistry* **2010**, *71*, 838–843. [CrossRef] [PubMed]
37. Mikulic-Petkovsek, M.; Stampar, F.; Veberic, R. Parameters of inner quality of the apple scab resistant and susceptible apple cultivars (Malus domestica Borkh.). *Sci. Hortic.* **2007**, *114*, 37–44. [CrossRef]
38. Vrhovsek, U.; Masuero, D.; Gasperotti, M.; Franceschi, P.; Caputi, L.; Viola, R.; Mattivi, F. A versatile targeted metabolomics method for the rapid quantification of multiple classes of phenolics in fruits and beverages. *J. Agric. Food Chem.* **2012**, *60*, 8831–8840. [CrossRef] [PubMed]
39. Vanzo, A.; Janeš, L.; Požgan, F.; Bolta, Š.V.; Sivilotti, P.; Lisjak, K. UHPLC-MS/MS determination of varietal thiol precursors in Sauvignon Blanc grapes. *Sci. Rep.* **2017**, *7*, 13122. [CrossRef]
40. Zupan, A.; Mikulic-Petkovsek, M.; Slantnar, A.; Stampar, F.; Veberic, R. Individual phenolic response and peroxidase activity in peel of differently sun-exposed apples in the period favorable for sunburn occurrence. *J. Plant Physiol.* **2014**, *171*, 1706–1712. [CrossRef]
41. Cebulj, A.; Halbwirth, H.; Mikulic-Petkovsek, M.; Veberič, R.; Slatnar, A. The impact of scald development on phenylpropanoid metabolism based on phenol content, enzyme activity, and gene expression analysis. *Hortic. Environ. Biotechnol.* **2020**, *61*, 849–858. [CrossRef]

Article

Cracking in Sweet Cherry Cultivars Early Bigi and Lapins: Correlation with Quality Attributes

Sandra Pereira [1,*], Vânia Silva [1], Eunice Bacelar [1], Francisco Guedes [2], Ana Paula Silva [1], Carlos Ribeiro [3] and Berta Gonçalves [1]

1   CITAB-Centre for the Research and Technology of Agro-Environmental and Biological Sciences, University of Trás-os-Montes e Alto Douro, Quinta de Prados, 5000-801 Vila Real, Portugal; vaniasilva@utad.pt (V.S.); areale@utad.pt (E.B.); asilva@utad.pt (A.P.S.); bertag@utad.pt (B.G.)
2   Cermouros—Cerejas de São Martinho de Mouros, Lda, Quinta da Ribeira, Bulhos, 4660-210 Resende, Portugal; fjmguedes@cermouros.pt
3   Department of Agronomy, University of Trás-os-Montes e Alto Douro, Quinta de Prados, 5000-801 Vila Real, Portugal; cribeiro@utad.pt
*   Correspondence: sirp@utad.pt

Received: 7 October 2020; Accepted: 8 November 2020; Published: 12 November 2020

Abstract: Fruit cracking is one of the main concerns in sweet cherry production and is caused by a heavy rainfall before and during the harvest. This physiological disorder leads to severe economic losses, which can be more or less effective depending on the cracked region of the fruit: in the cheeks (side cracks), in the stylar scar region, or in the stem cavity region. Sweet cherry cracking can be affected by several factors such as cultivar, growing conditions, rootstock, fruit size, flesh osmotic potential, cuticular characteristics of the skin, and stage of fruit development. In this sense, the objective of this work was to evaluate the cracking incidence in two sweet cherry cultivars (Early Bigi and Lapins grafted on "Saint Lucie 64" rootstock) and correlate the cracking index with other quality parameters. Fruits were harvested on 2 May (cv. Early Bigi) and on 27 May (cv. Lapins) 2019 at their commercial ripening stage. In the field, the total yield and the trunk cross-sectional area were determined for each tree in order to calculate the yield efficiency. In the laboratory, the cracking index was determined in 150 fruits without visual defects. In addition, fruit size and weight, wax content, flesh firmness, epidermis rupture force, total soluble solids, pH, titratable acidity, and maturity index of 30 fruits were also evaluated. In general, all the analyzed quality parameters were influenced by the cultivar, being that cv. Lapins presented larger, heavier, firmer, and sweeter fruits, with more acidity and higher maturation index. However, cv. Lapins also presented higher cracking index, which was positively correlated with all the parameters above-mentioned and negatively correlated with the wax content. In fact, cv. Early Bigi presented a high wax content and simultaneously a low cracking index. The stylar scar region cracks were the most prevalent in both cultivars. These results allowed us to conclude that, in the North Portugal region, the Lapins cherries presented better quality attributes than the Early Bigi cherries. However, the latter are still very valuable to the region due to its early ripening. Additionally, it was also possible to conclude that bigger, firmer, more mature, and with lower wax content cherries were more sensitive to cracking than the smaller fruits, soft-fleshed, less mature, and with higher wax content.

Keywords: cracking index; fruit quality; sweet cherry fruit; yield efficiency

---

1. Introduction

Sweet cherries (*Prunus avium* L.) are one of the most attractive and appreciated spring–summer fruits, especially due to its attractive appearance, color, taste, and sweetness [1].

According to FAOSTAT (Food and Agriculture Organization of the United Nations), in 2018, 2,547,944 tons of sweet cherry were produced in the world, in a global cultivated area of 432,314 ha [2]. In that year, the main producers in the world were Turkey, United States of America, Uzbekistan, Chile, and Iran [2]. Within Europe, Italy (114,798 tons), Spain (106,584 tons), Romania (90,837 tons), and Greece (90,290 tons) were the top sweet cherry producing countries, while Portugal produced, in the same year, 17,461 tons in a total cultivated area of 6056 ha [2].

Sweet cherries are extremely perishable fruits (short shelf life of 7–14 days in air cold storage) with a short harvest season. Its quality is highly affected by the environmental conditions, since excessive rainfall before and during the harvest can lead to the fruit cracking. This physiological disorder causes production and economic losses [3,4] and originates from the excess uptake of water by the fruit surface, which results in localized bursting of the skin [5,6]. Other authors support that fruit cracking is the result of skin shrinking after rapid cooling caused by rainfall or by a sharp temperature drop [7]. Cracking in sweet cherries can be divided in macro-cracks (extending into the epidermal and hypodermal cell layers and visible to the naked eye) and micro-cracks (induced by water and not detected by visual inspection) [6,8]. According to [9], three different types of macro-cracking can occur in sweet cherries: in the cheek region, in the stylar scar region (apical end), and in the stem cavity region.

The cracking susceptibility is difficult to quantify in the field, since the level, distribution, and duration of rainfall, fruit maturity stage, orchard factors, and environmental conditions are not standardized [10]. Indeed, fruit cracking can be affected by several factors such as cultivar, growing conditions, rootstock, fruit size, flesh osmotic potential, cuticular characteristics of the skin, and stage of fruit development [3,11]. Among the several factors affecting the cracking of sweet cherry, the effect of the cultivar on this disorder was evaluated in the present work. The cultivars studied were Early Bigi and Lapins.

The cv. Early Bigi, also known as Bigi Sol, is a self-sterile cherry cultivar, large in size but with less flavor than other cultivars such as Burlat and Lapins. However, it is of great agronomic interest, since it is one of the very early cultivars produced in Resende.

The cv. Lapins, also marketed as Cherokee, is a hybrid of Van (variety very used in cross pollinations for its excellent attributes) and Stella (first self-fertile variety) cultivars and it is currently the sweet cherry cultivar most planted around the world. This cultivar is a late-season sweet cherry cultivar, ripening about one month later than Early Bigi. Its fruit is of excellent quality, producing some of the largest and juiciest of the sweet cherries. Furthermore, this cultivar is self-fertile and do not require a pollinator.

The sensitivity of different cultivars to cracking is not completely understood. However, some authors support that by choosing the right cultivar, it is possible to minimize the cracking incidence. The authors in [12,13] also defended that good results in productivity, fruit size, firmness, and tolerance against cracking were the main criteria for cultivar selection.

In this sense, the main objective of this work was to evaluate the fruit cracking incidence and cracking characteristics of two sweet cherry cultivars Early Bigi and Lapins (early vs. late) and correlate it with other fruit quality parameters, namely, fruit size and weight, firmness, epidermis rupture force, wax content, total soluble solids, titratable acidity, maturity index, and pH.

2. Results and Discussion

2.1. Total Yield and Yield Efficiency

According to [14], high yield efficiency is an important characteristic that determines the commercial value of a given cultivar.

The results of the yield per tree evaluated in the field at harvesting time and the yield efficiency are presented in the Figure 1. No significant differences were observed in both parameters for the two studied cultivars ($p = 0.805$ and $p = 0.657$, respectively). Despite no significant differences were observed in the total yield per tree, 50.1% of the fruits of cv. Early Bigi were harvested already cracked, due to heavy rains during the maturity period (especially the month of April), while no cracked fruits were collected for cv. Lapins. This can be explained by the low rainfall during the month of May (only 0.3 mm of rain). In fact, high rainfall during the maturity period and near the harvest time usually leads to a higher cracking index.

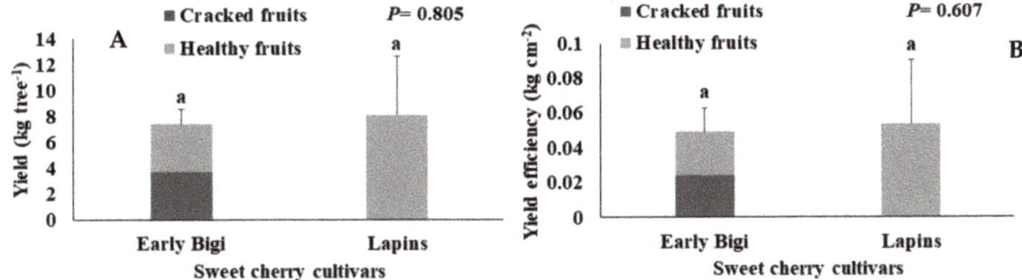

Figure 1. (**A**) Yield (kg tree^{-1}) and (**B**) yield efficiency (kg cm^{-2}) of two sweet cherry cultivars (Early Bigi. and Lapins). Data are expressed as mean ± standard deviation. Similar letters indicate no statistically significant differences ($p > 0.05$) between cultivars for each variable, according to Tukey's test.

The yield efficiency also did not present significant differences between cultivars. However, the highest values were observed in cv. Lapins. In a previous work developed by [13], in Slovenia, cv. Early Bigi presented a higher yield efficiency (0.25 kg cm^{-2}) than our trees, which can be explained by the different ages of the trees, different soils, and edaphoclimatic conditions.

2.2. Cracking Index and Crack Type Incidence

The CI (cracking index) of both sweet cherry cultivars after an immersion in distilled water during 6 h and the crack type incidence are presented in Figure 2. The cv. Lapins presented a significantly higher cracking index than cv. Early Bigi ($p = 0.001$), with a fold increase of 440%, despite being considered as a moderately resistant cultivar to cracking.

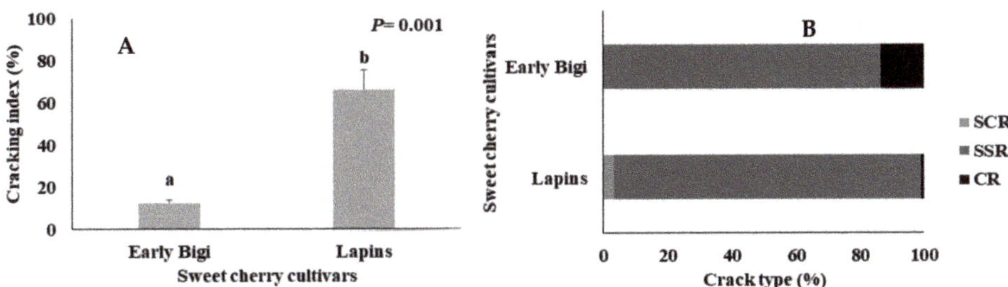

Figure 2. (**A**) Cracking index (%) and (**B**) crack type incidence (%) of two sweet cherry cultivars (Early Bigi and Lapins). SCR—stem cavity region, SSR—stylar scar region, and CR—cheek region. Data are expressed as mean ± standard deviation. Different letters indicate statistically significant differences ($p < 0.05$) between cultivars for each variable, according to Tukey's test.

Taking into account the different types of cracking, in the present work, the SSR cracks were the most frequent in both cultivars (86.41% in cv. Early Bigi and 96.12% in cv. Lapins). This can be explained by the higher osmotic concentration of solutes in this part of the fruit, which accounted for a more rapid water absorption through the skin, resulting in a quicker formation of cracks [15]. Furthermore, cv. Early Bigi also presented 13.59% of CR cracks and no SCR-type cracking was found in this cultivar, while cv. Lapins presented 3.1% of SCR cracks and only around 0.78% of CR cracks.

According to [15], the small cracks at the base and top end of the fruit (SCR and SSR cracks) often occurred at a very early stage, when the fruits were not yet mature. The same author [9] also suggested that cherries with these two types of cracks were tolerated by consumers since no fungal infection is present. However, fruit with CR cracks (more predominant in cv. Early Bigi than in cv. Lapins) are usually rejected by the consumers.

Despite the higher cracking index of cv. Lapins observed in the present work and determined in the laboratory, it is necessary to take in account that cracking of fruit attached to the tree and that of fruit detached and submerged differs significantly. In fact, according to [16], fruit attached to the tree cracked more slowly and required more water to crack.

2.3. Cracking Index and Biometric Attributes

Fruit size is considered as the main benchmark in commercial cherry grading, being a large factor in consumer preference, and is a huge determinant of both farm gate and market price.

The correlations between the CI and the fruit width and its weight are presented in Figure 3A,B. In Figure 3A, it is possible to observe that cv. Early Bigi presented simultaneously lower values of CI and width, which means that these two parameters were positively correlated ($R^2 = 0.9529$, $p = 0.001$). The same profile was observed in Figure 3B. Indeed, CI and fruit weight were also positively correlated ($R^2 = 0.9482$, $p = 0.001$), with the highest values of both parameters to be observed in cv. Lapins. In fact, the weight of fruits from cv. Lapins was significantly higher (11.12 g of average weight) than in cv. Early Bigi (6.10 g of average weight) ($p = 0.000$). According to several authors, the fruit size depends on cultivar, maturation stage, agricultural practices, and environmental conditions [17–19]. In this work, the higher CI observed in cv. Lapins may have been influenced by the big fruit size of this cultivar (29.00 mm of average width) in contrast to cv. Early Bigi (24.46 mm of average width), thus there were significant differences between cultivars ($p = 0.000$). In fact, according to [20] and [21], larger fruits are more prone to cracking than small ones. In fact, in big fruits, the physical stress on the enclosing membrane (the skin) is bigger, and consequently, cracks occur in areas of the fruit where stress is the greatest [4]. However, the first author also suggested that the correlation between the CI and the fruit size was greater within a cultivar than between cultivars, since it is necessary to take into account the specificities of each cultivar.

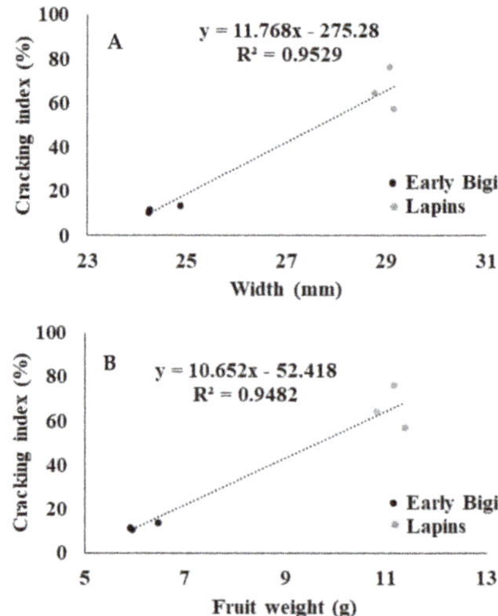

Figure 3. Correlation between (**A**) the cracking index (%) and the fruit width (mm) and (**B**) the cracking index (%) and the fruit weight (g) of two sweet cherry cultivars: Early Bigi and Lapins.

2.4. Cracking Index, Soluble Cuticular Wax Content, and Texture Parameters

The fruit cuticle is a hydrophobic and semi-permeable membrane consisting of two major lipid types: cutin and cuticular waxes [22]. Cuticular waxes play an important role in the water permeability of sweet cherries [23].

The correlation between the CI and the cuticular wax content is presented in Figure 4A. Cv. Early Bigi presented lower CI, but significantly higher wax content than cv. Lapins ($p = 0.000$). This means that these two parameters were negatively correlated ($R^2 = 0.8389$, $p = 0.012$). In general, the lower the wax content, the greater the water intake for the fruit, and consequently, the higher the cracking index. Indeed, according to [24], cultivars more tolerant to cracking have higher wax contents.

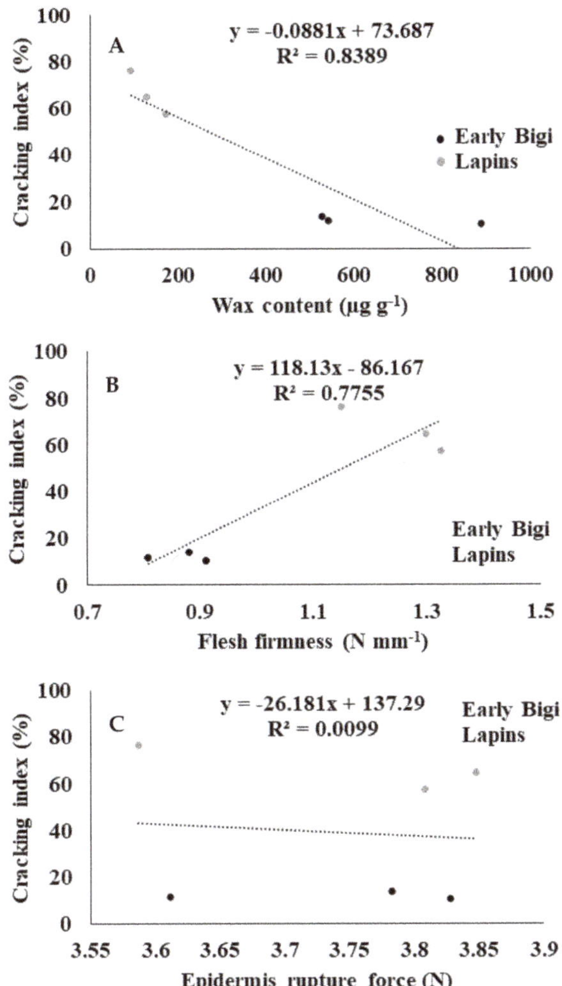

Figure 4. Correlations between (**A**) the cracking index (%) and the wax content ($\mu g\ g^{-1}$), (**B**) the cracking index (%) and the flesh firmness ($N\ mm^{-1}$), and (**C**) the cracking index (%) and the epidermis rupture force (N) of two sweet cherry cultivars: Early Bigi and Lapins.

Fruit firmness (FF) is an important quality attribute in sweet cherries, which is associated with a greater resistance to decay and mechanical damage, and consequently, to the increase of storage life [25].

In the present work, according to Figure 4B, the CI is positively correlated with the FF ($R^2 = 0.7755$, $p = 0.021$), being that cv. Lapins presented the highest values for both parameters. Indeed, cv. Lapins presented a significantly higher FF than cv. Early Bigi ($p = 0.000$). According to [26], sweet cherries with higher flesh firmness can have consequently fewer physiological disorders during handling, storage, and shipping, but according to [21], these firmer fruits are simultaneously more affected by cracking. In fact, it is usually assumed that cultivars with firm-fleshed fruits have more tendency to cracks, but the literature concerning this correlation is still very unclear [19]. It is likely that the epidermis elasticity is smaller in the hardest fruits.

In a previous work developed by [27,28], cv. Lapins presented an average fruit firmness of 2.65 N at harvest, which was lower than the results obtained in the present work for both cultivars (3.75 N for Lapins and 3.74 N for Early Bigi), perhaps due to the pre-harvest applied treatments and/or the different edaphoclimatic conditions.

No correlation was observed between the CI and the epidermis rupture force (ERF) ($R^2 = 0.0099$, $p = 0.870$) (Figure 4C). Furthermore, no significant differences were observed in ERF between cultivars ($p = 0.963$).

2.5. Cracking Index (CI) and Maturity Evaluation Parameters

The cv. Lapins presented higher values for all the analyzed routine parameters (Figure 5A–D). According to [29], total soluble solids (TSS) depend on cultivar, mainly due to glucose and fructose and less to the presence of sucrose and sorbitol, being related to flavor intensity. In this work, cv. Lapins presented a TSS value twice higher than that of cv. Early Bigi (averages of 19.37 and 9.80 °Brix, respectively), indicating a significantly higher sugar content in cv. Lapins ($p = 0.000$). Maturity index (MI), as the ratio between TSS and TA (titratable acidity), is also an indicator of sweetness and consequently of maturity [30,31], allowing us to conclude that cv. Lapins (MI = 2.77) was significantly more mature at harvesting time than cv. Early Bigi (MI = 1.73) ($p = 0.000$). As cv. Early Bigi is one of the earliest cultivars collected in the Resende region, it can be sold at a good price and it is therefore a very profitable cultivar in the region.

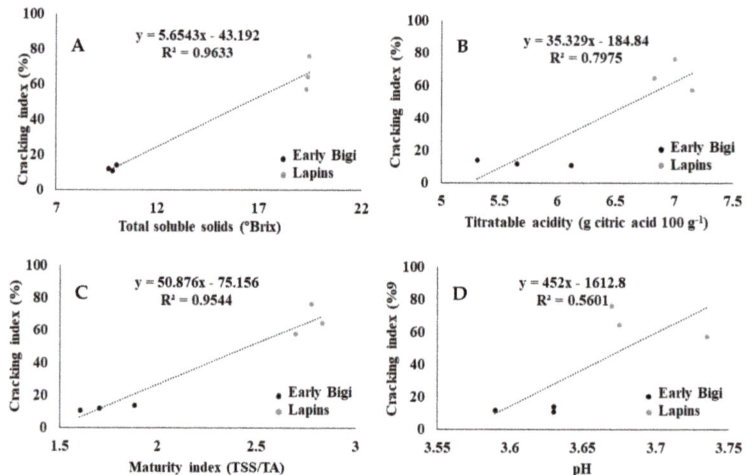

Figure 5. Correlations between (**A**) the cracking index (%) and the total soluble solids (°Brix), (**B**) the cracking index (%) and the titratable acidity (g citric acid 100 g^{-1}), (**C**) the cracking index (%) and the maturity index (TSS/TA), and (**D**) the cracking index (%) and the pH of two sweet cherry cultivars: Early Bigi and Lapins.

Positive correlations were observed between the CI and all the analyzed routine parameters, with the exception of pH (Figure 5A–D). This means that the higher the total soluble solids, the titratable acidity and the maturity index, the greater the CI. Stronger correlations were found between the CI and TSS (R^2 = 0.9633, p = 0.001) and the CI and the MI (R^2 = 0.9544, p = 0.001). These results were consistent with the findings of [32], who also reported a positive correlation between the TSS and fruit cracking in sweet cherry cultivars. [33] also demonstrated that the susceptibility to cracking increased with the maturation stage.

2.6. Principal Component Analysis (PCA)

To better understand the correlations between all the evaluated parameters, a chemometric analysis was performed integrating all the data (Figure 6). The PCA based on the correlation matrix standardizes the data and this analysis was performed using a correlation matrix (Corr-PCA). In this analysis, the first two factorial axes (PC1 and PC2) represent 95.04% of the total variance, with factor 1 the one that presents the higher weight (82.81%).

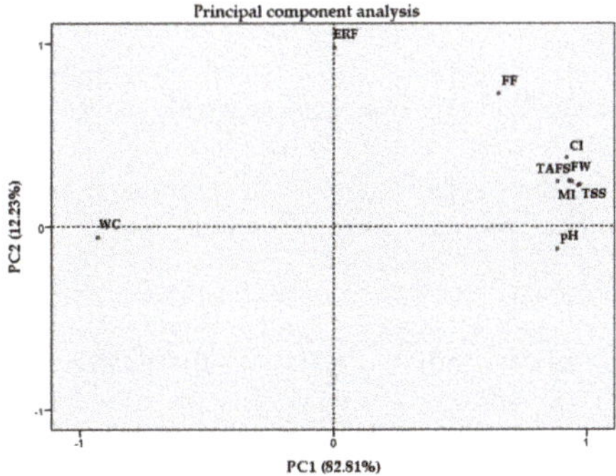

Figure 6. Principal component analysis using the whole dataset of two sweet cherry cultivars (Early Bigi and Lapins). Analyzed parameters: WC—wax content; ERF—epidermis rupture force; FF—flesh firmness; pH; CI—cracking index; FW—fruit weight; FS—fruit size (width); TA—titratable acidity; TSS—total soluble solids; MI—maturity index.

The CI as well as the most of the analyzed parameters (fruit size, fruit weight, total soluble solids, titratable acidity, and maturity index) were together in the right PCA quadrant as well as flesh firmness, epidermis rupture force, and pH, although the latter were further away from the remaining parameters. On the other hand, the wax content was spatially separated and placed in the left PCA quadrant, corroborating the negative correlation between this parameter and the CI.

3. Material and Methods

3.1. Experimental Design and Sweet Cherry Raw Material

The experiment was conducted in 2019 in a 10 year-old sweet cherry orchard located in São João de Fontoura, Resende (Viseu district, Northern Portugal; latitude 41°12′ N, longitude 7°93′ W, altitude 149 m).

Weather data were recorded by a weather station located near the experimental site. The daily minimum temperature in March, when the first flowers appeared in the trees, ranged between 3.7 and 12.2 °C and the daily maximum temperature ranged between 11.9 and 22.6 °C. The total precipitation in that month was 4.1 mm, with day 6 as the rainiest. April was a very rainy month, with a total rainfall of 8.1 mm throughout the month. The daily minimum temperature ranged between 2.0 and 14.3 °C and the daily maximum temperature varied between 10.9 and 26.0 °C. The hottest days were 29 and 30 April. May was the warmest month (daily minimum temperature ranged between 6.4 and 17.4 °C and daily maximum temperature fluctuated between 16.2 and 32.9 °C) and the least rainy month (only 0.3 mm of total precipitation).

For this study, ten trees of each sweet cherry cultivar (Early Bigi and Lapins, grafted on 'Saint Lucie 64' rootstock) were used. Trees were spaced 3.0 m between rows and 2.5 m from each other in the row. Fruits were hand-picked at their commercial ripening stage, which was on 2 May and 27 May 2019 for cv. Early Bigi and cv. Lapins, respectively. The commercial ripening stage for each cultivar was defined by the producer. In fact, the fruits to analyze were collected at the same time that the producer harvested it to send to the market. Within each cultivar, about 500 g samples of fruit were collected randomly from all trees and two fruits per tree were additionally collected to determine the wax content. In the field, collected fruits were kept on ice and placed in the refrigerator upon reaching the laboratory. There, fruits were randomly divided into two sub batches. The first sub batch (150 sweet cherries without visual defects) was assigned to the evaluation of the cracking index (CI), while the second sub batch, consisting of 30 sweet cherries, was assigned for fruit weight (FW), size (FS), epidermis rupture force (ERF), flesh firmness (FF), total soluble solids (TSS), titratable acidity (TA), maturity index (MI), and pH measurements.

3.2. Production and Yield Efficiency

At harvest time, the production per tree was assessed, evaluating the weight of healthy and cracked cherries, separately. The perimeter of the trunks was also measured to determine the trunk cross-sectional area (TCSA) and the yield efficiency was determined as the ratio between the yield and the TCSA (kg cm^{-2}).

3.3. Cracking Index

The cracking index (CI) was determined in the first sub batch of cherries according to the method of [34], and modified by [15]. For this, the 150 cherries were divided into three replicates of 50 fruits and immersed in 2 L containers filled with distilled water (20 ± 1 °C). After 2, 4, and 6 h, the fruits were observed for macroscopic cracks. At each time, cracked cherries were removed and counted, and fruits without cracks were re-incubated. The CI was calculated as follows:

$$CI = ((5a + 3b + c) * 100)/250 \tag{1}$$

where a, b, and c represent the number of cracked cherries after 2, 4, and 6 h of immersion, respectively.

The crack type (CT) incidence (%) was also determined and expressed as the number of cracks of a particular type (SCR—stem cavity region, SSR—stylar scar region or CR—cheek region) compared to the total number of cracks.

3.4. Soluble Cuticular Waxes Content

Total soluble epi- and intra-cuticular wax content of sweet cherries was determined using two fruits per tree, by the method of [35]. Two fruits without peduncle per tree were weighed and immersed in a mix of chloroform:methanol (3:1) for 2 min at 25 °C. After that, the fruits were discarded and the solution was filtered and placed in a glass previously identified and weighed. After about a week, when the glasses were dry, they were weighted again. The weight of the fruits and the difference in the weight of the glasses were used to calculate the wax content, which was expressed as µg g^{-1} of fresh weight.

3.5. Fruit Weight and Size

Fruit weight and size were determined for the 30 fruits of the second sub batch. Fruit weight (g) was evaluated using an electronic balance (EW2200-2NM, Kern, Germany) and fruit size was determined by measuring the width (mm) using electronic calipers (500-196-30, Mitutoyo, UK).

3.6. Epidermis Rupture Force and Flesh Firmness

The same 30 fruits from the second sub batch were used to determine the texture parameters, epidermis rupture force (N) and flesh firmness (N mm^{-1}) using a TA.XT.plus texture analyzer (Stable Micro Systems, Godalming, UK) employing a 50 N loading cell and a 2.0 mm diameter cylindrical probe. The maximum force when compressing 5 mm was measured at a speed of 1 mm s^{-1}.

3.7. Total Soluble Solids, Titratable Acidity, Maturity Index, and pH

After determining the texture, the same 30 fruits from the second sub batch were divided into three groups of 10 fruits. The juice from these 10 fruits was extracted with an electrical extractor (ZN350C70, Tefal Elea, China) for one minute. Total soluble solids (TSS, in °Brix) were determined using a digital refractometer (PR-101, Atago, Tokyo, Japan) and expressed as the percentage of soluble solids in juice (%). Then, pH was also measured using a pH meter (3310 Jenway). Titratable acidity (TA, in g citric acid 100 g^{-1} of fresh weight) was determined by diluting 10 mL of fruit juice with 10 mL of distilled water and titrating with 0.1 mol L^{-1} sodium hydroxide (NaHO) until reaching pH 8.2 using a Schott Easy Titroline automatic titrator. The maturity index was expressed as the ratio of TSS and TA.

3.8. Statistical Analysis

Statistical analysis was performed using SPSS V.25 software (SPSS-IBM, Corp., Armonk, New York, NY, USA). Statistical differences between the two cultivars were evaluated by one-way analysis of variance (ANOVA), followed by Tukey's post-hoc multiple range test ($p < 0.05$). In addition, Pearson's rank correlation was performed for the relationship between the cracking index and the other quality parameters (three average values for each cultivar). Principal component analysis (PCA) was also performed. PCA is mostly used as a tool in exploratory data analysis and for making predictive models. PCA was performed by eigenvalue decomposition of the data correlation (Corr-PCA) matrix after normalizing the data matrix for each attribute.

4. Conclusions

Nowadays, the big challenge of producers is to guarantee high production without affecting the quality of the fruit. For this, the selection of the best cultivars for each region is crucial.

In the present work, cv. Lapins cherries presented better quality attributes than the cv. Early Bigi cherries, having potentially greater commercial value and is an ideal cultivar for the Resende region. Lapins cherries simultaneously presented a higher cracking index, which was related to the bigger size, higher fruit firmness, more mature fruits, and less wax content. However, it is necessary to take into account that the cracking index is higher in detached fruits compared to cracking in fruits on the trees. Even so, Lapins cherries presented higher susceptibility for cracking in the laboratory assays, indicating a possible higher cracking incidence in the orchard if intensive rains occur near or during the harvest.

Although it has not shown itself as an ideal cultivar in terms of quality parameters, the cv. Early Bigi is still very valuable to producers in the Resende region due to its early ripening. The differences in terms of productivity and harvest time of these two cultivars are also important to the producers, especially due to the climate change and irregular rainfall that can affect one cultivar more than the other, according to its maturity stage.

Author Contributions: All authors contributed in a significant way to the activities of the paper. S.P., V.S., A.P.S., and B.G. had a major role in the design of the experiment and the coordination of the activities. S.P. and V.S. did the practical work in the field and in the laboratory. S.P. did the literature research and drafted the manuscript including the tables and figures; and E.B., A.P.S., and B.G. helped write the final manuscript. F.G. followed and helped in the field work and C.R. in methodology and data analysis. All authors have read and agreed to the published version of the manuscript.

Funding: This research was funded by "Fundo Europeu Agrícola de Desenvolvimento Rural (FEADER)" and by "Estado Português" in the context of "Ação 1.1 «Grupos Operacionais»", integrated in "Medida 1. «Inovação» do

PDR 2020—Programa de Desenvolvimento Rural do Continente"—Grupo Operacional para a valorização da produção da Cereja de Resende e posicionamento da subfileira nos mercados (iniciativa no. 362). The authors also acknowledge the support of National Funds by the FCT—Portuguese Foundation for Science and Technology, under the project UIDB/04033/2020.

Acknowledgments: The authors are grateful to José de Almeida from Cermouros for providing access to his orchard as well as to Sílvia Afonso and Ivo Oliveira for their support in the field work.

Conflicts of Interest: The authors declare no conflict of interest.

References

1.   Jakobek, L.; Seruga, M.; Voca, S.; Sindrak, Z.; Dobricevic, N. Flavonol and phenolic acid composition of sweet cherries (cv. Lapins) on six different vegetative rootstocks. *Sci. Hortic.* **2009**, *123*, 23–28. [CrossRef]
2.   FAOSTAT. Agricultural Production. Crops Primary–Cherries. Food and Agriculture Organization of the United Nations. 2020. Available online: http://www.fao.org/faostat/en/#data/QC (accessed on 10 July 2020).
3.   Lane, W.D.; Meheriuk, M.; McKenzie, D.-L. Fruit cracking of a susceptible, an intermediate, and a resistant sweet cherry cultivar. *HortScience* **2000**, *35*, 239–242. [CrossRef]
4.   Khadivi-Khub, A. Physiological and genetic factors influencing fruit cracking. *Acta Physiol. Plant.* **2015**, *37*, 1–14. [CrossRef]
5.   Knoche, M.; Peschel, S. Water on the surface aggravates microscopic cracking of the sweet cherry fruit cuticle. *J. Am. Soc. Hortic. Sci.* **2006**, *131*, 192–200. [CrossRef]
6.   Correia, S.; Schouten, R.; Silva, A.P.; Gonçalves, B. Sweet cherry fruit cracking mechanisms and prevention strategies: A review. *Sci. Hortic.* **2018**, *240*, 369–377. [CrossRef]
7.   Koumanov, S. On the mechanisms of the sweet cherry (*Prunus avium* L.) fruit cracking: Swelling or shrinking? *Sci. Hortic.* **2015**, *184*, 169–170. [CrossRef]
8.   Knoche, M.; Winkler, A. Chapter 7: Rain-induced cracking of sweet cherries. In *Cherries: Botany, Production and Uses*; Quero-García, J., Iezzoni, A., Puławska, J., Lang, G., Eds.; CABI: Wallingford, UK, 2017; pp. 140–165.
9.   Christensen, J.V. Rain-induced cracking of sweet cherries: Its causes and prevention. In *Cherries: Crop Physiology, Production and Uses*; Webster, A.D., Looney, N.E., Eds.; CAB International: Wallingford, UK, 1996; pp. 297–327.
10.   Measham, P.F.; Bound, S.A.; Gracie, A.J.; Wilson, S.J. Crop load manipulation and fruit cracking in sweet cherry (*Prunus avium* L.). *Adv. Hortic. Sci.* **2012**, *26*, 25–31.
11.   Moing, A.; Renaud, C.; Christmann, H.; Fouilhaux, L.; Tauzin, Y.; Zanetto, A. Is there a relation between changes in osmolarity of cherry fruit flesh or skin and fruit cracking susceptibility? *J. Am. Soc. Hortic. Sci.* **2004**, *129*, 635–641. [CrossRef]
12.   Rozpara, E. Growth and yield of eleven sweet cherry cultivars in Central Poland. *Acta Hortic.* **2008**, *795*, 571–576. [CrossRef]

13. Usenik, V.; Fajt, N. Sweet cherry cultivar testing in Slovenia. *Acta Hortic.* **2019**, *1235*, 265–270. [CrossRef]

14. Wermund, U.; Fearne, A.; Hornibroo, S.A. Consumer purchasing behaviour with respect to cherries in the United Kingdom. *Acta Hortic.* **2005**, *667*, 539–544. [CrossRef]

15. Christensen, J.V. Cracking in Cherries III. Determination of cracking susceptibility. *Acta Agric. Scand.* **1972**, *22*, 128–136. [CrossRef]

16. Winkler, A.; Blumenberg, I.; Schurmann, L.; Knoche, M. Rain cracking in sweet cherries is caused by surface wetness, not by water uptake. *Sci. Hortic.* **2020**, *269*, 109400. [CrossRef]

17. Gonçalves, B.; Silva, A.P.; Moutinho-Pereira, J.; Bacelar, E.; Rosa, E.; Meyer, A.S. Effect of ripeness and postharvest storage on the evolution of colour and anthocyanins in cherries (*Prunus avium* L.). *Food Chem.* **2007**, *103*, 976–984. [CrossRef]

18. Usenik, V.; Fabcic, J.; Stampar, F. Sugars, organic acids, phenolic composition and antioxidant activity of sweet cherry (*Prunus avium* L.). *Food Chem.* **2008**, *107*, 185–192. [CrossRef]

19. Usenik, V.; Fajt, N.; Mikulic Petkovsek, M.; Slatnar, A.; Stampar, F.; Veberic, R. Sweet cherry pomological and biochemical characteristics influenced by rootstock. *J. Agric. Food Chem.* **2010**, *58*, 4928–4933. [CrossRef]

20. Christensen, J.V. Cracking in cherries VII. Cracking susceptibility in relation to fruit size and firmness. *Acta Agric. Scand.* **1975**, *25*, 301–312. [CrossRef]

21. Yamaguchi, M.; Sato, I.; Ishiguro, M. Influences of epidermal cell sizes and flesh firmness on cracking susceptibility in sweet cherry (*Prunus avium* L.) cultivars and selections. *J. Jpn. Soc. Hortic. Sci.* **2002**, *71*, 738–746. [CrossRef]

22. Kunst, L.; Samuels, A.L. Plant cuticles shine: Advances in wax biosynthesis and export. *Curr. Opin. Plant Biol.* **2009**, *12*, 721–727. [CrossRef]

23. Balbontín, C.; Ayala, H.; Bastías, R.M.; Tapia, G.; Ellena, M.; Torres, C.; Yuri, J.A.; Quero-García, J.; Ríos, J.C.; Silva, H. Cracking in sweet cherries: A comprehensive review from a physiological, molecular, and genomic perspective. *Chil. J. Agric. Res.* **2013**, *73*, 66–72. [CrossRef]

24. Yamaguchi, M.; Sato, I.; Watanabe, A.; Ishiguro, M. Cultivar differences in exocarp cell growth at apex, equator, stalk cavity and suture during fruit development in sweet cherry (*Prunus avium* L.). *J. Jpn. Soc. Hortic. Sci.* **2003**, *72*, 465–472. [CrossRef]

25. Barret, D.M.; Gonzalez, C. Activity of softening enzymes during cherry maturation. *J. Food Sci.* **1994**, *59*, 574–577. [CrossRef]

26. Wang, Y.; Long, L.E. Respiration and quality responses of sweet cherry to different atmospheres during cold storage and shipping. *Postharvest Biol. Technol.* **2014**, *92*, 62–69. [CrossRef]

27. Li, M.; Zhi, H.H.; Dong, Y. Applications of glycine betaine on fruit quality attributes and storage disorders of 'Lapins' and 'Regina' cherries. *HortScience* **2019**, *54*, 1540–1545. [CrossRef]

28. Li, M.; Wang, Y.; Dong, Y. Pre-harvest application of harpin β protein improves fruit on-tree and storage quality attributes of 'Lapins' and 'Regina? Sweet cherry (*Prunus avium* L.). *Sci. Hortic.* **2020**, *263*, 109–115. [CrossRef]

29. Martínez-Romero, D.; Alburquerque, N.; Valverde, J.M.; Guillén, F.; Castillo, S.; Valero, D.; Serrano, M. Postharvest sweet cherry quality and safety maintenance by *Aloe vera* treatment: A new edible coating. *Postharvest Biol. Technol.* **2006**, *39*, 93–100. [CrossRef]

30. Romano, G.S.; Cittadini, E.D.; Pugh, B.; Schouten, R. Sweet cherry quality in the horticultural production chain. *Stewart Postharvest Rev.* **2006**, *6*, 1–8.

31. Chockchaisawasdee, S.; Golding, J.B.; Vuong, Q.V.; Papoutsis, K.; Stathopoulos, C.E. Sweet cherry: Composition, postharvest preservation, processing and trends for its future use. *Trends Food Sci. Technol.* **2016**, *55*, 72–83. [CrossRef]

32. Simon, G.; Hrotkó, K.; Magyar, L. Fruit quality of sweet cherry cultivars grafted on four different rootstocks. *Acta Hortic.* **2004**, *658*, 365–370. [CrossRef]

33. Christensen, J.V. Cracking in Cherries VI. Cracking susceptibility in relation to the growth rhythm of the fruit. *Acta Agric. Scand.* **1973**, *23*, 52–54. [CrossRef]

34. Verner, L.; Blodgett, E.C. Physiological studies of the cracking of sweet cherries. University of Idaho. *Agric. Exp. Stn. Bull.* **1931**, *184*, 1–15.
35. Hamilton, R.J. *Waxes: Chemistry, Molecular Biology and Functions*; Orly Press: Edinburgh, UK, 1995.

Publisher's Note: MDPI stays neutral with regard to jurisdictional claims in published maps and institutional affiliations.

 *plants*

Article

Effects of Different Processing Treatments on Almond (*Prunus dulcis*) Bioactive Compounds, Antioxidant Activities, Fatty Acids, and Sensorial Characteristics

Ivo Oliveira [1],*, Anne S. Meyer [2], Sílvia Afonso [1], Alex Sequeira [3], Alice Vilela [4], Piebiep Goufo [1], Henrique Trindade [1] and Berta Gonçalves [1]

[1] Centre for the Research and Technology of Agro-Environmental and Biological Sciences—CITAB, Universidade de Trás-os-Montes e Alto Douro (UTAD), Quinta de Prados, 5000-801 Vila Real, Portugal; safonso@utad.pt (S.A.); pgoufo@utad.pt (P.G.); htrindad@utad.pt (H.T.); bertag@utad.pt (B.G.)

[2] Department of Biotechnology and Biomedicine, Technical University of Denmark, DTU Building 221, DK-2800 Kgs, 2800 Lyngby, Denmark; asme@dtu.dk

[3] Universidade de Trás-os-Montes e Alto Douro (UTAD), Quinta de Prados, 5000-801 Vila Real, Portugal; alexasequeira94@gmail.com

[4] Biology and Environment Department, CQ-VR, Chemistry Research Centre–Vila Real, Food and Wine Sensory Lab, University of Trás-os-Montes and Alto Douro, 5001-801 Vila Real, Portugal; avimoura@utad.pt

* Correspondence: ivo.vaz.oliveira@utad.pt

Received: 26 October 2020; Accepted: 19 November 2020; Published: 23 November 2020

Abstract: Almond is one of the most commonly consumed nuts worldwide, with health benefits associated with availability of bioactive compounds and fatty acids. Almond is often eaten raw or after some processing steps. However, the latter can positively or negatively influence chemical and sensorial attributes of almonds. This work was carried out to assess the effects of two processing treatments, namely; roasting and blanching on (i) contents of bioactive compounds, (ii) contents of fatty acids (3) antioxidant activities (4), sensorial characteristics of four neglected Portuguese almond cultivars (Casanova, Molar, Pegarinhos and Refêgo) and two foreign cultivars (Ferragnès and Glorieta). Results showed that in general, levels of bioactive compounds and antioxidant activities increased with roasting and decreased with blanching. Fatty acid profiles of raw kernels of all cultivars were generally identical although Refêgo exhibited a high content of α-linolenic acid. Following roasting and blanching, content of polyunsaturated fatty acids increased while saturated fatty acids, monounsaturated fatty acids and several health lipid indices decreased. Roasting positively affected perception of skin color and sweetness of Ferragnès and Glorieta as well as skin roughness of Molar and Pegarinhos. Blanching on the other hand led to positive changes in textural properties of Refêgo and Pegarinhos. This study reveals the nutritive benefits of consuming neglected almond cultivars in Portugal, and the novel data reported here could be of interest to growers, processing companies and consumers.

Keywords: *Prunus dulcis*; processing; sensorial analysis; fatty acids; antioxidant

1. Introduction

The consumption of nuts, including almonds, is associated with some positive health benefits such as antioxidant capacities, anticancer and antiatherogenic actions, as well as the regulation of immune and inflammatory responses [1]. The health benefits of almond are related to the availability of unsaturated fatty acids [2] and polyphenols which are known to improve human health [3]. Although almonds are mainly eaten in the raw state, sliced, or roasted, almonds can also be processed to obtain products such as marzipan, butter, milk, and oil [4]. Besides their direct consumption, almonds are added to several sweet or savory dishes and food products for special purposes, such as improve complexion or texture. Two major methods used in the processing of almonds are roasting and blanching [5]. Roasting and blanching can significantly alter the physical, chemical, and nutritive properties of the almond kernel, thus resulting in desired changes in texture, color, flavor, aroma, and taste [6]. Positive changes are particularly evident in the case of the brittle roasted almond, whose pleasant color and aroma results from the decrease in moisture level observed after roasting [7]. Modifications at a microstructural level due to processing can also lead to unwanted changes such as lipid oxidation and nutrient loss [8]. Almond processing can also result in unpleasant odor, flavor, and color [9] which influences sensorial acceptance by consumers [10]. The chemical composition of almonds is greatly influenced by pre-harvest factors such as geographical location, physiological events, cultural practices, harvesting plans (e.g., the maturity stage of the fruit at harvest) and genotype often regarded as the most important [11,12]. Thus, it can be hypothesized that the effect of processing on the composition of almond will vary based on the cultivars. In the north of Portugal, there is a large area of almond cultivation which has been assigned a Protected Designation of Origin (PDO) status: Amêndoa Douro. The cultivars grown in Amêndoa Douro have been the subject of few scientific studies [13–15]. Preliminary data indicate that these cultivars are low-yielding due to unsuitable soils and environments as well as poor management practices [13]. Currently, new orchards are being planted with foreign almond cultivars, including the French cultivar Ferragnès and the Spanish cultivar Glorieta [16]. Lack of knowledge about the nutritive value of Portuguese cultivars has led to the assumption that these cultivars are of low quality. As a result, most growers and processing companies have lost interest in the Portuguese cultivars. The overall goal of this study was to determine and compare raw, roasted, and blanched Portuguese (cultivars Casanova, Molar, Pegarinhos, and Refêgo) and foreign cultivars (Ferragnès and Glorieta) concerning their nutritive value (fatty acid composition), eating qualities (sensorial characteristics) and bioactivities (content of bioactive compounds and antioxidant activities).

2. Results and Discussion

2.1. Content of Bioactive Compounds in Almond Cultivars

Significant cultivar differences were observed in the total phenolic and total flavonoid contents of raw kernels (Table 1). The total phenolic content ranged from 0.048 in Glorieta to 0.189 mg gallic acid equivalent (GAE)/g in Pegarinhos. The above range is similar to the range reported for other Portuguese cultivars (0.09–1.63 mg GAE/g; [17], but lower than the values reported for California (1.27–2.41 mg GAE/g; [18]) almonds. The values obtained for the total phenolic content in the present study are considerably different from the values obtained in our previous work using the same cultivars harvested in 2017 [14–16]. The above observation may be attributed to climate variability among the different growing seasons. Averagely, Portuguese cultivars had higher levels of phenolics (0.10 mg GAE/g) than the foreign cultivars (0.06 mg GAE/g in Ferragnès and 0.05 mg GAE/g in Glorieta). The total flavonoid content ranged from 0.35 in Pegarinhos to 1.86 mg catechin equivalents (CE)/g in Refêgo; the above values were considerably lower than those reported for the same cultivars harvested in previous years in Portugal [14]. Besides differences linked to the genotype, a strong influence of the harvest time is already known, suggesting that the synthesis of antioxidant compounds can occur in the last stage of ripening has already been reported. In the present work, all samples were harvested at commercial maturity, but slight variations on the ripening stage might be also linked to the differences between years.

2.2. Effects of Roasting and Blanching on the Content of Bioactive Compounds in Almond

Processing had a significant effect on the content of bioactive compounds in all almond cultivars (Table 1). Analysis of the results showed higher levels of phenolics in roasted kernels relative to raw kernels as observed in previous studies [19–21]. The above observation may be related to the potential increase in phenol extractability after the destruction of cellular structures in roasted kernels. In blanched almonds, the total phenolic content was generally reduced when compared to raw almonds although no significant effect was found for Casanova, Ferragnès, and Glorieta. Since blanched samples were also roasted, it may be assumed that the observed reductions in the total phenolic content in blanched kernels are due to skin removal. Indeed, 70–100% of almond phenolics is present in the skin [22] which is removed during blanching. Similarly, blanching led to reduced levels of flavonoids in the kernels. Accordingly, the total flavonoid content was similar for all cultivars after blanching but lower than contents in both raw and roasted samples. In general, roasting had no effect on the total flavonoid content of almond cultivars except for a 58% decrease for Molar and a 72% decrease for Refêgo. The above results are in contradiction to the increase in total phenolic content observed in all cultivars after roasting and might be an indication of the degradation of more complex phenolic structures such as polymerized proanthocyanidins and glycosylated flavonoids at high temperatures [19–21].

Table 1. Total phenolic content, total flavonoid content, and antioxidant activities of raw and processed almond kernels (mean f.w., n = 3).

	Cultivar	Raw	Roasted	Blanched	p Value
	Casanova	0.09B a,b	0.49A b	0.04B b	0.001
	Ferragnès	0.06B b	0.58A b	0.05B a,b	0.001
Phenolics	Glorieta	0.05B b	1.33A a,b	0.01B c	0.000
(mg GAE/g FW)	Molar	0.09B a,b	1.16A a,b	0.02C b,c	0.007
	Pegarinhos	0.19B a B	0.88A a,b	0.08C a	0.003
	Refêgo	0.02B b	2.66A a	0.01C c	0.013
	p value	0.002	0.019	0.023	
	Casanova	0.76A b,c	1.23A a	0.09B	0.002
	Ferragnès	0.59A c	0.85A a,b	0.16B	0.033
Flavonoids	Glorieta	0.77A b,c	0.62A a,b	0.08B	0.020
(mg CE/g FW)	Molar	1.38A a,b	0.58A b	0.06C	0.000
	Pegarinhos	0.35A c	0.44A b	0.14B	0.000
	Refêgo	1.86A a	0.53A b	0.11B	0.001
	p value	0.000	0.014	0.578	
	Casanova	4.02B b	9.48A b	0.48C a,b	0.000
	Ferragnès	2.82B c	12.96A a	0.42C b	0.001
DPPH	Glorieta	1.54B d	4.86A b	0.70C a	0.016
(µg Trolox/g)	Molar	3.37A b,c	4.51A b	0.49B a,b	0.016
	Pegarinhos	6.42A a	7.60A b	0.64B a	0.004
	Refêgo	1.01A d	0.33B c	0.01C c	0.000
	p value	0.000	0.000	0.014	
	Casanova	8.81B a,b	13.96A a	0.47C b	0.000
	Ferragnès	5.07B c,d	14.23A a	0.56C a,b	0.000
ABTS	Glorieta	2.51B d,e	8.92A b	0.44C b	0.000
(µg Trolox/g)	Molar	7.27A b,c	8.05A b	0.52B a,b	0.000
	Pegarinhos	11.59A a	11.14A a,b	0.68B a	0.000
	Refêgo	1.56A e	2.64A c	0.41B b	0.009
	p value	0.000	0.000	0.004	
	Casanova	92.77A a,b	78.50B	71.41C	0.000
	Ferragnès	95.85A a,b	78.25B	70.79C	0.000
ß carotene	Glorieta	96.22A a	80.65B	67.16C	0.000
bleaching assay	Molar	88.60A b,c	76.08B	69.75C	0.005
(% inhibition)	Pegarinhos	84.89A 6c	71.33B	62.265C	0.000
	Refêgo	89.62A b,c	80.77A	67.91B	0.005
	p value	0.001	0.119	0.157	

Different small letters in front of mean within a column indicate significant differences among cultivars for the same treatment. Different capital letters in front of mean within a row indicate significant differences among treatments for the same cultivar ($p < 0.05$, ANOVA Tukey's test).

2.3. Antioxidant Activities of Almond Raw Extracts

The antioxidant potential of the six almond cultivars was evaluated using three different methods (ABTS, DPPH and β-carotene) which enabled elucidation of the mechanism of action. Raw extract data presented in Table 1 show that highest ABTS and DPPH activities were obtained with Pegarinhos, followed by Casanova. The highest percentage inhibition of lipid peroxidation (β-carotene-linoleic acid bleaching assay) on the other hand was obtained with Glorieta, Casanova and Ferragnès. Overall, the lowest antioxidant activities were recorded in Refêgo.

No correlations were found between the total flavonoid content and antioxidant activities. Nevertheless, positive correlations were found between the total phenolic content and ABTS ($R^2 = 0.7057$, $y = 50.67x + 1.7828$) and between the total phenolic content and DPPH ($R^2 = 0.7892$, $y = 0.0298x - 0.0093$). Such correlations have already been reported in previous works using the same almond cultivars [14–16] (and in studies using walnut [23].

2.4. Effects of Roasting and Blanching on the Antioxidant Activities of Almond Extracts

Even if not significant, the average DPPH and ABTS activities of all the extracts increased after roasting except for the DPPH for Refêgo which was reduced by 67%. As shown in Table 1, Ferragnès had the highest antioxidant activities after roasting followed by Casanova and Pegarinhos. Refêgo on the other hand exhibited the lowest activities. Additionally, Refêgo exhibited the lowest antioxidant activities after blanching. Blanching in general led to huge drops in DPPH and ABTS activities for all almond extracts. Indeed, the highest mean DPPH activity of 0.70 µg Trolox/g observed for Glorieta (statistically similar to Casanova, Molar, and Pegarinhos) blanched kernels was two times lower than that of the raw kernels; the highest mean ABTS activity of 0.68 µg Trolox/g observed for Pegarinhos (statistically similar to Ferragnès and Molar) blanched kernels was 17 times lower than that of the raw kernels. Results of the β-carotene-linoleic acid assay showed that roasting reduced the potential of almond samples to inhibit lipid peroxidation. The ability for almond to inhibit peroxidation was further reduced after blanching (Table 1).

Correlation analyses using data from roasted samples showed that ABTS ($R^2 = 0.708$, $y = -3.76362x + 14.27525$) and DPPH ($R^2 = 0.545$, $y = -2.48231x + 16.18607$) negatively correlated with the total phenolic content, a finding inconsistent with what was observed for raw samples. There were no significant correlations found with data from blanched samples. Negative correlations between phenolic levels and antioxidant activities were reported in few studies [24,25] linked these negative correlations to the unspecific nature of the Folin–Ciocalteu assay since the reagent targets all compounds with phenolic units with no consideration for the number of hydroxyl groups in a compound [26] or interfering compounds [27]. In the present study, negative correlations were likely due to different magnitudes of change due to roasting. For example, the 11000% increase in the total phenolic content of Refêgo after roasting was accompanied by only a 69% increase in the ABTS activity.

2.5. Fatty Acid Composition of Almond Cultivars

A total of 23 fatty acids were identified in the six almond cultivars. This number is similar to what was previously reported by Oliveira et al. (2019) [15] and Beyhan et al. (2011) [11]. Lower number of fatty acids in almonds have been reported in several studies [28,29]. These differences can be attributed mainly to genotype [30]. In the present study, 13 fatty acids exhibited high abundance and/or significant changes due to processing as shown in Table 2. In raw kernels, the most abundant fatty acid was C18:1 (elaidic + oleic acids), followed by C18:2 (linoleic + linolelaidic acids) except for Refêgo which had α-linolenic acid as the second abundant fatty acid. Other major fatty acids were α-linolenic and palmitic acids (Table 2). Most studies have reported oleic, linoleic, palmitic, and stearic acids [15,28,31–34] as the most abundant fatty acids in almond. The content of α-linolenic acid in almond is usually reported to be around 1%. Thus, the 16.21% of α-linolenic acid found in the present study for Refêgo can be considered high even when compared to the 11.00 % and 9.45% reported by Askin et al.

(2007) [35] and Oliveira et al. (2019) [15], respectively. This finding is extremely relevant given the important role that α-linolenic acid plays in the reduction of diabetes and coronary heart diseases [36]. α-linolenic acid is also the precursor of long-chain polyunsaturated fatty acids in the human body (e.g., eicosapentaenoic and docosahexaenoic acids) which have positive impacts on prothrombotic risk factors [37]. However, oils rich in linoleic and linolenic acids have reduced oxidative stability and hence are more prone to rancidity; this consequently results in the production of undesirable volatile compounds and off-flavors [38]. Oils rich in oleic acid on the other hand have better resistance to oxidation, longer shelf life and higher nutritional values [39]. Differences in the fatty acid composition of the six almond cultivars can be related to several factors such as the genotype and growing temperature which are the most important [11,40].

2.6. Effects of Roasting and Blanching on the Fatty Acid Composition of Almonds

Although not always the case, both processing treatments (roasting and blanching) had similar effects on the fatty acid profile of almonds; however, the effect of blanching was stronger than that of roasting. After roasting, the major fatty acid in all cultivars remained C18:1 (elaidic + oleic acids). However, the second most abundant fatty acid in raw kernels (C18:2) was substituted in roasted kernels with α-linolenic acid. Indeed, there was a sharp increase in the level of α-linolenic acid in all cultivars after roasting except for Refêgo. Other major fatty acids identified in roasted kernels were erucic acid (third most abundant fatty acid in Casanova, Glorieta, and Pegarinhos), nervonic acid (third most abundant fatty acid in Ferragnès) and palmitic acid (second most abundant fatty acid in Refêgo and third in Molar). C18:1 (elaidic + oleic acids) and α-linolenic acid were also the most abundant fatty acids after blanching except for Molar which had erucic acid as the second most abundant compound. Other fatty acids found in high contents after blanching were palmitic acid (third major fatty acid in Casanova), erucic acid (second major fatty acid in Molar and third in Ferragnès, Pegarinhos, and Refêgo) and nervonic acid (third major fatty acid in Molar).

Both roasting and blanching led to decreases in the contents of C18:1 (elaidic + oleic acids) and C18:2 (linoleic + linolelaidic acids) with the exception of Refêgo where an increase in the content of C18:1 was found after roasting. In fact, C18:2 (linoleic + linolelaidic acids) was undetectable in Ferragnès after roasting. Several studies [6,37,41,42] have identified both increases and decreases in C18:1 levels after subjecting almonds to different processing treatments, although the changes observed in these studies were much lower in magnitude than those found in the present study. The content of palmitic acid generally decreased in almond samples after roasting and blanching (except for an increase for Molar and Refêgo after roasting and no change for Casanova, Glorieta and Molar after blanching). Research shows that roasting lead to increased levels of palmitic acid in almond [6,43]; the effect however, depends on the roasting temperature and time. In fact, the content of palmitic acid can be gradually reduced with prolonged exposure of kernels to high temperatures [41]. The content of stearic acid in almonds tended to decrease after roasting (except for an increase for Pegarinhos and Refêgo) and to increase after blanching (except for a decrease for Casanova and Refêgo).

Table 2. Contents of the main (most abundant and/or most affected) fatty acids in almond oil extracted from raw, roasted, and blanched kernels (%, mean, n = 3). Different small letters in front of mean within a row indicate significant differences among cultivars for the same treatment. Different capital letters in front of mean within a row indicate significant differences among treatments for the same cultivar ($p < 0.05$, ANOVA Tukey's test). n.d.–not detected.

Cultivar	Casanova			Ferragnès			Glorieta			Molar			Pegarinhos			Refêgo		
	Raw	Roasted	Blanched	Raw	Roasted	Blanched	Raw	Roasted	Blanched	Raw	Roasted	Blanched	Raw	Roasted	Blanched	Raw	Roasted	Blanched
Palmitic	6.94A a	2.32B c	6.27A a	3.93A c	3.02AB c	2.71B c	2.92c	2.52c	2.96c	2.63B c	4.95A b	2.79B c	6.54A ab	2.14C c	3.57B bc	5.45B b	7.10A a	4.04C b
Stearic	0.132A a	n.d.	0.071B bc	n.d.	n.d.	0.36a	0.13a	n.d.	0.259ab	0.03B c	n.d.	0.33A a	0.11C b	0.17B a	0.23A ab	n.d.	0.09b	n.d.
Elaidic + Oleic	70.02A ab	34.05B d	68.26A a	61.51A c	48.28B b	45.97B c	77.03A a	40.86C c	54.11B b	65.08A bc	50.06B b	40.64C d	65.90A bc	26.87C e	42.67B cd	66.07B bc	82.19A a	52.17C b
Linoleic + Linolelaidic	13.54A ab	3.04B b	0.15C e	8.89A b	n.d.	1.05B a	10.34B ab	2.86C b	0.84A b	13.09A ab	0.36B c	0.54B cd	14.18A a	3.53B a	0.41C d	1.08A c	0.26C c	0.59B c
γ-Linolenic	0.17B	0.52A b	n.d.	0.104B	0.52B b	3.31A a	0.096C	0.75B a	2.31A b	0.08C	0.88B a	2.32A b	0.12C	0.48B b	1.66A c	n.d.	n.d.	0.943d
α-Linolenic	4.14C cd	33.89A b	18.89B b	4.99C bc	33.54A b	9.94B c	4.72B bc	32.73A b	6.18B cd	1.92B d	30.59A b	2.48B d	6.74C b	46.97A a	23.85B a	16.21A a	4.77B c	6.10B cd
cis-11,14-Eicosadienoic	n.d.	0.33b	n.d.	n.d.	n.d.	1.39a	n.d.	0.49B a	0.67A b	n.d.	n.d.	0.97b	n.d.	0.40B ab	0.67A b	n.d.	n.d.	0.69b
cis-8,11,14-Eicosatrienoic	n.d.	n.d.	0.10e	n.d.	n.d.	0.64c	n.d.	n.d.	1.42a	n.d.	0.27B b	0.83A b	n.d.	0.58A a	0.37B d	n.d.	n.d.	0.32d
cis-5,8,11,14,17-Eicosapentaenoic	n.d.	n.d.	0.10d	n.d.	n.d.	0.35c	n.d.	n.d.	0.79a	n.d.	0.29B b	0.62A ab	n.d.	0.48a	0.42bc	n.d.	n.d.	0.30cd
Erucic	1.97B ab	5.64A a	2.25B d	2.72B a	3.62B b	7.34A b	0.98B cd	5.73A a	5.73A c	0.51C d	3.44B b	9.35A a	1.41C bc	4.29B ab	5.81C c	2.47B a	2.59B b	6.32A bc
Nervonic	0.71B d	5.58A a	0.73B c	4.26A a	4.77A ab	3.31C b	1.06C c	4.18B bc	6.22A a	0.47C d	3.36B c	5.52A a	1.20C c	3.70B bc	5.18A a	2.31B b	0.69C d	5.29A a

Following roasting or blanching, the content of α-linolenic acid considerably increased in all cultivars, except for Refêgo, which showed an opposite response. Few studies found significant changes in the content of α-linolenic acid after processing of almonds: Ghazzawi and Al-Ismail (2017) [37] recorded a slight increase in the content of linolenic acid after frying, but not after roasting; Schlörmann et al. (2015) [42] did not detect this fatty acid in raw and roasted almond samples. In fact, the low level of linolenic acid in several cultivars is likely responsible for the lack of data regarding the effects of processing on almond lipids. In walnut, increases or decreases in the content of α-linolenic were found, depending on the roasting conditions [42,44]; In cashew, pistachio and pine nuts, these changes depended on the processing method [37]. In all cultivars, the amount of γ-linolenic acid rose or tended to increase after roasting (except for Refêgo in which it was not detected) and blanching (except for a decrease in Casanova). Additional fatty acids whose contents rose in all cultivars after roasting or blanching were erucic and nervonic acids. In the case of nervonic acid however, two exceptions were found: for a decrease in Refêgo after roasting, and a decrease in Ferragnès after blanching.

Interestingly, three long-chain polyunsaturated fatty acids are reported in this study that were undetected in raw almond kernels, and only emerge after processing. These included after roasting cis-11,14-eicosadienoic acid in. Casanova, Glorieta and Pegarinhos, cis-8,11,14-eicosatrienoic in Molar and Pegarinhos, andcis-5,8,11,14,17-eicosapentaenoic acid in Molar and Pegarinhos. With the exception of cis-11,14-eicosadienoic acid in Casanova, the three long-chain polyunsaturated fatty acids were detected in all cultivars following blanching.

2.7. Health Lipid Indices of Almond Cultivars

The content of saturated fatty acids (SFA) ranged from 4.36 % in Glorieta to 18.04 % in Molar (Table 3). SFA values were close to the threshold of 10 % [40] and similar to those reported for the same cultivars by Oliveira et al. (2019) [15]. Research shows that high-fat fruit rich in SFA are less susceptible to lipid oxidation and rapid deterioration than low-fat fruit. Similarly, Molar's ability to resist oxidation than the other cultivars can be deduced from its high SFA content. Significant differences in cultivars relative to the content of monounsaturated fatty acids (MUFA) were also observed and values ranged from 65.58 % to 78.09 %, in agreement with previous studies [15,45]. The content of polyunsaturated fatty acids (PUFA) ranged from 16.37 % in Molar to 23.81 % (Table 3). It is reported that nuts with high levels of MUFA (oleic acid in particular) are more stable and less susceptible to oxidative rancidity than those with high levels of PUFA [29,39]. In the present study, PUFA/MUFA values were similar for all cultivars and lower than 1, indicating good oil stability. The UFA/SFA ratio is another parameter related to the shelf life of food products; the lower the UFA/SFA ratio, the higher the prospective shelf life of almonds [46]. In this study, the lowest UFA/SFA value of 4.57 % was calculated for Molar, which indicates its ability to withstand long storage periods. It is important to note that despite their relevance to the oxidative stability of almonds, SFA at high levels are harmful to the cardiovascular system [29].

The atherogenicity index (AI) is defined as the relationship between the main saturated (pro-atherogenic) and unsaturated (anti-atherogenic) fatty acids. The lower the AI values, the less likely the cardiovascular risk [47]. AI values were similar for all cultivars with the exception of Molar which obtained the highest value due to the abundance of SFA (Table 3). The thrombogenicity index (TI) indicates the propensity of lipids to form clots in blood vessels and is defined as the relationship between saturated (pro-thrombogenetic) and unsaturated (MUFAs, PUFAs—n6, and PUFAs—n3; anti-thrombogenetic) fatty acids. The lower the TI values, the healthier the oil or the fat contained in a food (Ulbricht and Southgate, 1991). The "cultivar" factor significantly affected TI and the highest value of 0.16 was calculated for Molar. The hypocholesterolemic/hypercholesterolemic (h/H) ratio estimates the functional role of fatty acids in the metabolism of lipoproteins involved in the transport of plasmatic cholesterol. Thus, the h/H can be used as an indicator for the risk level of cardiovascular disease incidence [48] (Santos-Silva et al., 2002). Glorieta obtained the highest h/H value indicating its ability to contribute to improved cardiovascular health.

Table 3. Contents of the main (most abundant and/or most affected) fatty acids in almond oil extracted from raw, roasted, and blanched kernels (%, mean, n = 3) Different small letters in front of mean within a row indicate significant differences among cultivars for the same treatment. Different capital letters in front of mean within a row indicate significant differences among treatments for the same cultivar ($p < 0.05$, ANOVA Tukey's test). n.d.–not detected. SFA–saturated fatty acids; MUFA–monounsaturated fatty acids; PUFA–polyunsaturated fatty acids; AI–atherogenic index; TI–thrombogenic index; h/H–hypocholesterolemic/hypercholesterolemic index.

Cultivar	Casanova			Ferragnès			Glorieta			Molar			Pegarinhos			Refêgo		
	Raw	Roasted	Blanched	Raw	Roasted	Blanched	Raw	Roasted	Blanched	Raw	Roasted	Blanched	Raw	Roasted	Blanched	Raw	Roasted	Blanched
SFA	8.01B c	9.95A a	7.68B d	12.18A b	5.19C d	10.53B b	4.36C d	6.67B c	9.04A c	18.04A a	6.55C c	11.04B b	8.78A c	6.14B cd	8.74A cd	9.24B c	8.07C a	12.62A a
MUFA	70.73A ab	41.71B c	69.19A a	66.41A b	54.03B b	51.63B c	78.09A a	46.05C c	61.59B b	65.58A b	54.33B b	47.56C d	67.41A b	31.55C d	50.45B cd	68.58B b	82.88A a	61.32C b
PUFA	21.26B abc	48.34A b	23.14B c	20.88C abc	40.78A cd	37.84B a	17.55C bc	47.28A bc	29.36B b	16.37C c	39.12B d	41.39A a	23.81C a	62.1A a	40.80B a	22.17B ab	9.04C e	26.06A bc
PUFA/MUFA	0.30B ab	1.18A b	0.34B d	0.32B ab	0.76A c	0.733A b	0.225C b	1.03A bc	0.476B c	0.25C ab	0.72B c	0.872A a	0.35C a	1.98A a	0.81B ab	0.32B ab	0.11C d	0.425A cd
UFA/SFA	11.60A b	9.05B d	12.02A a	7.17C c	18.38A a	8.51B c	21.96A a	14.07B bc	10.08C b	4.57C d	14.38A bc	8.07B cd	10.39B b	15.28A ab	10.47B b	9.85B b	11.39A cd	6.929C d
AI	0.17A b	0.02C b	0.14A a	0.09A b	0.04B b	0.04B c	0.09A b	0.03B b	0.06A bc	0.73A a	0.06B b	0.04B c	0.14A b	0.01C b	0.05B c	0.13B b	0.43A a	0.08C b
TI	0.11A b	0.01C c	0.04B ab	0.06A c	0.02B c	0.03B b	0.05A c	0.02B c	0.04A b	0.16A a	0.03B b	0.03B b	0.09A b	0.01C c	0.03B b	0.05B c	0.09A a	0.05B a
hH	13.19B cd	34.22A ab	14.45B b	19.12b B	29.39b A	27.31A a	32.03A a	34.02A ab	27.29B a	9.08C d	17.83B c	24.82A ab	13.56C cd	39.91A a	22.38B ab	15.91B bc	12.70C c	18.46A ab

2.8. Effects of Roasting and Blanching on Almond Health Lipid Indices

With some exceptions, blanching similarly affected health lipid indices of all almond cultivars (Table 3). Casanova, Ferragnès, Glorieta, Molar, and Pegarinhos exhibited similar responses after roasting, in contrast to Refêgo. Besides Casanova and Glorieta in which increases were observed, there was a decrease in the content of SFA in almond cultivars after roasting. Furthermore, almond cultivars showed a decreasing trend in SFA content after blanching with the exception of Glorieta and Refêgo. The content of MUFA generally decreased in almond samples after roasting and blanching except for Refêgo after roasting; the inverse was true for the PUFA content. The most common responses observed after nut processing are increases in SFA [37,41,42] and decreases in PUFA and MUFA [37] levels. In the study by Valdés et al. (2015) [6], the contents of SFA and MUFA increased while that of PUFA decreased in almond after processing. In the present study, both increases and decreases were found in relation to the contents of SFA, MUFA, and PUFA. Beside the "cultivar" effect, these variations are most likely related to processing conditions and basal levels of all fatty acids, but also to the fact that minor fatty acids are not usually quantified by investigators.

Similar to the PUFA content, an increasing trend was observed in relation to the PUFA/MUFA ratio of all cultivar after processing with the exception of that of Refêgo after roasting. The lowest PUFA/MUFA value obtained after roasting was found for Refêgo. In the case of blanching, both Refêgo and Casanova achieved the lowest PUFA/MUFA. The above observation indicates that these cultivars are most likely to be less prone to oxidation after processing. After roasting, the UFA/SFA ratio increased in four cultivars (Ferragnès, Molar, Pegarinhos, and Refêgo) and decreased in two cultivars (Casanova and Glorieta). Both Casanova and Refêgo attained the lowest UFA/SFA values, as also observed with the PUFA/MUFA values. Blanching differentially affected the UFA/SFA ratio: The UFA/SFA ratio calculated for Refêgo and Glorieta decreased whereas an increase was recorded for Ferragnès and Molar. Casanova and Pegarinhos on the other hand showed no significant changes relative to the UFA/SFA ratio.

Health lipid indices were significantly affected by both processing treatments. The AI and TI values decreased or tended to decrease in all cultivars with the exception of Refêgo in which an increase was observed after roasting. The h/H ratio on the other hand generally increased after roasting and blanching although some exceptions were observed. For instance, a decrease in the h/H ratio was observed for Refêgo and Glorieta after roasting and blanching respectively. In roasted kernels, AI and TI values were similar for all cultivars with the exception of Refêgo which obtained high values. These high values of AI and TI observed for Refêgo can be associated with increased risks to cardiovascular problems; indeed, Refêgo exhibited the lowest h/H value after roasting. Overall, the highest h/H value was calculated for the oil extracted from roasted kernels of Pegarinhos.

2.9. Sensorial Analysis of Raw Almond Samples

The sensory profiles of raw kernels from all almond cultivars studied were similar to that shown in Figure 1A. Significant differences were observed relative to skin color, bitter almond flavor and bitter taste. Skin color was found to be darker in the foreign cultivars (Glorieta and Ferragnès) than the Portuguese cultivars. Bitter almond flavor and bitter taste were more associated with Molar and Pegarinhos than the rest of the cultivars. The primary taste characteristics easily identified by consumers are sweetness and astringency followed by bitterness or sourness [49]. The bitter taste derived from eating raw kernels of Molar and Pegarinhos might be due to a high concentration of benzaldehyde in these cultivars. Indeed, benzaldehyde is a chemical with a bitter taste and a low odor threshold [50]. Several Portuguese almond cultivars including Pegarinhos have high levels of benzaldehyde [14].

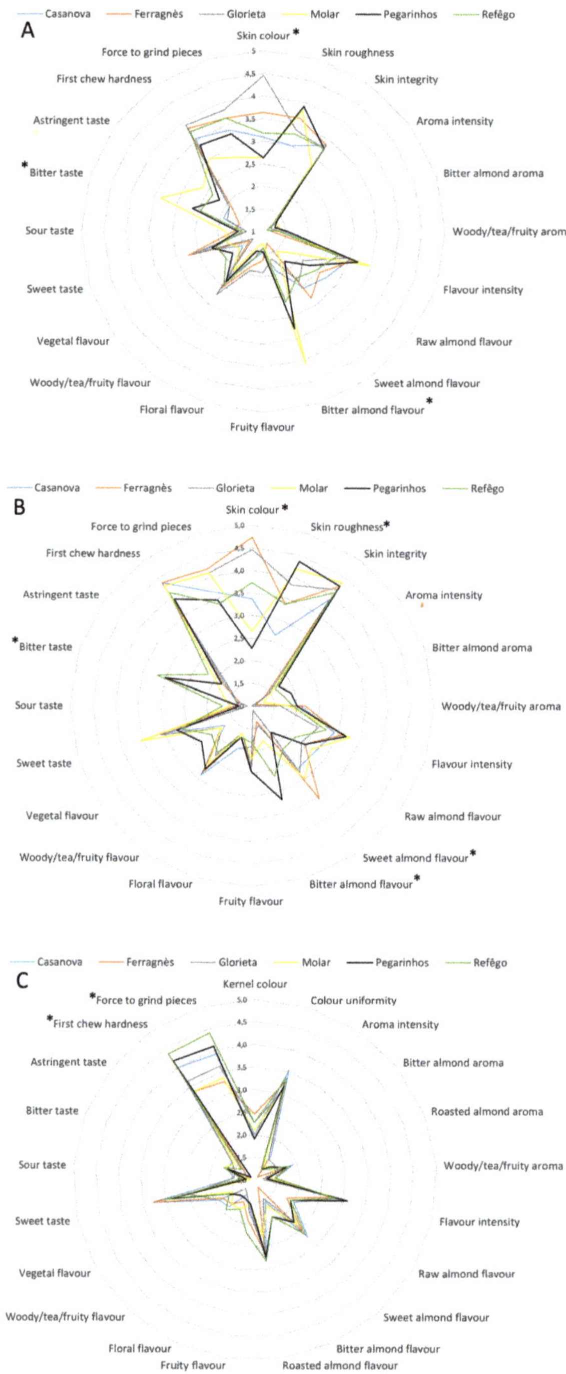

Figure 1. Spider plot of the sensory profile of raw (**A**), roasted (**B**) and blanched (**C**) almond kernles. Asterisks (*) indicate represent significant differences among cultivars $p < 0.05$, ANOVA Tukey's test.

2.10. Effects of Roasting and Blanching on Sensory Characteristics of Almonds

The main sensory attributes differentiating the six almond cultivars after roasting were found to be skin color, bitter almond flavor, bitter taste, skin roughness, and sweet almond flavor (Figure 1B). Skin color rated highest for the foreign cultivars (Ferragnès and Glorieta) and lowest for Molar and Pegarinhos. Relative to skin roughness, the skin of Molar and Pegarinhos were found to be rough whereas that of Casanova was found to be smooth. The sweet almond flavor was strongly associated with Ferragnès and barely perceived for Pegarinhos. Pegarinhos and Refêgo achieved the highest rating for bitter almond flavor and bitter taste whereas Glorieta was rated lowest for bitter almond flavor. Roasting is known to increase the levels of compounds such as pyrazines, furans, and pyrrols in food products. The above compound represents three groups of chemicals with nutty and roasted aromas [50,51] that are formed through non-enzymatic Maillard browning reactions [52]; these compounds are also known to enhance the aromas of roasted almonds [40]. In our previous work [14], pyrazines were not detected in any of the six cultivars studied; it was, however, detected in the Portuguese cultivar Amendoão. Xiao et al. (2014) [53] reported decreases in the levels of benzaldehyde and several alcohols in food products after roasting. For the six cultivars object of the present study an increase in the amount of benzaldehyde was observed after roasting. The above observation may partly account for the significant differences among cultivars relative to bitter sensorial characteristics.

In the case of blanched kernels, differences in sensorial characteristics among cultivars were only recorded for textural parameters, namely first chew hardness and force to grind pieces (Figure 1C). Both characteristics rated highest for Refêgo and Pegarinhos and lowest for Ferragnès and Molar. These changes in textural parameters are likely due to the low water content of blanched kernels which were subjected to two processing steps both involving heating. Indeed, water is known to directly influence the hardness, crispness, crunchiness and toothpack attributes of almonds [54]. Consumers usually show a preference for crispy, crunchy [54,55], and sweet almonds [55]. Hence, good ratings by the panelist for blanched kernels relative to textural properties could have positive implications for the almond industry. Indeed, hard, fracturable and crunchy almonds are used as cooking ingredients to improve food texture while moist and chewy almonds are suitable for beverages [56].

2.11. Multivariate Analyses of Bioactive Compounds, Antioxidant Activities and Fatty Acids Data

All data obtained (with the exception of those from sensorial analyses) were submitted to linear discriminant analysis (LDA) to identify parameters that could be used to differentiate both the almond cultivars and the processing methods. Seventeen variables which differed significantly between cultivars and processing treatments were selected using LDA and subjected to Principal component analysis (PCA) (Figure 2). PCA generated a 17-component model that explained 100% of the total variance in the data. The first and second components accounted for 43.398% and 17.992% (for a total of 61.318%) of the total variance respectively (Figure 2a). All cultivars and processing methods were well separated on the PCA except for roasted and blanched kernels of Refêgo and Casanova respectively. Indeed, the loading plot in Figure 2B showed that palmitic acid, MUFA, AI and TI contributed the most to the separation of Refêgo from other cultivars. Blanched kernels of Casanova on the other hand contained high levels of C18:1 (elaidic + oleic acid), palmitic acid, MUFA and AI and low levels of α-linolenic acid.

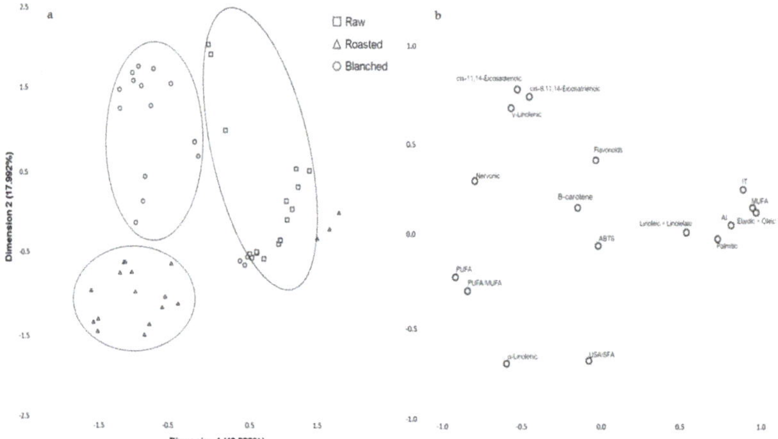

Figure 2. Principal component analysis of bioactive compounds, antioxidant activities and fatty acids data from raw, roasted and blanched almond kernels: scores plot of the first and second principal components (**a**) showing the clustering of cultivars and treatments; loadings plot (**b**) reflecting the influence of parameters on the separation of samples.

3. Materials and Methods

3.1. Almond Samples and Processing Treatments

Almond samples were obtained in 2019 from growers in the municipality of Torre de Moncorvo (Northeastern Portugal, 41°10′26″ N 7°3′0″ W), and were constituted of fruit of the Portuguese traditional cultivars (Casanova, Molar, Pegarinhos and Refêgo), French cultivar Ferragnès and Spanish cultivar Glorieta. Samples from each cultivar were harvested at commercial maturity and subdivided into three parts to obtain three replicates. Before processing, almond fruit were deshelled to obtain raw kernels which included the skin. Medium roasting of the kernels was conducted at 138 °C for 33 min [53]. Blanching was performed using the method described by Milbury et al. [18] with some modifications; kernels were immersed in boiling water for 30 s and the skins were removed by hand. Deskinned kernels were left to dry at room temperature and this was followed by roasting at 138 °C for 33 min. These treatments were selected due to their extensive use in industrial processing of almonds. Raw, roasted, and blanched kernels were finely ground to obtain a fresh flour for chemical analyses.

3.2. Bioactive Compounds: Total Phenolic and Total Flavonoid Contents

An "antioxidant extract" was prepared by vortex-mixing 40 mg of the sample with 1 mL of 70 % methanol. The mixture obtained was heated for 30 min at 70 °C, and then centrifuged (Eppendorf Centrifuge 5804 R, Hamburg, Germany) at 25,200 rcf for 15 min at 1 °C. The supernatant which constituted the "antioxidant extract" was filtered (Spartan filters 0.2 mm) into HPLC amber vials and used for the determination of the total phenolic content, total flavonoid content, and antioxidant activities. A slight modification of the methodology of Singleton and Rossi [26] (1965) was used for quantification of total phenolics as follows: 20 μL of the "antioxidant extract" was mixed with 100 μL of Folin-Ciocalteu phenol reagent (1:10 v/v/in bidistilled H2O) and 80 μL of 7.5% Na2CO3 in a 96-well microplate (Multiskan™ FC Microplate Photometer, Waltham, MA, USA). The microplate was incubated in the dark for 15 min at 45 °C. Afterward, the absorbance values against the blank value were recorded at 765 nm in a microplate reader (Multiskan GO Microplate Spectrophotometer, Thermo Scientific, Vantaa, Finland). Simultaneously, gallic acid solutions of different concentrations were analyzed and a calibration curve was constructed for quantitation purposes. The total phenolic content was expressed as mg gallic acid equivalent (GAE)/g f.w (fresh weight). The total flavonoid content was determined using the colorimetric method described in Dewanto et al. [21] with some modifications. In a 96-well microplate, 25 μL of "antioxidant extract" was combined with 100 μL of water and 10 μL of NaNO2. The microplate was placed in the dark at room temperature. After 5 min, 15 μL of 10% AlCl3 was added to the wells and the microplate was incubated again at room temperature in the dark for 6 min. Then, 50 μL of NaOH 1M and 50 μL of water were added. The absorbance values against the blank value were recorded at 510 nm. Simultaneously, catechin solutions of different concentrations were analyzed. The total flavonoid content was quantified using the calibration curve of catechin and expressed as mg catechin equivalent (CE)/g f.w.

3.3. Antioxidant Activities

The 2,2'-azino-bis(3-ethylbenzothiazoline-6-sulfonic) acid (ABTS) radical scavenging activity was evaluated in a 96-well microplate using the method of Re et al. [57]. An ABTS radical solution was prepared by mixing 7 mM of ABTS at pH 7.4 (5 mM NaH2PO4, 5 mM Na2HPO4, and 154 mM NaCl) with 2.5 mM K2S2O8. After incubation in the dark at room temperature for 16 h, the ABTS solution was diluted with ethanol until an absorbance of 0.70 ± 0.02 units at 734 nm obtained. In each microplate well, 15 μL of the "antioxidant extract" was combined with 285 μL of freshly prepared ABTS solution. The mixtures were incubated at room temperature in the dark for 10 min and absorbance values were measured at 734 nm. ABTS activity was expressed using the linear calibration curve of trolox as g trolox equivalent/g f.w. The 2,2-diphenyl-1-picrylhydrazyl (DPPH) antioxidant activity assay was performed by spectrophotometry as described by Siddhraju & Becker [58]. Briefly, 20 μL of the "antioxidant extract" and 280 μL of freshly prepared methanolic radical DPPH solution ($6 \times 10{-}5$ mol/L) were introduced into a 96-well microplate. The microplates were covered with a foil paper and left for 30 min at room temperature. The reduction in absorbance was measured at 517 nm. The DPPH activity was expressed using the linear calibration curve of trolox as g trolox equivalent/g f.w. To assess the ability of almond extracts to inhibit lipid oxidation, the β-carotene-linoleic acid bleaching assay was performed using the method described by Salleh et al. [59] with minor modifications as follows: a mixture of β-carotene and linoleic acid was prepared by adding 0.5 mg β-carotene with 1 mL chloroform (HPLC grade), 25 μL linoleic acid and 200 mg Tween 40. After complete evaporation of the chloroform under vacuum, 100 mL of water was added to the residue which was gently stirred to form a yellowish emulsion. To 50 μL of the "antioxidant abstract" were added 0.25 μL of the yellowish emulsion. After thorough mixing, the mixture was incubated in a water bath at 50 °C for 2 h followed by the measurement of absorbance values at 470 nm against a blank. Percentage inhibition (I%) of lipid peroxidation was calculated using the following equation:

$$I\% = (A\beta\text{-carotene after 2 h}/A\text{initial }\beta\text{-carotene}) \times 100; \tag{1}$$

where Aβ-carotene after 2 h is the absorbance value of β-carotene after 2 h of incubation and Ainitial β-carotene is the absorbance value of β-carotene before incubation.

3.4. Fatty Acids

Total oil was extracted from 5 g almond sample with petroleum ether in a Soxhlet apparatus [30] operating at 135 °C for 2 h. The extracted oil was used to prepare methyl esters of the corresponding fatty acids (FAME) according to EEC (1991) [60]. The resulting FAMEs were analyzed with a Shimadzu GC-2010 Plus gas chromatograph (Shimadzu, Kyoto, Japan) equipped with a flame ionization detector (FID-2010 Plus). Component peak separation was obtained on a DB-225MS capillary column (0.25 μm, 30 m × 0.25 mm i.d., Agilent Technologies, Wilmington, DE, USA). Helium was used as the carrier gas (200 kPa, constant flow, 1 mL/min). The temperature of the column was maintained for 10 min at 200 °C and then increased to 220 °C at 5°C per min. The inlet and detector temperatures were set at 270 °C. The split ratio was 5:1 and the injected volume was 1.0 µL. The constituent FAMEs were identified by comparison with standard FAME mixtures (FAME 37, Supelco, Bellefonte, PA, USA). The amount of each FAME was expressed as a weight percentage (%) of the total FAMEs represented in the chromatogram, with the assumption that no other major lipids and other substances were present in the almond oils. In several samples, the peaks for elaidic and oleic acids co-eluted, as well as the peaks for linoleic and linolelaidic acids. Thus, peak areas corresponding to these compounds were summed in the calculations. Fatty acid data was used for the calculation of several health lipid indices. The atherogenic index (AI) and the thrombogenic index (TI) were calculated using the equations proposed by Ulbricht and Southgate [47].

The hypocholesterolemic/hypercholesterolemic (h/H) index indicates the sample's potential to provide good over bad cholesterol and was calculated using the equation proposed by Santos-Silva et al. [48].

3.5. Sensorial Analysis

Raw, roasted, and blanched almond kernels were evaluated by a panel of 12 tasters from Departamento de Biologia e Ambiente (DeBA-ECVA), Universidade de Trás-os-Montes e Alto Douro in Portugal. The panelist was well trained and have participated in sensory tests from previous studies. The test took place between 4:00 pm and 5:00 pm in a room with regulated temperature and air pressure/flow. Almond kernels were presented to the tasters in white pyrex dishes. The testing environment and procedure/equipment were all following ISO 8589:2007. Three tasting sessions were carried out; the first, second and third sessions were done using raw, roasted, and blanched almonds respectively. A Quantitative Descriptive Analysis (QDA) was performed using 20 descriptors adapted from Civille et al. [49] and a structured scale from 1 (least intense) to 5 (most intense) for each descriptor (ISO 4121:2003).

3.6. Statistical Analyses

Data are presented as mean (f.w) of three replicates. Differences among means were determined by analysis of variance (ANOVA) using SPSS (Statistical Package for Social Sciences) 19.0 (IBM Corporation, New York, NY, USA). The fulfillment of ANOVA requirements, namely the normal distribution of the residuals and the homogeneity of variance was evaluated using the Shapiro–Wilk's test (n < 50) and the Levene's test, respectively. The Tukey test was used for the comparison of means which were considered different at a 5% significance level. A stepwise linear discriminant analysis (LDA) was performed to find the linear combination of parameters that best characterized raw, roasted, or blanched samples. The LDA involved the use of a combination of forward selection and backward elimination procedures for variable separation. Before selecting a new variable to be included in the model, it was ascertained that all previously selected variables were still significant which enabled identification of all possible significant variables. The Wilk's lambda test was applied for variable selection through verification of the significance of each canonical discriminant functions using the probabilities $F = 3.84$ to add and $F = 2.71$ to remove. To avoid overoptimistic data modulation, the model performance was assessed using a leave-one-out cross validation procedure. Principal component analysis (PCA) was performed by plotting all data in a multidimensional space using LDA significant variables as dimensions (factor scores). With this approach, the number of variables was reduced to a smaller number of newly derived variables (principal component or factors) that adequately summarized the original information, and highlighted underlying patterns in the data collected. Scree plots were used for retaining the most useful factors taking into consideration eigenvalues greater than one and internal consistencies of high Cronbach α values.

4. Conclusions

This study evaluated the effects of processing (roasting and blanching) on the levels of phenolics, flavonoids and fatty acids in four Portuguese (Casanova, Molar, Pegarinhos and Refêgo) and two foreign (Ferragnès and Glorieta) almond cultivars, as well as on the antioxidant activities and sensorial characteristics of kernels. Antioxidant activities and levels of bioactive compounds were generally enhanced following roasting but reduced following blanching. The increased antioxidant activity upon roasting may at first seem unexpected, as heat could be envisaged to decrease (oxidize) phenolics. However, we interpret that roasting may either induce cell wall disruption allowing better antioxidant extraction, or may alternatively cause chemical alterations that result in heat-induced production of hyper-antiradical scavenging species. This might be linked to browning reaction s compounds, known to affect total phenolics quantification and antioxidant activities. Roasting and blanching reduced the ability of all cultivars to inhibit lipid oxidation. The fatty acid profiles of all cultivars were similar although raw kernels of Refêgo exhibited a high content of α-linolenic acid. Glorieta and Molar were characterized by a low and a high content of SFA respectively. Both roasting and blanching led to significant changes in the fatty acid profiles of almonds, similar for all cultivars with the exception of Refêgo, and the effect of blanching was stronger than that of roasting. Relative to health lipid indices, Pegarinhos and Molar had better responses to roasting and blanching than the other cultivars. Very few significant differences in cultivars and treatments relative to sensorial characteristics were found. The negative features bitter taste and bitter almond flavor were noticed for raw kernels of Molar and Pegarinhos. Roasting of Ferragnès led to kernels with a strong sweet almond flavor, while blanching positively affected the textural properties of Refêgo and Pegarinhos. The findings of this study shed light on the nutritive and eating qualities of raw and processed kernels from neglected Portuguese almond cultivars, and highlight the potential use of these cultivars in various food industries.

Author Contributions: Conceptualization, I.O.; laboratory work, I.O., S.A., P.G. and A.S. writing—Original draft preparation, I.O. writing—Review and editing, I.O., A.S.M., S.A., A.V., P.G., H.T. and B.G.; supervision, B.G., A.S.M; All authors have read and agreed to the published version of the manuscript.

Funding: Ivo Oliveira is grateful to FCT—Fundação para a Ciência e a Tecnologia (FCT) for the financial support by national funds FCT/MCTES in the Post-doctoral Fellowship SFRH/BPD/111005/2015. This work is supported by National Funds by FCT—Portuguese Foundation for Science and Technology, under the project UIDB/04033/2020.

Conflicts of Interest: The authors declare no conflict of interest.

References

1. Rabadán, A.; Álvarez-Ortí, M.; Pardo, J. A comparison of the effect of genotype and weather conditions on the nutritional composition of most important commercial nuts. *Sci. Horti.* **2019**, *244*, 218–224. [CrossRef]

2. Berryman, C.; Preston, A.; Karmally, W.; Deckelbaum, R.; Kris–Etherton, P. Effects of almond consumption on the reduction of LDL–cholesterol: A discussion of potential mechanisms and future research directions. *Nutr. Rev.* **2011**, *69*, 171–185. [CrossRef] [PubMed]

3. Garrido, I.; Monagas, M.; Gómez–Cordovés, C.; Bartolomé, B. Polyphenols and antioxidant properties of almond skins: Influence of industrial processing. *J. Food Sci.* **2008**, *73*, C106–C115. [CrossRef] [PubMed]

4. Gradziel, T.M. Origin and dissemination of almond. In *Horticultural Reviews*; Janick, J., Ed.; John Wiley & Sons, Inc.: Hoboken, NJ, USA, 2011; pp. 23–81.

5. Grundy, M.; Lapsley, K.; Ellis, P. A review of the impact of processing on nutrient bioaccessibility and digestion of almonds. *Int. J. Food Sci. Technol.* **2016**, *51*, 1937–1946. [CrossRef]

6. Valdés, A.; Beltrán, A.; Karabagias, I.; Badeka, A.; Kontominas, M.; Garrigós, M. Monitoring the oxidative stability and volatiles in blanched, roasted and fried almonds under normal and accelerated storage conditions by DSC, thermogravimetric analysis and ATR-FTIR. *Eur. J. Lipid Sci. Technol.* **2015**, *117*, 1199–1213. [CrossRef]

7. Severini, C.; Gomes, T.; de Pilli, T.; Romani, S.; Massini, R. Autoxidation of packed almonds as affected by Maillard reaction volatile compounds derived from roasting. *J. Agric. Food Chem.* **2000**, *48*, 4635–4640. [CrossRef]

8. Alamprese, C.; Ratti, S.; Rossi, M. Effects of roasting conditions on hazelnut characteristics in a two-step process. *J. Food Eng.* **2009**, *95*, 272–279. [CrossRef]

9. McClements, D.; Decker, E. Lı´pidos. In *Fennema Quı́mica de los Alimentos*, 3rd ed.; Samodaran, S., Parkin, K., Fennema, O., Eds.; Editorial Acribia: Zaragoza, Spain, 2010; pp. 155–214.

10. Agila, A.; Barringer, S. Effect of roasting conditions on color and volatile profile including HMF level in sweet almonds (*Prunus dulcis*). *J. Food Sci.* **2012**, *77*, C461–C468. [CrossRef]

11. Beyhan, Ö.; Aktas, M.; Yilmaz, N.; Simsek, N.; Gerçekçioglu, R. Determination of fatty acid compositions of some important almond (*Prunus amygdalus* L.) varieties selected from Tokat province and Eagean region of Turkey. *J. Med. Plants Res.* **2011**, *5*, 4907–4911.

12. Zacheo, G.; Cappello, M.S.; Gallo, A.; Santino, A.; Cappello, A.R. Changes associated with postharvest ageing in almond seeds. LWT. *Food Sci. Tech.* **2000**, *33*, 415–423.

13. Oliveira, I.; Meyer, A.; Afonso, S.; Ribeiro, C.; Gonçalves, B. Morphological, mechanical and antioxidant properties of Portuguese almond cultivars. *J. Food Sci. Technol.* **2018**, *55*, 467–478. [CrossRef] [PubMed]

14. Oliveira, I.; Malheiro, R.; Meyer, A.S.; Pereira, J.A.; Gonçalves, B. Application of chemometric tools for the comparison of volatile profile from raw and roasted regional and foreign almond cultivars (*Prunus dulcis*). *J. Food Sci. Technol.* **2019**, *56*, 3764–3776. [CrossRef] [PubMed]

15. Oliveira, I.; Meyer, A.S.; Afonso, S.; Aires, A.; Goufo, P.; Trindade, H.; Gonçalves, B. Phenolic and fatty acid profiles, α–tocopherol and sucrose contents, and antioxidant capacities of understudied Portuguese almond cultivars. *J. Food Biochem.* **2019**, *43*, e12887. [CrossRef] [PubMed]

16. Cabrita, L.; Apostolova, H.; Neves, A.; Leitão, J. Genetic diversity assessment of the almond (*Prunus dulcis* (Mill.) D.A. Webb) traditional germplasm of Algarve, Portugal, using molecular markers. *Plant Genet. Resour.* **2014**, *12*, 164–167. [CrossRef]

17. Barreira, J.; Ferreira, I.; Oliveira, M.; Pereira, J. Antioxidant activity and bioactive compounds of ten Portuguese regional and commercial almond cultivars. *Food Chem. Toxicol.* **2008**, *46*, 2230–2235. [CrossRef]

18. Milbury, P.; Chen, C.; Dolnikowski, G.; Blumberg, J. Determination of flavonoids and phenolics and their distribution in almonds. *J. Agric. Food Chem.* **2006**, *54*, 5027–5033. [CrossRef]

19. Yu, J.; Ahmedna, M.; Goktepe, I.; Dai, J. Peanut skin procyanidins: Composition and antioxidant activities as affected by processing. *J. Food Compos. Anal.* **2006**, *19*, 364–371. [CrossRef]

20. Rohn, S.; Buchner, N.; Driemel, G.; Rauser, M.; Kroh, L.W. Thermal degradation of onion quercetin glucosides under roasting conditions. *J. Agric. Food Chem.* **2007**, *55*, 1568–1573. [CrossRef]

21. Dewanto, V.; Wu, X.; Liu, R. Processed sweet corn has higher antioxidant activity. *J. Agric. Food Chem.* **2002**, *50*, 4959–4964. [CrossRef]

22. Bolling, B.; McKay, D.; Blumberg, J. The phytochemical composition and antioxidant actions of tree nuts. *Asia Pacific J. Clin. Nutr.* **2010**, *19*, 117.

23. Pereira, J.; Oliveira, I.; Sousa, A.; Ferreira, I.; Bento, A.; Estevinho, L. Bioactive properties and chemical composition of six walnut (*Juglans regia* L.) cultivars. *Food Chem. Tox.* **2008**, *46*, 2103–2111. [CrossRef] [PubMed]

24. Jesus, F.; Gonçalves, A.; Alves, G.; Silva, L. Exploring the phenolic profile, antioxidant, antidiabetic and antihemolytic potential of *Prunus avium*, vegetal parts. *Food Res. Int.* **2019**, *116*, 600–610. [CrossRef] [PubMed]

25. Tawaha, K.; Alali, F.Q.; Gharaibeh, M.; Mohammad, M.; El–Elimat, T. Antioxidant activity and total phenolic content of selected Jordanian plant species. *Food Chem.* **2007**, *104*, 1372–1378. [CrossRef]

26. Singleton, V.; Rossi, J. Colorimetry of total phenolics with phosphomolybdic–phosphotungstic acid reagents. *Am. J. Enology Vitic.* **1965**, *16*, 144–158.

27. Prior, R.L.; Wu, X.; Schaich, K. Standardized methods for the determination of antioxidant capacity and phenolics in foods and dietary supplements. *J. Agric. Food Chem.* **2005**, *53*, 4290–4302. [CrossRef] [PubMed]

28. Čolić, S.; Akšić, M.; Lazarević, K.; Zec, G.; Gašić, U.; Zagorac, D.; Natić, M. Fatty acid and phenolic profiles of almond grown in Serbia. *Food Chem.* **2017**, *234*, 455–463. [CrossRef] [PubMed]

29. Kodad, O.; Socias i Company R. Variability of oil content and of major fatty acid composition in almond (*Prunus amygdalus* Batsch) and its relationship with kernel quality. *J. Agric. Food Chem.* **2008**, *56*, 4096–4101. [CrossRef]

30. Kodad, O.; Estopanán, G.; Juan, T.; Alonso, J.M.; Espiau, M.T.; Socias i Company, R. Oil content, fatty acid composition and tocopherol concentration in the Spanish almond genebank collection. *Sci. Hort.* **2014**, *177*, 99–107. [CrossRef]

31. Barreira, J.; Casal, S.; Ferreira, I.; Peres, A.; Pereira, J.; Oliveira, M. Supervised chemical pattern recognition in almond (*Prunus dulcis*) portuguese PDO cultivars: PCA-and LDA-based triennial study. *J. Agric. Food Chem.* **2012**, *60*, 9697–9704. [CrossRef]

32. Rabadán, A.; Álvarez-Ortí, M.; Gómez, R.; Pardo–Giménez, A.; Pardo, J. Suitability of Spanish almond cultivars for the industrial production of almond oil and defatted flour. *Scientia Hort.* **2017**, *225*, 539–546. [CrossRef]

33. Zamany, A.; Samadi, G.; Kim, D.; Keum, Y.; Saini, R. Comparative Study of Tocopherol Contents and Fatty Acids Composition in Twenty Almond Cultivars of Afghanistan. *J. Am. Oil Chem. Soc.* **2017**, *94*, 805–817. [CrossRef]

34. Zhu, Y.; Wilkinson, K.; Wirthensohn, M. Lipophilic antioxidant content of almonds (*Prunus dulcis*): A regional and varietal study. *J. Food Compos. Anal.* **2015**, *39*, 120–127. [CrossRef]

35. Askin, M.; Balta, M.; Tekintas, F.; Kazankaya, A.; Balta, F. Fatty acid composition affected by kernel weight in almond [*Prunus dulcis* (Mill.) DA Webb.] genetic resources. *J. Food Compos. Anal.* **2007**, *20*, 7–12. [CrossRef]

36. Virtanen, J.; Mursu, J.; Voutilainen, S.; Uusitupa, M.; Tuomainen, T. Serum omega-3 polyunsaturated fatty acids and risk of incident type 2 diabetes in men: The Kuopio Ischemic Heart Disease Risk Factor study. *Diabetes Care* **2014**, *37*, 189–196. [CrossRef]

37. Ghazzawi, H.A.; Al-Ismail, K. A comprehensive study on the effect of roasting and frying on fatty acids profiles and antioxidant capacity of almonds, pine, cashew, and pistachio. *J. Food Quality* **2017**, *2017*, 9038257. [CrossRef]

38. Buranasompob, A.; Tang, J.; Powers, J.R.; Reyes, J.; Clark, S.; Swanson, B.G. Lipoxygenase activity in walnuts and almonds. LWT. *Food Sci. Technol.* **2007**, *40*, 893–899.

39. Gama, T.; Wallace, H.; Trueman, S.; Hosseini–Bai, S. Quality and shelf life of tree nuts: A review. *Sci. Hort.* **2018**, *242*, 116–126. [CrossRef]

40. Yada, S.; Lapsley, K.; Huang, G. A review of composition studies of cultivated almonds: Macronutrients and micronutrients. *J. Food Compos. Analysis* **2011**, *24*, 469–480. [CrossRef]

41. Lin, J.T.; Liu, S.C.; Hu, C.C.; Shyu, Y.S.; Hsu, C.Y.; Yang, D.J. Effects of roasting temperature and duration on fatty acid composition, phenolic composition, Maillard reaction degree and antioxidant attribute of almond (*Prunus dulcis*) kernel. *Food Chem.* **2016**, *190*, 520–528. [CrossRef]

42. Schlörmann, W.; Birringer, M.; Böhm, V.; Löber, K.; Jahreis, G.; Lorkowski, S.; Müller, A.; Schöne, F.; Glei, M. Influence of roasting conditions on health–related compounds in different nuts. *Food Chem.* **2015**, *180*, 77–85. [CrossRef]

43. Hojjati, M.; Lipan, L.; Carbonell–Barrachina, Á.A. Effect of roasting on physicochemical properties of wild almonds (*Amygdalus scoparia*). *J. Am. Oil Chem. Soc.* **2016**, *93*, 1211–1220. [CrossRef]

44. Ghafoor, K.; Al-Juhaimi, F.; Geçgel, Ü.; Babiker, E.E.; Özcan, M.M. Influence of Roasting on Oil Content, Bioactive Components of Different Walnut Kernel. *J. Oleo Sci.* **2020**, *69*, 423–428. [CrossRef] [PubMed]

45. Čolić, S.; Zec, G.; Natić, M.; Fotirić–Akšić, M. Almond (*Prunus dulcis*) oil. In *Fruit Oils: Chemistry and Functionality*; Ramadan, M., Ed.; Springer: Berlin/Heidelberg, Germany, 2019; pp. 149–180.

46. Fokou, E.; Achul, M.; Kanscil, G.; Ponkal, R.; Fotso, M.; Techouanguep, F. Chemical properties of some cucurbitacease oils from Cameroon. *Pak. J. Nut.* **2009**, *8*, 1325–1334. [CrossRef]

47. Ulbricht, T.L.V.; Southgate, D.A.T. Coronary heart disease: Seven dietary factors. *Lancet* **1991**, *338*, 985–992. [CrossRef]

48. Santos Silva, J.; Bessa, R.J.B.; Santos–Silva, F. Effect of genotype, feeding system and slaughter weight on the quality of light lambs: II. Fatty acid composition of meat. *Livest. Prod. Sci.* **2002**, *77*, 187–194. [CrossRef]

49. Civille, G.V.; Lapsley, K.; Huang, G.; Yada, S.; Seltsam, J. Development of an almond lexicon to assess the sensory properties of almond varieties. *J. Sens. Stud.* **2010**, *25*, 146–162. [CrossRef]

50. Franklin, L.M.; Mitchell, A.E. Review of the Sensory and Chemical Characteristics of Almond (*Prunus dulcis*) Flavor. *J. Agric. Food Chem.* **2019**, *67*, 2743–2753. [CrossRef]

51. Erten, E.; Cadwallader, K. Identification of predominant aroma components of raw, dry roasted and oil roasted almonds. *Food Chem.* **2017**, *217*, 244–253. [CrossRef]

52. Vázquez–Araújo, L.; Verdú, A.; Navarro, P.; Martínez–Sánchez, F.; Carbonell–Barrachina, Á.A. Changes in volatile compounds and sensory quality during toasting of Spanish almonds. *Int. J. Food Sci. Technol.* **2009**, *44*, 2225–2233. [CrossRef]

53. Xiao, L.; Lee, J.; Zhang, G.; Ebeler, S.E.; Wickramasinghe, N.; Seiber, J.; Mitchell, A.E. HS-SPME GC/MS characterization of volatiles in raw and dry-roasted almonds (*Prunus dulcis*). *Food Chem.* **2014**, *151*, 31–39. [CrossRef]

54. Vickers, Z.; Peck, A.; Labuza, T.; Huang, G. Impact of almond form and moisture content on texture attributes and acceptability. *J. Food Sci.* **2014**, *79*, S1399–S1406. [CrossRef] [PubMed]
55. Cheely, A.N.; Pegg, R.B.; Kerr, W.L.; Swanson, R.B.; Huang, G.; Parrish, D.R.; Kerrihard, A.L. Modeling sensory and instrumental texture changes of dry-roasted almonds under different storage conditions. *LWT* **2018**, *91*, 498–504. [CrossRef]
56. King, E.S.; Chapman, D.M.; Luo, K.; Ferris, S.; Huang, G.; Mitchell, A.E. Defining the Sensory Profiles of Raw Almond (*Prunus dulcis*) Varieties and the Contribution of Key Chemical Compounds and Physical Properties. *J. Agric. Food Chem.* **2019**, *67*, 3229–3241. [CrossRef]
57. Re, R.; Pellegrini, N.; Proteggente, A.; Pannala, A.; Yang, M.; Rice–Evans, C. Antioxidant activity applying an improved ABTS radical cation decolorization assay. *Free Rad. Biol. Med.* **1999**, *26*, 1231–1237. [CrossRef]
58. Siddhraju, P.; Becker, K. Antioxidant properties of various solvents extracts of total phenolic constituents from three different agroclimatic origins of drumstick tree (Moringa oleifera Lam) leaves. *J. Agric. Food Chem.* **2003**, *51*, 2144–2155. [CrossRef]
59. Salleh, W.; Ahmad, F.; Yen, K.; Sirat, H. Chemical compositions, antioxidant and antimicrobial activity of the essential oils of *Piper officinarum* (Piperaceae). *Nat. Prod. Commun.* **2003**, *7*, 1659–1662. [CrossRef]
60. EEC European Commision. *EU Official Method (EEC Regulation 2568/91, Corresponding to AOCS Method Ch 2-91)*; European Commision: Brussels, Belgium, 1991.

Publisher's Note: MDPI stays neutral with regard to jurisdictional claims in published maps and institutional affiliations.

plants

Article

Effect of Ozone Treatment on the Quality of Sea Buckthorn (*Hippophae rhamnoides* L.)

Anita Zapałowska [1,*], Natalia Matłok [2], Miłosz Zardzewiały [2], Tomasz Piechowiak [3] and Maciej Balawejder [3]

1   Department of Land Management and Environmental Protection, University of Rzeszów, St. Ćwiklińskiej 1a, 35-601 Rzeszów, Poland
2   Department of Food and Agriculture Production Engineering, University of Rzeszow, St. Zelwerowicza 4, 35-601 Rzeszów, Poland; nmatlok@ur.edu.pl (N.M.); mzardzewialy@ur.edu.pl (M.Z.)
3   Department of Chemistry and Food Toxicology, University of Rzeszow, St. Ćwiklińskiej 1a, 35-601 Rzeszów, Poland; tom_piech@interia.pl (T.P.); maciejb@ur.edu.pl (M.B.)
*   Correspondence: azapalowska@ur.edu.pl

Abstract: The aim of this research was to show the effect of the ozonation process on the quality of sea buckthorn (*Hippophae rhamnoides* L.). The quality of the ozonated berries of sea buckthorn was assessed. Prior to and after the ozone treatment, a number of parameters, including the mechanical properties, moisture content, microbial load, content of bioactive compounds, and composition of volatile compounds, were determined. The influence of the ozonation process on the composition of volatile compounds and mechanical properties was demonstrated. The ozonation had negligible impact on the weight and moisture of the samples immediately following the treatment. Significant differences in water content were recorded after 7 days of storage. It was shown that the highest dose of ozone (concentration and process time) amounting to 100 ppm for 30 min significantly reduced the water loss. The microbiological analyses showed the effect of ozone on the total count of aerobic bacteria, yeast, and mold. The applied process conditions resulted in the reduction of the number of aerobic bacteria colonies by 3 log cfu g^{-1} compared to the control (non-ozonated) sample, whereas the number of yeast and mold colonies decreased by 1 log cfu g^{-1} after the application of 100 ppm ozone gas for 30 min. As a consequence, ozone treatment enhanced the plant quality and extended plant's storage life.

Keywords: ozone; sea buckthorn; mechanical properties; microbial load; bioactive components

Citation: Zapałowska, A.; Matłok, N.; Zardzewiały, M.; Piechowiak, T.; Balawejder, M. Effect of Ozone Treatment on the Quality of Sea Buckthorn (*Hippophae rhamnoides* L.). *Plants* **2021**, *10*, 847. https://doi.org/10.3390/plants10050847

Academic Editor: Ivo Vaz de Oliveira

Received: 30 March 2021
Accepted: 19 April 2021
Published: 22 April 2021

Publisher's Note: MDPI stays neutral with regard to jurisdictional claims in published maps and institutional affiliations.

1. Introduction

The sea buckthorn (*Hippophae rhamnoides* L.) is a spiny, deciduous shrub which belongs to the oleaster family (*Elaeagnaceae Juss.*) The fruits are rich in a variety of phytochemicals with physiological properties, such as lipids, carotenoids, ascorbic acid, tocopherols, and flavonoids [1,2]. A characteristic feature of *Hippophae rhamnoides* L. is intense orange berries that densely overgrow the shoots. A peculiar smell is caused by the presence of chemical compounds such as alcohols, aldehydes, ketones, and terpenes [3]. Sea buckthorn contains carotenoids, flavonoids, phospholipids, tannins, vitamins, and macro- and microelements [4]. Due to its abundance of the biologically active compounds, the sea buckthorn is widely used as a human health-promoter. This plant is a valuable material, particularly in pharmaceutical, cosmetic, and food industries [5]. Ozone, a highly reactive triatomic form of oxygen, is a powerful agent able to degrade organic substances. Gaseous ozone, due to its properties, has been approved by the Food and Drug Administration for direct contact with food, and can have a significant impact on the quality of the fruit. Numerous studies have confirmed its beneficial effect on the quality of products, evidenced in the extension of their shelf life [6,7]. Due to ozonation, select parameters of plant materials, such as antioxidant potential, polyphenol content, or enzymatic activity, are modified. Many studies have confirmed that ozone treatment inhibits the growth of microorganisms responsible for fruit spoilage, and reduces the loss of nutritional and sensory value of fruit during storage [8–17]. These studies have shown that the use of ozonated water is only effective in combination with blanching [8,13]. Using ozone gas is more effective, but it is necessary to change the atmosphere with other gases [9,12]. A plant composition change may occur as a result of the ozonation process [14,15]. This gas, as confirmed by numerous studies undertaken by the Environmental Protection Agency, is considerably safer than chlorine or other sanitizers [18]. In contact with solid materials, it decomposes into molecular oxygen and does not cause additional fouling.

The studies undertaken to date on the utilization of ozone treatment in fruit storage have principally concerned the effect of ozonation on the level of microbial contamination, biological activity of the fruit, and fruits' sensory characteristics [14,15,19]. However, the available data on changes in the bioactive compounds of ozonated fruits are variable and difficult to compare because they are related to different factors, forms, and concentrations of ozone, in addition to different fruits and environmental conditions in which they are produced [20].

No reports have been published about the effect of ozone treatment on the quality of sea buckthorn (*Hippophae rhamnoides* L.). To obtain factual information, we conducted a survey, the results of which are shown below. The aim of the research was to show the effect of the ozonation process on the quality of sea buckthorn (*Hippophae rhamnoides* L.).

2. Results and Discussion

2.1. Moisture Content

Sea buckthorn berries consist of more than 80% water, which fluctuates greatly during storage. The moisture content depends largely on the storage conditions. Fruit stored in the same conditions should therefore maintain similar quality and strength parameters, which are dependent on the moisture content. The fruit shelf life can be influenced by the metabolic system being activated. This can be performed by adjusting the storage to proper conditions (appropriate storage temperature) or by making use of both biotic and abiotic factors (in the form of nutrition, cutting, light, water, etc.). One such factor (abiotic) is the ozone gaseous form, which may be used to slow the ripening process. The findings of our experiment show the effects on the ozone concentration and its duration on the moisture content.

The applied ozone treatment had a direct influence on the sea buckthorn shelf life. The observed changes in moisture content depended on ozone concentrations and exposure time (Figure 1). The largest water loss, which amounted to 5.20%, was observed in the control sample (0 ppm, 0 min) on the 7th day of storage, and the lowest water loss of 3.65% was recorded in the samples subjected to ozone treatment at a rate of 100 ppm, for the duration of 30 min. The performed statistical analysis showed a significant influence of ozone concentration and ozone exposure time ($LSD_{0.05}$ (least significant difference) 0.47) on the moisture content of the sea buckhorn berries. This corresponds to the observation of the lowest microbial load for fruit subjected to ozonation under these conditions. It is most likely that this was caused by a decrease in the activity of microorganisms on the surface of this fruit. Ozone, as an agent that disinfects fruit surfaces, affects the activity of microorganisms and the storage process. Microbial inactivation by ozone, mainly due to the rupture of its cellular membranes, prevents damage of the fruit surface, leading to a reduction in water loss and increasing the membranes' mechanical strength. The influence of the ozonation process on the reduction of water loss in stored plant material has been also confirmed in other studies. Zardzewiały et al. [21] observed a similar effect of ozone treatment applied to rhubarb petioles. The authors recorded the lowest water loss for the exposure of 30 min with an ozone concentration of 100 ppm. The influence of the ozonation process on improving quality parameters and reducing water loss in stored ground cucumbers was also confirmed by Gorzelany et al. [22]. Antos et al. [23], using ozone on stored apples, observed a similar relationship.

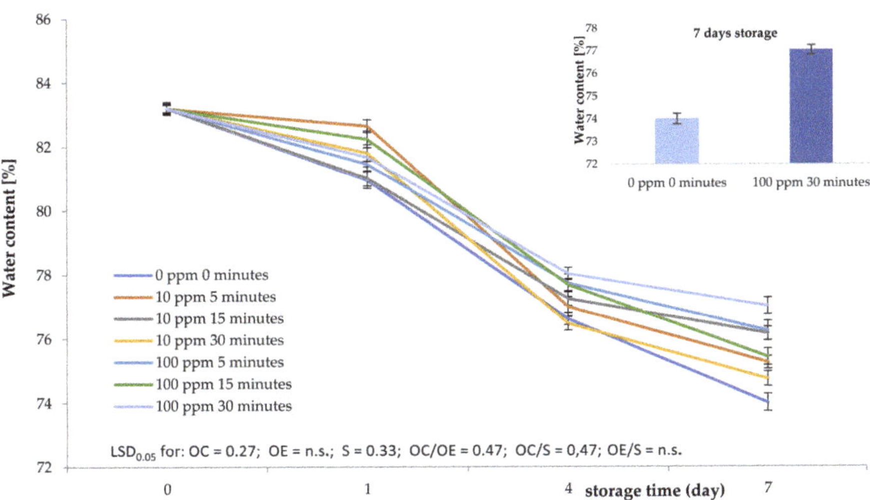

LSD$_{0.05}$ for: OC = 0.27; OE = n.s.; S = 0.33; OC/OE = 0.47; OC/S = 0,47; OE/S = n.s.

Figure 1. Effect of the ozone treatment concentration (OC), ozone exposure time (OE), and storage time (S) on moisture content of *Hippophae rhamnoides* L. fruit.

2.2. The Mechanical Properties

The mechanical parameters determine the quality of the raw material plants, which is important in terms of the storage ability. During the storage time, all ozonated sea buckthorn berries showed a higher resistance to the destructive force (Fmax) compared to non-ozonated fruits (Table 1). The mean values of indentation force in the control sample amounted to 6.84 N. To correlate moisture content with selected mechanical properties, these parameters were measured at the same time of storage (on the 1st, 4th, and 7th days). The results of the test show that water loss from cellular structures directly affects the parameters of the mechanical properties. The value of destructive force (Fmax) subjected to these conditions after the 1st day of storage was 10.1% higher compared to non-ozonated fruits. Significantly better results were recorded on the 4th and the 7th day of storage, when the value of Fmax was 36.3% and 58.6% higher, respectively, compared to the control sample (non-ozonated). The result shows that, as in the case of moisture content, the highest measured parameter was recorded on the 7th day following treatment at the ozone rate of 100 ppm for 30 min (Fmax = 9.69 N). This was likely the result of less damage to the ozonated surface of the fruit caused by microbes. Increasing the resistance of ozonated sea buckthorn berries to the destructive force has been associated with the alteration of the fruit surface due to ozone gas. Similar relationships were observed on the recorded deformation of sea buckthorn berries to the point of destruction (dl to Fmax) and destructive force (Fmax/Lmax). These parameters were significantly higher in the case of fruit exposed to gaseous ozone. The performed statistical analysis showed a significant influence of ozone treatment concentration, ozone exposure time, and storage time on the mechanical parameters (Table 1). The impact of the ozonation process on selected mechanical parameters of the plant material was also confirmed in other studies. Zardzewiały et al. [21] observed a similar effect of the ozonation process on the mechanical resistance of rhubarb in the compression test. The best results were found in the case of the material subjected to the process of ozone treatment for 15 min with a concentration of 10 ppm. Gorzelany et al. [22], using gaseous ozone for post-harvest ground cucumber, obtained a material with improved mechanical and sensory properties compared to the control. The increase in the parameters of the destructive power of ozonated apples was noted by Antos et al. [23]. Using cyclic gaseous ozone on apples, the authors observed a similar relationship between the conditions of the ozonation process and the mechanical parameters of the stored fruit.

Table 1. Effect of ozone treatment concentration (OC), ozone exposure time (OE), and storage time (S) on the mechanical parameters of *Hippophae rhamnoides* L. fruit.

Parameter	Treat		Storage [Days]		
	Ozone Concentration [ppm]	Ozone Exposure [min]	1	4	7
F_{max} [N]	0	0	6.82	6.20	6.11
	10	5	6.06 [de]	6.26 [de]	5.86 [e]
		15	7.31 [bcde]	8.05 [abc]	8.17 [ab]
		30	7.45 [bcde]	6.99 [cde]	8.30 [a]
	100	5	6.46 [cde]	6.28 [e]	6.57 [cde]
		15	7.69 [bcd]	7.36 [abcde]	7.71 [bd]
		30	7.53 [bc]	8.45 [b]	9.69 [a]
	LSD$_{0.05}$ for: OC = 0.13; OE = 0.23; S = 0.17; OC/OE = 0.30; OC/S = n.s; * OE/S = 0.30				
dl to F_{max} [mm]	0	0	2.32	2.30	2.06
	10	5	2.65 [abc]	2.37 [abcd]	2.44 [ab]
		15	2.30 [abcd]	2.34 [cd]	2.02 [d]
		30	2.80 [ab]	2.41 [abcd]	2.60 [abc]
	100	5	2.57 [abc]	2.58 [abc]	2.26 [bcd]
		15	2.82 [ab]	2.59 [abc]	2.42 [abc]
		30	2.69 [abc]	2.87 [a]	2.45 [abc]
	LSD$_{0.05}$ for: OC = 0.02; OE = 0.04; S = 0.05; OC/OE = 0.05, OC/S = 0.06; OE/S = n.s				
W to F_{max} [J]	0	0	5.22	4.67	4.32
	10	5	6.48 [abc]	6.51 [abc]	6.22 [abc]
		15	4.80 [bcd]	4.48 [cd]	3.91 [d]
		30	7.29 [a]	6.17 [abc]	7.05 [a]
	100	5	5.76 [abcd]	5.39 [abcd]	5.63 [abcd]
		15	6.88 [ab]	6.21 [abc]	6.09 [abc]
		30	6.78 [abc]	6.60 [a]	6.61 [abc]
	LSD$_{0.05}$ for: OC = 0.10; OE = 0.24; S = 0.24; OC/OE = 0.29; OC/S = n.s; OE/S = 0.41				
F_{max}/L_{max} [N/mm^2]	0	0	11.93	16.65	15.29
	10	5	11.30 [i]	26.06 [ab]	18.75 [def]
		15	13.23 [ghi]	30.07 [a]	18.63 [defg]
		30	9.65 [i]	18.48 [defg]	19.86 [cdeg]
	100	5	11.97 [i]	12.20 [hi]	13.17 [fghi]
		15	11.97 [hi]	17.17 [efgh]	23.72 [bc]
		30	14.38 [efghi]	23.09 [bc]	19.50 [def]
	LSD$_{0.05}$ for: OC = 0.55; OE = 0.62; S = 0.68; OC/OE = 0.90; OC/S = 0.96; OE/S = 1.15				

Note: * n.s.—non-significant difference; Fmax—destructive force (N), dL to Fmax—deflection at the moment of destruction (mm), W to Fmax—energy required for destruction (J), Fmax/Lmax—destructive force measurement (N/mm^2). [a,b,c,d,e,f,g,h,i]—statistically significant differences for the effect: ozone concentration (ppm) × ozone exposure (min) × storage (days).

2.3. Content of Bioactive Compounds

The results of the present study show increased total polyphenol content after 24 h (Figure 2A). This was determined by the duration of the process and the ozone concentration. The best results were observed when the plant was exposed to ozone for 5 min with an ozone concentration of 10 ppm. In this case, the change in the total polyphenol content was 15.9% higher compared to the control sample (non-ozonated fruit). When higher ozone concentrations of 100 ppm were used, the best results were achieved after 15 min. In this case, an increase in total polyphenol growth of more than 12.8% was found in the ozonated plant. A strong relationship with ozone dose was observed in the case of antioxidant potential (Figure 2B). The antioxidant capacity of the fruit 24 h after the ozone treatment of 100 ppm for 15 min was higher by 11.5% compared to the control sample (non-ozonated fruit). A quantity of 100 ppm of ozone significantly disturbed the fruit balance, which caused oscillation of the content of components, particularly polyphenols. Such oscillations were not observed for the antioxidant potential. Statistical analysis showed the impact of the ozone concentration and ozone exposure time on the total content of polyphenols in sea buckthorn berries ($LSD_{0.05}$ = 53.07), and their antioxidant properties ($LSD_{0.05}$ = 145.77). In addition, the antioxidant potential depended significantly on the time of the ozonation ($LSD_{0.05}$ = 103.84) and ozone concentrations ($LSD_{0.05}$ = 133.95). The increase in the concentration of secondary metabolites in ozonated sea buckthorn berries after 24 h of storage can be attributed to metabolic changes due to the occurrence of oxidative stress caused by ozone. The effect of ozone on the total polyphenol content and antioxidant potential of the ozonated fruit has been confirmed by other scientists. Rodoni et al. [24] recorded a 50% increase in the total polyphenol content compared to the non-ozonated fruit, on the sixth day after the ozone treatment, while examining the effect of 10 min ozone exposure at a dose of 10 μL L^{-1} on tomato fruits. The increase in total polyphenol content was also observed in the case of the ozonated fruits of *Carica papaya* L. [25]. Alothman et al. [26], using ozone at a flow of 8 $\pm$ 0.2 mL s^{-1} for 0, 10, 20 and 30 min to ozonate pineapple and banana fruits, found an increase in the total polyphenols in ozonated fruit. The best results were achieved when the ozonation process was conducted for 30 min. Furthermore, the total polyphenol content of the ozonated pineapple and bananas was higher by 9.5% and 23.5%, respectively, compared to non-treated fruit. The process of ozonation under properly selected conditions may influence an increase in ascorbic acid content in the plant material [21,27]. Ozone treatment on the buckthorn berries resulted in changes in ascorbic acid content after 24 h of the conducted process (Figure 2C). However, this impact was different depending on the concentration of the ozone used and the duration of the process. The best results were obtained for 30 min using ozone at a concentration of 10 ppm. At that time, the ascorbic acid content in the tested material was higher by 12.7% compared to the control sample (non-ozonated fruit). However, when the ozone concentration was increased to 100 ppm (30 min), a significant decrease in the ascorbic acid content was observed. It is likely that the observed effects of the oxidation of ascorbic acid was the result of the activation of ascorbate oxidase being maintained by stress related to the action of ozone gas. This enzyme degrades ascorbic acid to dehydroascorbic acid [28]. The degradation of ascorbic acid under the influence of inappropriate selection of process conditions for the type of ozonated plant material was also confirmed by others. Alothman et al. [26], treating guava fruit with ozone with a flow rate of 8 $\pm$ 0.2 mL s^{-1} for 30 min, noted a decrease in ascorbic acid by 67.1% compared to fruit not exposed to this gas (control test). The increase in the content of ascorbic acid in the plant material subjected to the ozone treatment in properly selected conditions (concentration and time of ozone exposure) was also confirmed in other studies. Piechowiak et al. [15], while ozonating raspberry fruits with ozone at a concentration of 8–10 ppm for 30 min, observed an increase in ascorbic acid content in fruits subjected to the ozonation process. The performed statistical analysis showed a significant influence of the interaction of ozone concentration and ozonation time on the ascorbic acid content in sea buckthorn berries ($LSD_{0.05}$ = 12.46).

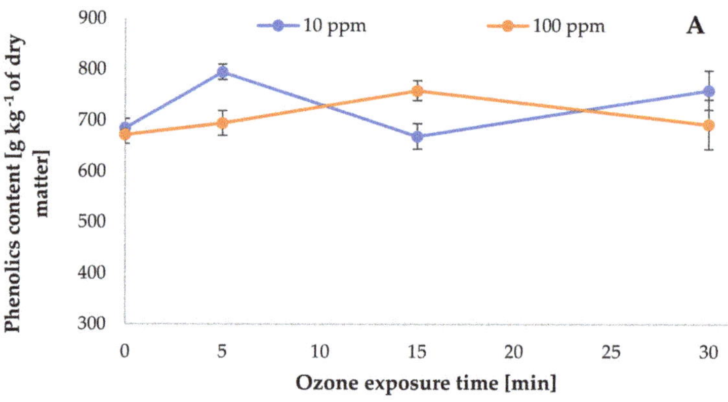

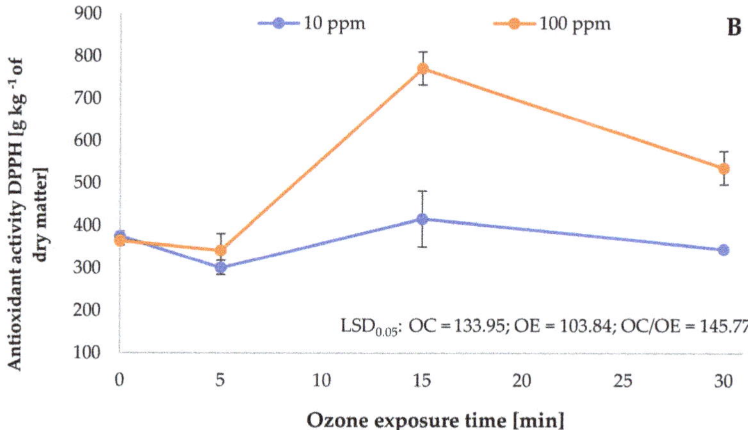

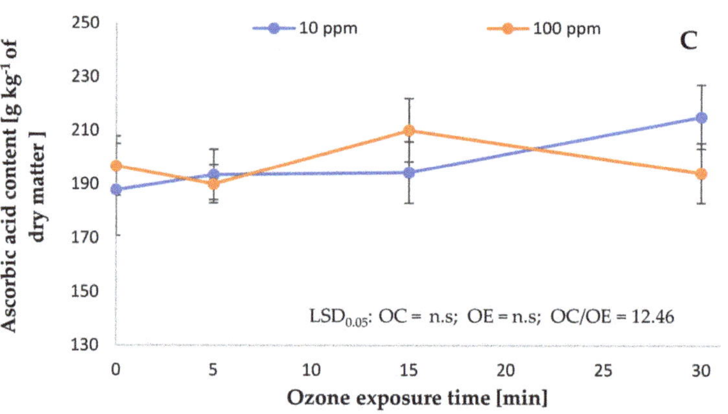

Figure 2. The impact of ozonation process on the polyphenolic content (**A**), antioxidant activity against DPPH radical (**B**), and total ascorbic acid content (**C**) in *Hippophae rhamnoides* L. fruit (*n* = 3). Note: LSD for α = 0.05 for the impact of ozone treatment concentration (OC), ozone exposure time (OE), and the interaction between tested parameters (OC)/(OE).

2.4. Microbial Load in the Raw Material

The ozone treatment resulted in a reduction of the microbial load in the plant material after 24 h of the conducted process (Table 2). The results of microbiological and statistical analyses showed the impact of ozone on the total number of aerobic bacteria colonies, in addition to yeasts and molds. Exposure of sea buckthorn berries to the effect of gaseous ozone at a concentration of 100 ppm for 30 min resulted in the lowest number of colonies of aerobic bacteria (cfu g^{-1}). The applied process conditions resulted in the reduction of the number of aerobic bacteria colonies by 3 log cfu g^{-1} compared to the control (non-ozonated) sample. Similar effects of gaseous ozone were observed in the case of reducing the number of yeast and mold colonies on the surface of ozonated sea buckthorn berries. The number of yeast and mold colonies after the application of 100 ppm gaseous ozone for 30 min decreased by 1 log cfu g^{-1} compared to non-ozone fruit. The reduction of the microbial load was due to the antimicrobial effect of gaseous ozone, which results from its strong oxidizing properties [24]. The influence of the ozonation process on the reduction of the microbial load of fruits has been confirmed in the studies of other authors. Piechowiak et al. [19] found a decrease in the total number of mesophilic aerobic bacteria, by 1.18 log cfu g^{-1}, at 48 h after the first ozone treatment of raspberry fruit (8–10 ppm for 30 min every 12 h), compared to the control.

Table 2. Microbiological load of sea buckthorn berries 24 h after the ozonation process ($n = 3$).

Parameter	Treat		Storage
	Ozone Concentration [ppm]	Ozone Exposure [min]	1 Day
Count of yeast and mould [cfu g^{-1}]	0	0	4.82
	10	5	4.07 [c]
		15	3.91 [d]
		30	4.26 [a]
	100	5	4.14 [b]
		15	3.55 [e]
		30	3.55 [e]
	LSD$_{0.05}$: OC = 2.64; OE =2.98; OC/OE= 2.92		
Count of aerobic bacteria [cfu g^{-1}]	0	0	6.28
	10	5	6.71 [a]
		15	5.10 [c]
		30	4.61 [c]
	100	5	5.84 [b]
		15	3.64 [c]
		30	3.63 [c]
	LSD$_{0.05}$: OC = 4.96; OE = 5.07; OC/OE= 5.15		

Note: LSD for $\alpha = 0.05$ for the impact of ozone treatment concentration (OC), ozone exposure time (OE), and the interaction between tested parameters (OC)/(OE). [a,b,c,d,e] statistically significant differences for the effect: ozone concentration (ppm) × ozone exposure (min).

2.5. Profile of Volatile Compounds

The ozonation process affected the chemical composition of the sea buckthorn berries. The group of compounds that are particularly susceptible to decomposition under the influence of this process are volatile compounds (Table 3). In sea buckthorn fruits, these compounds belong to main two groups: organic acid esters (compounds with a fruity smell) and terpene derivatives. The first group of compounds are aliphatic derivatives that are resistant to the oxidation process. HS-SPME analysis showed that the percentages of these substances vary depending on the ozonation conditions used, but their mutual proportions are similar. The differences in the percentage composition are due to the loss of other compounds that were most likely susceptible to the ozone decomposition process. The greatest differences were observed for carvacrol presence. Carvacrol is a phenol derivative that was identified only in the control sample (non-ozonated fruit). This is an interesting observation because the aromatic compounds show a moderate tendency to ozonolysis [29]. The process carried out with the use of gaseous ozone mainly affects the surface of the ozonated raw material. The secondary metabolites are often found on the surface of the fruit in tissues capable of storing hydrophobic substances. They are often components of the cuticle, which is a barrier rich in hydrophobic compounds. Terpenes are hydrophobic substances that are often found in the cuticle. One of the components of the volatile fraction of sea buckthorn berries is carvacrol. Carvacrol is a component with a low boiling point and a relatively high vapor pressure, which makes the compound susceptible to the surface action of ozone. The HS-SPME technique used makes it possible to identify volatile compounds present in the headland phase. The lack of carvacrol in this phase in the case of ozonated fruit (Figure 3) indicates that it has been removed from the top layers of sea buckthorn berries, but this does not exclude its presence in deeper structures. It should be noted, however, that the lack of this substance in the headspace will likely affect the subjective smell of ozonated fruits. A cursory sensory analysis did not reveal the absence of foreign odors. Ozonolysis can lead to the formation of carbonyl derivatives, which often exhibit strong aromatizing properties. These compounds can arise as oxidation products of unsaturated fatty acids in the deeper layers of plant tissues [30,31]. However, they arise from long-chain unsaturated hydrocarbon derivatives. In the case of sea buckthorn berries, the aromatic hydrocarbon derivative (carvacrol) was degraded, and presumably degraded into highly volatile compounds (glyoxal), the presence of which was not detected. No significant losses of the esters, which are mainly responsible for the fruity fragrance, were observed during the ozonation process. As shown by Slynko et al. [32], these esters are the main ingredients in the scent of sea buckthorn berries and represent 71.45% of the ingredients of Hippophae scents. In the case of ozonated fruit, the proportion of these esters ranges from 97% (100 ppm, 30 min) to 60.98% for non-ozonated fruit.

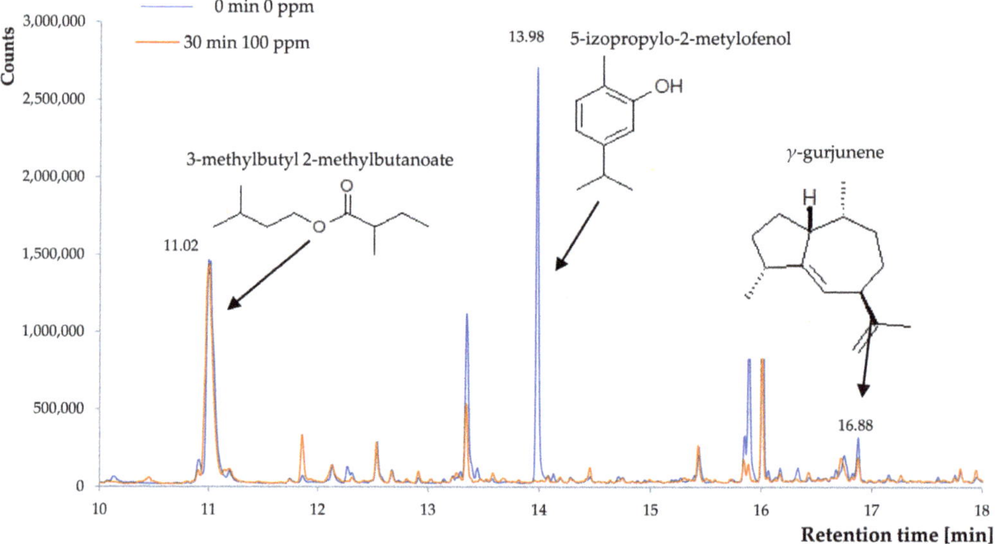

Figure 3. Chromatogram SPME-GC for the volatile fraction of *Hippophae rhamnoides* L. fruit.

Table 3. Effect of ozone treatment concentration and ozone exposure time on chemical composition of headspace fractions of *Hippophae rhamnoides* L. fruit after one day of storage ($n = 3$).

No.	RT [min]	Peak Share in the Chromatogram [%]							Ordinary Substance Name	Systematic Substance Name	No CAS
		0 ppm 0 min	10 ppm 5 min	10 ppm 15 min	10 ppm 30 min	100 ppm 5 min	100 ppm 15 min	100 ppm 30 min			
1	8.93	trace	trace	4.49 [a]	trace	3.46 [b]	trace	trace	ethyl caproate	ethyl hexanoate	123-66-0
1	10.91	2.05 [b]	3.56 [b]	3.66 [b]	2.32 [b]	5.30 [a]	trace	trace	(*E*)-sabinene hydrate	(2R,5R)-2-methyl-5-propan-2-ylbicyclo (3.1.0) hexan	17699-16-0
2	10.99	trace	trace	trace	trace	trace	trace	trace	2-methylbutyl pentanoate	2-methyl butyl valerate	55590-83-5
3	11.02	32.77 [b]	43.77 [b]	39.84 [b]	51.83 [b]	47.80 [b]	43.88 [b]	64.02 [a]	isoamyl 2-methyl butyrate	3-methylbutyl 2-methylbutanoate	27625-35-0
4	11.19	0.51 [a]	trace	0.77 [a]	trace	trace	2.64 [a]	trace	ipsenol	2-methyl-6-methylideneoct-7-en-4-ol	35628-05-8
5	12.13	1.06 [b]	trace	4.72 [a]	trace	1.73 [b]	9.01 [b]	trace	ethyl benzoate	ethyl benzoate	93-89-0
6	12.27	1.55 [a]	trace	trace	trace	trace	8.58	trace	4-carvomenthenol	4-methyl-1-propan-2-ylcyclohex-3-en-1-ol	562-74-3
7	12.53	2.52 [b]	3.57 [b]	5.83 [a]	2.65 [b]	4.77 [a]		5.24 [a]	ethyl caprylate	ethyl octanoate	106-32-1
9	13.35	10.25 [a]	10.89 [a]	13.85 [a]	13.06 [a]	10.37 [a]	9.66 [a]	8.91 [a]	isoamyl caprylate	isoamyl octanoate	2035-99-6
9	13.98	20.37	nd	nd	nd	nd	nd	nd	carvacrol	5-izopropylo-2-metylofenol	499-75-2
10	15.44	1.75 [b]	2.44 [b]	3.23 [b]	3.52 [b]	2.26 [b]	tarce	5.42 [a]	benzyl valerate	benzyl pentanoate	10361-39-4
11	15.85	2.18 [b]	3.29 [b]	2.10 [b]	4.18 [a]	2.76 [b]	trace	trace	α-gurjunene	(2S,4R,7R,8R)-3,3,7,11-tetramethyltricyclo[6.3.0.02.4]undec-1(11)-ene	489-40-7
12	15.89	8.54 [a]	7.69 [a]	3.86 [b]	5.64 [b]	5.97 [b]	trace	trace	caryophyllene	(1R,4E,9S)-bicyclo[7.2.0]undec-4-ene, 4,11,11-trimethyl-8-methylene-,	87-44-5
13	16.01	11.64 [a]	15.48 [a]	12.30 [a]	15.79 [a]	14.67 [a]	16.94 [a]	12.78 [a]	isoamyl benzoate	3-methylbutyl benzoate	94-46-2
14	16.75	1.79 [a]	3.58 [a]	trace	trace	trace	trace	trace	β-selinene	(3S,4aR,8aS)-8a-methyl-5-methylidene-3-prop-1-en-2-yl-1,2,3,4,4a,6,7,8-octahydronaphthalen	17066-67-0
15	16.88	2.01 [a]	2.08 [a]	3.97 [a]	trace	trace	7.85 [a]	2.29 [a]	(1R,4R,7R)-1,4-dimethyl-7-prop-1-en-2-yl-1,2,3,3a,4,5,6,7-octahydroazulene	γ-gurjunene	22567-17-5
TOTAL		98.99	96.35	98.62	98.99	98.60	98.56	98.66			

Note: [a, b]—difference at significant level $p < 0.05$.

3. Materials and Methods

3.1. Plant Material

The research material was acquired from a local producer in a neighborhood of Rzeszów (southern Poland) after harvest. The plant material was analyzed at the Department of Food Chemistry and Toxicology University of Rzeszów.

3.2. Ozone Exposure

Apparatus for the ozone treatment of plant material was fed with ozone generated by a TS 30 (Ozone Solution, Hull, MA, USA) ozone generator, and the ozone concentration was measured by a 106 M UV Ozone Solution detector (Ozone Solution, Hull, MA, USA), with the range of 0–1000 ppm [33]. The plant material was placed into the reactor chamber. Samples (500 g) of the sea buckthorn were exposed to ozone at the concentration of 10 ppm and 100 ppm for 10, 15, and 30 min respectively with the gas flow rate of 4 m^3 h^{-1} at room temperature (20 °C). The experiments were conducted in triplicate.

3.3. Determination of the Moisture Content

The moisture content in the plant material was determined using a SLW 115 SMART moisture analyzer (POL-EKO-APARATURA, Wodzisław Śląski Poland). The maximal temperature during this procedure was 105 °C.

3.4. Measurement of Mechanical Properties

Hippophae rhamnoides L. samples were subjected to the puncture strength test with a pen puncture probe with a diameter of: a~8 mm, b~6 mm, using Zwick/ Roell Z010 machine (Zwick Roell Polska Sp. z o.o. Sp. K., Wrocław, Poland). Measurements of the mechanical properties were conducted at the preliminary power F = 1 N (initial force), and the speed of traverse of the beam load cell V = 0.3 mm·mins^{-1} (crosshead return speed). The measurements were conducted in 36 repetitions for each of the analyzed varieties, on the fresh and the ozonated fruit after the first, the fourth, and the seventh day of storage at a temperature of 4 °C.

3.5. Content of Bioactive Compounds

The content of polyphenols in *Hippophae rhamnoides* L. fruit was measured in accordance with the methodology described by Matłok et al. [34] utilizing the Folin–Ciocalteu method. The total ascorbic acid content and the antioxidant activity DPPH were determined according to the methodology described by Oszmiański et al. [35] and Piechowiak et al. [19]. The analysis was performed in three replicates.

3.6. Microbiological Analysis

Microbiological analyses consisting of the indication of the total number of yeasts (cfu g^{-1}) and mold and aerobic bacteria (cfu g^{-1}) were carried out in accordance with the methods described by Matlok et al. [36]. The analyses were performed in three replications.

3.7. Head Space-Solid Phase Microextraction (HS-SPME) and Chromatographic Analysis

Head Space–Solid Phase Microextraction (HS-SPME) and chromatographic analysis was conducted in accordance with the method described by Matłok et al. [36]. The analyses were performed in three replications.

3.8. Statistical Analysis

The significance of the differences was checked for each term of analysis. The results were statistically processed using the analysis of variance in a two-factor random block design. Furthermore, the effect of storage time on the specified quality parameter of the control and ozonated fruit in terms of the value of the specified mechanical quality parameter and moisture content was determined using the analysis of variance in a three-factor random blocks design. The significance of differences with LSD of 0.05 was evaluated using Tukey's multiple test. Statistical analysis of results was carried out using the ANALWAR—5.3 FR programme by Franciszek Rudnicki (University of Life Science and Technology in Bydgoszcz, Poland)

4. Conclusions

The results of the present study show the effect of ozone treatment on the quality of sea buckthorn (*Hippophae rhamnoides* L.) The study shows that the application of gaseous ozone leads to reduced microbial load and loss of water, and improved mechanical properties in the plant material. Furthermore, the findings show increased ascorbic acid content after ozone treatment (for 10 ppm at 30 min and 100 ppm at 15 min), and higher total content of polyphenols. Moreover, the observed increase was dependent on the applied dose of ozone. The current findings show that ozone treatment may effectively be used to extend the shelf life of sea buckthorn.

Author Contributions: Conceptualization, methodology, formal analysis and writing—original draft: A.Z., N.M., M.B. investigation— M.Z., T.P. All authors have read and agreed to the published version of the manuscript.

Funding: The project is financed by the program of the Minister of Science and Higher Education named "Regional Initiative of Excellence" in the years 2019–2022, project number 026/RID/2018/19, the amount of financing PLN 9 542 500.00.

Informed Consent Statement: Informed consent was obtained from all subjects involved in the study.

Conflicts of Interest: The authors declare no conflict of interest.

References

1. Pop, R.M.; Weesepoel, Y.; Socaciu, C.; Pintea, A.; Vincken, J.-P.; Gruppen, H. Carotenoid composition of berries and leaves from six Romanian sea buckthorn (*Hippophae rhamnoides* L.) varieties. *Food Chem.* **2014**, *147*, 1–9. [CrossRef] [PubMed]
2. Teleszko, M.; Wojdyło, A.; Rudzińska, M.; Oszmiański, J.; Golis, T. Analysis of Lipophilic and Hydrophilic Bioactive Com-pounds Content in Sea Buckthorn (*Hippophaë rhamnoides* L.) Berries. *J. Agric. Food Chem.* **2015**, *63*, 4120–4129. [CrossRef] [PubMed]
3. Tiitinen, K.; Hakala, M.; Kallio, H. Headspace volatiles from frozen berries of sea buckthorn (*Hippophaë rhamnoides* L.) varieties. *Eur. Food Res. Technol.* **2006**, *223*, 455–460. [CrossRef]
4. Czaplicji, S.; Nowak-Polakowska, H.; Zadernowski, R. Aktywność biologiczna lipofilnej fakcji owoców rokitnika w zależności od zawartości karotenoidów i tokoferoli. *Bromat. Chem. Toksykol.* **2005**, *38*, 185–188.
5. Tang, X.R.; Kalviainen, N.; Tuorila, H. Sensory and hedonic characteristics of juice of sea buckthorn (*Hippophae rhamnoides* L.) origins and hybrids. *Lebensm. Wiss. Technol.* **2001**, *34*, 102–110. [CrossRef]
6. Pérez, A.G.; Sanz, C.; Ríos, J.J.; Olías, R.; Olías, J.M. Effects of ozone treatment on postharvest strawberry quality. *J. Agric. Food Chem.* **1999**, *47*, 1652–1656. [CrossRef]
7. Najafi, M.B.H.; Khodaparast, M.H. Efficacy of ozone to reduce microbial populations in date fruits. *Food Control* **2009**, *20*, 27–30. [CrossRef]
8. Alexandre, E.M.; Santos-Pedro, D.M.; Brandão, T.R.; Silva, C.L. Influence of aqueous ozone, blanching and combined treatments on microbial load of red bell peppers, strawberries and watercress. *J. Food Eng.* **2011**, *105*, 277–282. [CrossRef]
9. Carbone, K.; Mencarelli, F. Influence of Short-Term Postharvest Ozone Treatments in Nitrogen or Air Atmosphere on the Metabolic Response of White Wine Grapes. *Food Bioprocess Technol.* **2015**, *8*, 1739–1749. [CrossRef]
10. Chen, C.; Zhang, H.; Dong, C.; Ji, H.; Zhang, X.; Li, L.; Ban, Z.; Zhang, N.; Xue, W. Effect of ozone treatment on the phenyl-propanoid biosynthesis of postharvest strawberries. *RSC Adv.* **2019**, *9*, 25429–25438. [CrossRef]
11. Contigiani, E.V.; Jaramillo-Sánchez, G.; Castro, M.A.; Gómez, P.L.; Alzamora, S.M. Postharvest Quality of Strawberry Fruit (Fragaria x Ananassa Duch cv. Albion) as Affected by Ozone Washing: Fungal Spoilage, Mechanical Properties, and Structure. *Food Bioprocess Technol.* **2018**, *11*, 1639–1650. [CrossRef]
12. Horvitz, S.; Cantalejo, M.J. Application of Ozone for the Postharvest Treatment of Fruits and Vegetables. *Crit. Rev. Food Sci. Nutr.* **2013**, *54*, 312–339. [CrossRef]
13. Jaramillo-Sánchez, G.; Contigiani, E.V.; Castro, M.A.; Hodara, K.; Alzamora, S.M.; Loredo, A.G.; Nieto, A.B. Freshness Mainte-nance of Blueberries (*Vaccinium corymbosum* L.) During Postharvest Using Ozone in Aqueous Phase: Microbiological, Structure, and Mechanical issues. *Food Bioprocess Technol.* **2019**, *12*, 2136–2147. [CrossRef]
14. Piechowiak, T.; Grzelak-Błaszczyk, K.; Sójka, M.; Balawejder, M. Changes in phenolic compounds profile and glutathione status in raspberry fruit during storage in ozone-enriched atmosphere. *Postharvest Biol. Technol.* **2020**, *168*, 111277. [CrossRef]
15. Piechowiak, T.; Skóra, B.; Balawejder, M. Ozone Treatment Induces Changes in Antioxidative Defense System in Blueberry Fruit During Storage. *Food Bioprocess Technol.* **2020**, *13*, 1240–1245. [CrossRef]

16. Pinto, L.; Palma, A.; Cefola, M.; Pace, B.; D'Aquino, S.; Carboni, C.; Baruzzi, F. Effect of modified atmosphere packaging (MAP) and gaseous ozone pre-packaging treatment on the physico-chemical, microbiological and sensory quality of small berry fruit. *Food Packag. Shelf Life* **2020**, *26*, 100573. [CrossRef]
17. Yaseen, T.; Ricelli, A.; Turan, B.; Albanese, P.; D'Onghia, A.M. Ozone for postharvest treatment of apple fruits. *Phytopathol. Mediterr.* **2015**, *54*, 94–103.
18. Environmental Protection Agency. *Alternative Disinfectants and Oxidants Guidance Manual*; United States Environmental Protection Agency: New York, NY, USA, 1999.
19. Piechowiak, T.; Antos, P.; Kosowski, P.; Skrobacz, K.; Józefczyk, R.; Balawejder, M. Impact of ozonation process on the microbiological and antioxidant status of raspberry (*Rubus ideaeus* L.) fruit during storage at room temperature. *Agric. Food Sci.* **2019**, *28*, 35–44. [CrossRef]
20. Materska, M. Bioactive phenolics of fresh and freeze-dried sweet and semi-spicy pepper fruits (*Capsicum annuum* L.). *J. Funct. Foods* **2014**, *7*, 269–277. [CrossRef]
21. Zardzewiały, M.; Matlok, N.; Piechowiak, T.; Gorzelany, J.; Balawejder, M. Ozone Treatment as a Process of Quality Improvement Method of Rhubarb (*Rheum rhaponticum* L.) Petioles during Storage. *Appl. Sci.* **2020**, *10*, 8282. [CrossRef]
22. Gorzelany, J.; Migut, D.; Matłok, N.; Antos, P.; Kuźniar, P.; Balawejder, M. Impact of Pre-Ozonation on Mechanical Properties of Selected Genotypes of Cucumber Fruits During the Souring Process. *Ozone Sci. Eng.* **2017**, *39*, 188–195. [CrossRef]
23. Antos, P.; Piechowicz, B.; Gorzelany, J.; Matłok, N.; Migut, D.; Józefczyk, R.; Balawejder, M. Effect of Ozone on Fruit Quality and Fungicide Residue Degradation in Apples during Cold Storage. *Ozone Sci. Eng.* **2018**, *40*, 482–486. [CrossRef]
24. Rodoni, L.; Casadei, N.; Concellón, A.; Chaves Alicia, A.R.C.; Vicente, A.R. Effect of Short-Term Ozone Treatments on Tomato (Solanum lycopersicumL.) Fruit Quality and Cell Wall Degradation. *J. Agric. Food Chem.* **2010**, *58*, 594–599. [CrossRef] [PubMed]
25. Ong, M.K.; Ali, A.; Alderson, P.G.; Forney, C.F. Effect of different concentrations of ozone on physiological changes associated to gas exchange, fruit ripening, fruit surface quality and defence-related enzymes levels in papaya fruit during ambient storage. *Sci. Hortic.* **2014**, *179*, 163–169. [CrossRef]
26. Alothman, M.; Kaur, B.; Fazilah, A.; Bhat, R.; Karim, A.A. Ozone-induced changes of antioxidant capacity of fresh-cut tropical fruits. *Innov. Food Sci. Emerg. Technol.* **2010**, *11*, 666–671. [CrossRef]
27. Onopiuk, A.; Półtorak, A.; Moczkowska, M.; Szpicer, A.; Wierzbicka, A. The impact of ozone on health-promoting, microbiological, and colour properties of Rubus ideaus raspberries. *J. Food* **2017**, *15*, 563–573.
28. Lee, S.K.; Kader, A.A. Preharvest and postharvest factors influencingvitamin C content of horticultural crops. *Postharvest Biol. Technol.* **2000**, *20*, 207–220. [CrossRef]
29. Kazi, M.; Chatzopoulou, P.; Lykas, C. Effect of Ozonation on the Essential Oil Composition of Dried Aromatic Plants. In Proceedings of the Hellenic Association for Information and Communication Technologies in Agriculture, Food and Environment, Chania, Crete, Greece, 21–24 September 2017; pp. 772–780.
30. De Oliveira, J.M.; De Alencar, E.R.; Blum, L.E.B.; Ferreira, W.F.D.S.; Botelho, S.D.C.C.; Racanicci, A.M.C.; Leandro, E.D.S.; Mendonça, M.A.; Moscon, E.S.; Bizerra, L.V.A.D.S.; et al. Ozonation of Brazil nuts: Decomposition kinetics, control of Aspergillus flavus and the effect on color and on raw oil quality. *LWT* **2020**, *123*, 109106. [CrossRef]
31. Sert, D.; Mercan, E.; Kara, Ü. Butter production from ozone-treated cream: Effects on characteristics of physicochemical, microbiological, thermal and oxidative stability. *LWT* **2020**, *131*, 109722. [CrossRef]
32. Slynko, N.M.; Kuibida, L.V.; Tatarova, L.E.; Galitsyn, G.U.; Goryachkovskaya, T.N.; Peltek, S.E. Essential Oils from Different Parts of the Sea Buckthorn *Hippophae rhamnoides* L. *Adv. Biosci. Biotechnol.* **2019**, *10*, 233–243. [CrossRef]
33. Balawejder, M.; Szpyrka, E.; Antos, P.; Józefczyk, R.; Piechowicz, B.; Sadło, S. Method for Reduction of Pesticide Residue Levels in Raspberry and Blackcurrant Based on Utilization of Ozone. *Ochrona Srodowiska i Zasobów Naturalnych* **2014**, *25*, 1–5. [CrossRef]
34. Matłok, N.; Stępień, A.E.; Gorzelany, J.; Wojnarowska-Nowak, R.; Balawejder, M. Effects of Organic and Mineral Fertiliza-tion on Yield and Selected Quality Parameters for Dried Herbs of Two Varieties of Oregano (*Origanum vulgare* L.). *Appl. Sci.* **2020**, *10*, 5503. [CrossRef]
35. Oszmiański, J.; Kolniak-Ostek, J.; Lachowicz, S.; Gorzelany, J.; Matłok, N. Phytochemical Compounds and Antioxidant Ac-tivity in Different Cultivars of Cranberry (*Vaccinium macrocarpon* L). *J. Food Sci.* **2017**, *82*, 2569–2575. [CrossRef]
36. Matłok, N.; Piechowiak, T.; Zardzewiały, M.; Gorzelany, J.; Balawejder, M. Effects of Ozone Treatment on Microbial Status and the Contents of Selected Bioactive Compounds in *Origanum majorana* L. Plants. *Plants* **2020**, *9*, 1637. [CrossRef]

plants

MDPI

Article

Flaxseed and Camelina Meals as Potential Sources of Health-Beneficial Compounds

Silvia Tavarini [1,2,†], Marinella De Leo [2,3,†], Roberto Matteo [4], Luca Lazzeri [4], Alessandra Braca [2,3,*] and Luciana G. Angelini [1,2]

[1] Department of Agricultural, Food and Environment, University of Pisa, 56124 Pisa, Italy; silvia.tavarini@unipi.it (S.T.); luciana.angelini@unipi.it (L.G.A.)
[2] Research Centre for Nutraceutical and Healthy Foods "NUTRAFOOD", University of Pisa, 56124 Pisa, Italy; marinella.deleo@unipi.it
[3] Department of Pharmacy, University of Pisa, 56126 Pisa, Italy
[4] Council for Agricultural Research and Economics (CREA), Research Centre for Cereal and Industrial Crops, 40129 Bologna, Italy; roberto.matteo@crea.gov.it (R.M.); luca.lazzeri@crea.gov.it (L.L.)
* Correspondence: alessandra.braca@unipi.it; Tel.: +39-050-2219688
† These authors equally contributed.

Abstract: Seed meals and cakes, deriving from minor oilseed crops, represent interesting co-products for the presence of a high content of proteins and bioactive compounds that could be successfully explored as valuable plant-derived feedstocks for food and non-food purposes. In this contest, flaxseed (*Linum usitatissimum* L.) and camelina (*Camelina sativa* (L.) Crantz) are becoming increasingly important in the health food market as functional foods and cosmetic ingredients. Thus, this study aimed to evaluate the effect of genetic characteristics and cultivation sites on the chemical features of seed meals deriving from two flaxseed varieties (Sideral and Buenos Aires) and a camelina cultivar (Italia), cultivated in Central and Northern Italy (Pisa and Bologna). The content of total phenols and flavonoids, seed oil, proteins and fatty acids have been evaluated, together with the chemical profiles of flaxseed and camelina meals. In addition, radical-scavenging activity has been investigated. All the examined seed meals resulted as rich in bioactive compounds. In particular, flaxseed meal is a good source of the lignan secoisolariciresinol diglucoside (SDG) and hydroxycinnamic acid glucosides, while camelina meal contains glucosinolates and quercetin glycosides. Furthermore, all extracts exhibited a very strong radical-scavenging activity, that make these plant-derived products interesting sources for food or cosmetic ingredients with health outcomes.

Keywords: *Linum usitatissimum*; *Camelina sativa*; antioxidant capacity; bioactive compounds; glucosinolates; lignans; phenols; co-products valorization

Citation: Tavarini, S.; De Leo, M.; Matteo, R.; Lazzeri, L.; Braca, A.; Angelini, L.G. Flaxseed and Camelina Meals as Potential Sources of Health-Beneficial Compounds. *Plants* **2021**, *10*, 156. https://doi.org/10.3390/plants10010156

Received: 23 December 2020
Accepted: 12 January 2021
Published: 14 January 2021

Publisher's Note: MDPI stays neutral with regard to jurisdictional claims in published maps and institutional affiliations.

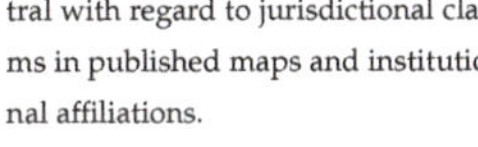

1. Introduction

In recent years, new perspectives for oilseed crops have revealed them to be renewable and valuable feedstocks for biorefinery processes, responding to the urgent need to transition toward a circular economy model based on the zero-waste concept [1]. These crops, in fact, are particularly suitable for obtaining, through a cascading use of total biomass, highest added-value products (pharmaceuticals, nutraceuticals, fine chemicals, cosmetics, agrochemicals, biomaterials), over bioenergy. In particular, oilseed meals and cakes, deriving from seed oil extraction, represent interesting co-products due to their high protein content, also for the presence of bioactive substances, such as phenolic acids, flavonoids, lignans and other antioxidant compounds [2], which could be used as food additives, supplements or cosmeceutical additives for foods and human health protection. In addition to the main oilseed crops, such as soybean and rapeseed used for food, feed and biofuel production, there is a growing interest in other minor oilseed crops suitable for marginal land, which could have a positive impact on the sustainability and resilience of agroecosystems. In this contest, flaxseed (*Linum usitatissimum* L., family Linaceae) and camelina (*Camelina sativa* (L.) Crantz, family Brassicaceae) are becoming more and more important in the health food market as a functional food and cosmetic ingredients [3,4]. Both crops, compared to the traditional oilseeds, display several agronomic advantages, such as great adaptability and phenotypic plasticity, low water and nutrient requirements, as well as good tolerance to pests and pathogens [5,6]. These positive agronomic attributes make these oilseed crops promising to be introduced in Mediterranean agroecosystems, where they might represent useful tools for enhancing biodiversity and cropping system diversification. These crops are not only characterized by positive agronomic traits, but they also have interesting chemical and functional features due to their products and co-products compositions. The nutritional importance of flaxseed is due to the high content of proteins (22%), lipids (43%) and minerals (3%). Its oil represents an important source of omega-3 fatty acids, especially linolenic acid (ALA) (more than 50% of the total fatty acids) [7]. Furthermore, flaxseed seeds, oil and cake are the richest source of the lignan secoisolariciresinol diglucoside (SDG), a natural cancer chemopreventive agent [8], in form of high molecular oligomers. At the same time, the high potential of camelina for nutritional applications is attributed to the distinctive fatty acid composition of its oil, rich in alpha-linolenic (18:3) and linoleic (18:2) acids [9,10]. Being an essential omega-3, alpha-linolenic acid has beneficial health effects on humans [11]. The presence of eicosenoic acid (11–19%) and tocopherols in relatively large amounts, and the low content of anti-nutritionals such as erucic acid, are additional distinctive differences of camelina in comparison with other commonly used vegetable oils [9,12]. All these compounds have antioxidant and free-radical scavenging activities and can play an important role in preventing several human diseases thanks their potential anti-tumoral, antiviral, antibacterial, and anti-mutagenic abilities [13]. For the aforementioned properties, flaxseed and camelina meals have interesting usable potential as ingredients for food and non-food purposes. Cake/meals composition is known to have a wide range of variability, depending on genetic and environmental growing conditions, and the extraction method. However, the current state of knowledge about the chemical composition of camelina and flaxseed meals, depending on the variety/cultivar and environmental conditions in which the plant is grown, is scarce. Therefore, the present study aimed to evaluate the role of environment and genotype in defining the chemical features of flaxseed and camelina meals. Consequently, the seed meals, obtained after solvent oil extraction, deriving from two flaxseed varieties (Sideral and Buenos Aires) and a camelina cultivar (Italia), cultivated in two environments of central and northern Italy (Pisa and Bologna), were analyzed for their phytochemical content and tested for their radical-scavenging activity. At the same time, the seed yield, oil and protein content and oil yield as well as fatty acid profile, were investigated for all the tested varieties in both environments, providing useful information about camelina and flaxseed yield potential under the climate conditions of Mediterranean region.

2. Results

2.1. Seed Yield and Qualitative Characteristics

In Figures 1 and 2, the main agronomic (seed and oil yield) and qualitative (seed oil and protein contents and fatty acid profile) traits, for all oilseed crops and sites were reported. Regarding flaxseed, the highest seed yield was obtained for Sideral in Pisa, while Buenos Aires in Bologna was characterized by the lowest crop yield. The same trend was observed for oil seed yield, with the highest value in Pisa for Sideral and the lowest one, once again, for Buenos Aires, grown in Bologna. On the contrary, no significant differences were found for oil content (%), indicating that the seed oil yield depended on seed yield rather than seed oil content. The seeds of Buenos Aires cultivated in Pisa were characterized by the lowest crude protein content (Figure 1).

Regarding camelina, the crop grown in Pisa showed the highest seed yield as well as the highest seed oil content, oil yield, and protein amount (Figure 1).

In Figure 2, the fatty acid composition of flaxseed and camelina seeds has been shown. Data highlighted that, for both flaxseed varieties, fatty acid composition did not vary depending on the cultivation site, while differences were observed between the two cultivars. In particular, Buenos Aires seeds were characterized by a lower content of linoleic acid and a higher content of α-linolenic one, in comparison with Sideral. On the contrary, for camelina, the cultivation site seemed to have a significant effect on the acidic composition, with higher contents of oleic and linoleic acids and a lower amount of α-linolenic acid in seeds obtained from the Pisa crop, compared with the fatty acid profile of camelina seed obtained in Bologna.

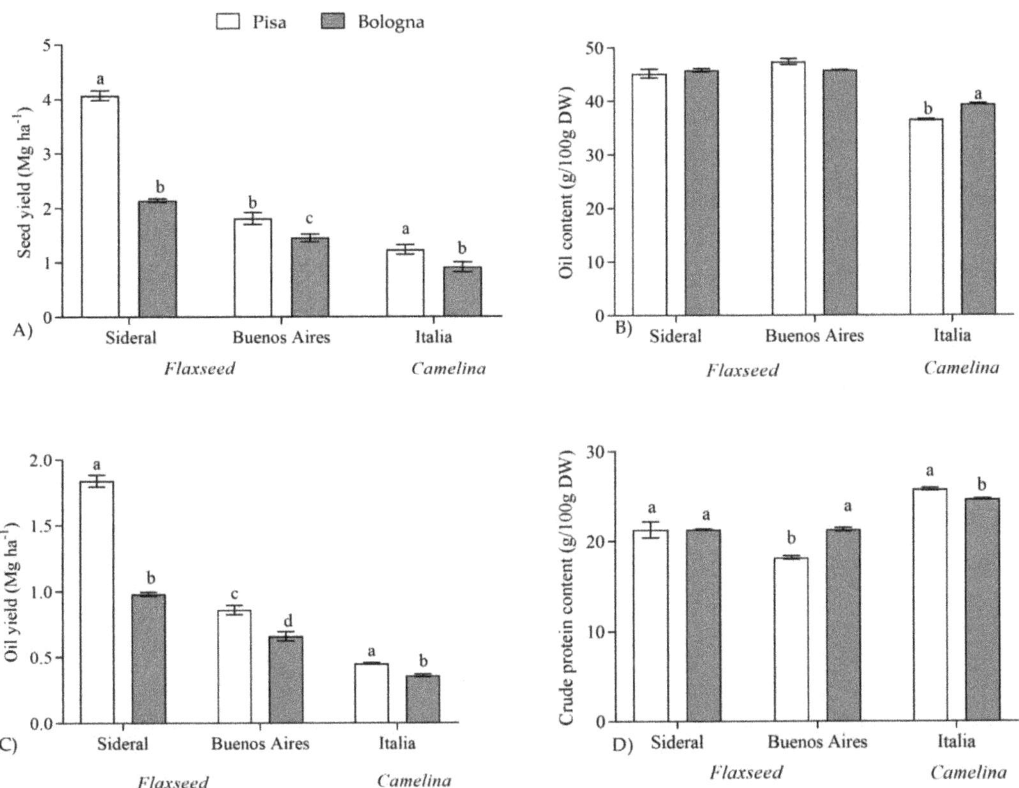

Figure 1. Seed yield (**A**), oil content (**B**), oil yield (**C**) and crude protein content (**D**) of the two oilseed crops (mean[†] $\pm$ SD) grown in two cultivation sites (Pisa and Bologna). [†] Values are the means of four replicates. Means followed by different letters are significantly different according to $LSD_{0.05}$ or Student's *t*-test.

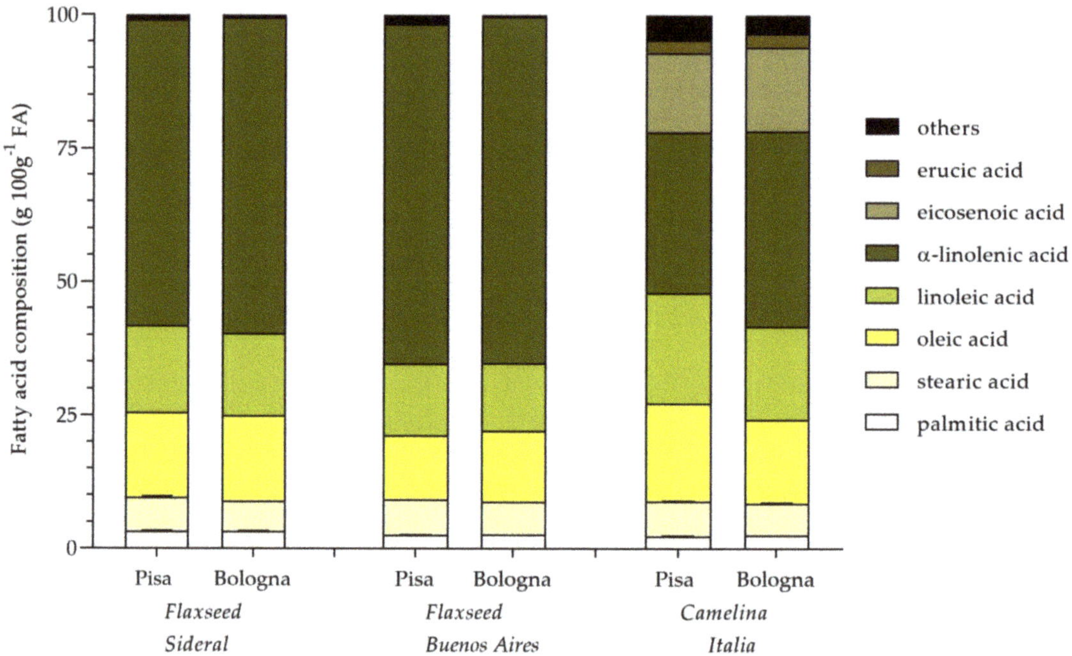

Figure 2. Fatty acid composition of oils of flaxseed (Sideral and Buenos Aires) and camelina (Italia) grown in two cultivation sites (Pisa and Bologna).

2.2. Phytochemical Screening and Anti-Radical Activity of Seed Meals

In Tables 1 and 2, the content of total phenols and flavonoids and anti-radical activity of flaxseed and camelina seed meals are shown. Regarding the phytochemical characteristics of flaxseed meal, ANOVA analysis showed no significant effect of variety and cultivation site on both the TPC (total phenolic content) and TFC (total flavonoids content) (Table 1).

A similar trend was also observed for anti-radical activity (Table 2), even if significant differences have been observed between varieties and sites for EC_{50} estimated by the DPPH assay. In particular, lower values, indicating a major anti-radical activity, were registered for meal obtained from Buenos Aires seeds and in general, for flaxseed meals deriving from the seeds produced in Pisa (Table 2). In camelina, the highest value of TPC was found in the defatted meal deriving from the seeds produced in Pisa (Table 1), while no differences between the two cultivation sites were observed for the flavonoid content. According to the phenol content, seed meals deriving from camelina produced in Pisa were characterized by the lowest anti-radical activity. The lower the EC_{50} value is, the higher the extract ability to scavenge radicals is, particularly peroxy radicals, which are the propagators of the autoxidation of lipid molecules and thereby break the free radical chain reaction (Table 2).

Table 1. Total phenols and total flavonoids (mean[†] ± SD) in the defatted seed meal of flaxseed (Sideral and Buenos Aires) and camelina (Italia) from two cultivation sites (Pisa and Bologna).

Crop/Variety	Total Phenolic Content (mg GAE/gDW)			Total Flavonoids (mg CAE/gDW)		
	Pisa	Bologna	*Mean Variety*	Pisa	Bologna	*Mean Variety*
Flaxseed Sideral	2.57 ± 0.28	2.74 ± 0.39	2.66	1.27 ± 0.05	1.36 ± 0.19	1.32
Flaxseed B.Aires	2.86 ± 0.15	2.74 ± 0.17	2.80	1.16 ± 0.23	1.23 ± 0.16	1.20
Mean Site	2.71	2.74		1.22	1.30	
Significance		*Variety (V) = n.s.* *Site (S) = n.s.* *VxS = n.s.*			*Variety (V) = n.s.* *Site (S) = n.s.* *VxS = n.s.*	
Camelina Italia	7.00 ± 0.18 a	6.26 ± 0.28 b	6.63	6.15 ± 0.79	5.48 ± 0.45	5.82
Significance		*Site (S) = ** *			*Site (S) = n.s.*	

[†] Values are the means of four replicates. Significance of variability factors according to F-test is reported as follows: n.s., not significant; **, significant at $p \leq 0.01$ level. For camelina's total phenolic content, means followed by different letters are significantly different at $p \leq 0.05$ based on the Student's *t*-test. CAE: catechin equivalent; DW: dry weight; GAE: gallic acid equivalent; SD: standard deviation.

Table 2. Anti-radical activity (expressed as EC_{50} and evaluated by DPPH and ABTS assays) in the defatted seed meal of flaxseed (Sideral and Buenos Aires) and camelina (Italia) in two cultivation sites (Pisa and Bologna). EC_{50} values of Trolox and BHA (butylated hydroxyanisole) were also shown as reference. Data are expressed as the mean[†] ± SD.

Crop/Variety	EC_{50} DPPH (mg mL^{-1})			EC_{50} ABTS (mg mL^{-1})		
	Pisa	Bologna	*Mean Variety*	Pisa	Bologna	*Mean Variety*
Flaxseed Sideral	3.60 ± 0.20	4.30 ± 0.30	3.95 A	3.10 ± 0.30	2.80 ± 0.20	2.95
Flaxseed B.Aires	3.10 ± 0.20	3.80 ± 0.20	3.45 B	3.00 ± 0.30	3.20 ± 0.30	3.10
Mean Site	3.35 B	4.05 A		3.05	3.00	
Significance		*Variety (V) = ** * *Site (S) = *** * *VxS = n.s.*			*Variety (V) = n.s.* *Site (S) = n.s.* *VxS = n.s.*	
Camelina Italia	1.50 ± 0.10 b	1.90 ± 0.10 a	1.70	2.10 ± 0.30 b	2.90 ± 0.30 a	2.50
Significance		*Site (S) = ** *			*Site (S) = ** *	
Trolox		0.052 ± 0.001			0.046 ± 0.001	
BHA		0.031 ± 0.001			0.029 ± 0.001	

[†] Values are the means of four replicates. SD: standard deviation. Significance of variability factors according to *F*-test is reported as follows: n.s., not significant; **, significant at $p \leq 0.01$ level; ***, significant at $p < 0.001$ level. Means followed by different letters are statistically different at $p \leq 0.05$ based on an LSD test or Student's *t*-test.

2.3. LC–PDA/UV–ESI–MS Profiles

2.3.1. Lignan Content of Flaxseed Meal

Flaxseed is known as a major source of lignan SDG, which is present in the form of high molecular oligomers due to ester bonds with 3-hydroxy-3-methylglutaric acid (HMGA) and glycosidic linkages with phenolic compounds, such as hydroxycinnamic acid derivatives and herbacetin diglucoside [14]. Both alkaline and acid hydrolysis of SDG oligomers is commonly used to analyze the lignan content of flaxseed [15]. In the present study, the chemical characterization of flaxseed lignans from the four analyzed extracts (defatted seed meal of flaxseed Sideral and Buenos Aires in the two cultivation sites Pisa and Bologna) was performed on the alkaline hydrolysates of SDG oligomers by the means of HPLC coupled to a PDA/UV detector and an electrospray ionization mass spectrometer (ESI–MS). The PDA/UV chromatograms (Figure 3) were acquired at 280 nm, which is the maximum absorption of SDG. All the extracts showed very similar profiles, with the presence of phenolic compounds due to the breaking of oligomer ester linkages. Indeed, the alkaline hydrolysis led to the formation of phenolic acid glucosides, such as *p*-coumaric acid glucoside (two isomeric forms, peaks **1** and **2**) and ferulic acid glucoside (two isomeric forms, peaks **3** and **4**), the flavonoid herbacetin diglucoside (peak **5**), the lignan SDG (two isomeric forms, peaks **6** and **7**), and ferulic acid (peak **8**), according to previous studies [15]. The tentative identification of all compounds was carried out comparing their elution order, ESI–MS/MS and PDA/UV data (Table 3) with those previously reported [16].

Masses of identified phenolics were detected in negative ion mode, originating deprotonated $[M - H]^-$ molecules and except for compound **6**, formiate $[M + HCOO]^-$ and acetate $[M + CH_3COO]^-$ adducts, leading to establish the molecular weight of detected substances. MS/MS of $[M + CH_3COO]^-$ ions for compounds **1/2** (*m/z* 385) and **3/4** (*m/z* 415) showed losses of a hexosyl moiety ($[M-162]^-$) due to the cleavage of the *O*-sugar bond, generating aglycon portions attributable to *p*-coumaric acid and ferulic acid, respectively. Thus, compounds **1/2** and **3/4** were identified as two isomeric forms of *p*-coumaric acid glucoside and ferulic acid glucoside, respectively, that cannot be distinguished on the basis of UV and MS data. The alkaline hydrolysates showed also the presence of ferulic acid (**8**), with λ_{max} 237 and 323 nm and diagnostic product ions (*m/z* 178, 149, and 134) generated in the MS/MS of deprotonated molecule $[M - H]^-$ at *m/z* 193. The full mass spectrum of compound **5** showed a deprotonated molecule $[M - H]^-$ at *m/z* 625, while MS/MS displayed two diagnostic fragment ions at *m/z* 463 and 301 generated by the subsequent losses of two hexosyl moieties. Thus, compound **5** was identified as herbacetin diglucoside, a flavonol considered part of the lignan macromolecule [17]. Compounds **6** and **7** were identified as two isomeric forms of SDG, the most abundant lignan present in linseed. The MS/MS experiment for the acetate adduct $[M + CH_3COO]^-$ at *m/z* 745 provided a product ion at *m/z* 583, due to the loss of a hexosyl moiety, according to the presence of a glucosyl residue. A few minor peaks remained unidentified.

The percentage composition of SDG oligomers in terms of detected phenols **1–8** after alkaline hydrolysis, calculated by integrating the peak areas at 280 nm in all camelina extracts, is shown in Figure 4. The most representative compound was *p*-coumaric acid glucoside (isomers **1** and **2**) with a percentage amount ranging from 48.8 to 60.0%, followed by SDG (isomers **6** and **7**; 20.2–26.4%), ferulic acid glucoside (isomers **3** and **4**, 7.6–16.0%), ferulic acid (**8**. 4.2–9.4%), and herbacetin diglucoside (**5**. 1.0–2.1%). Based on these results, meal extract from flaxseed Sideral cultivated in Pisa was the richest in *p*-coumaric acid glucoside, while meal extract from flaxseed Buenos Aires cultivated in Bologna was the richest in SDG and ferulic acid glucoside.

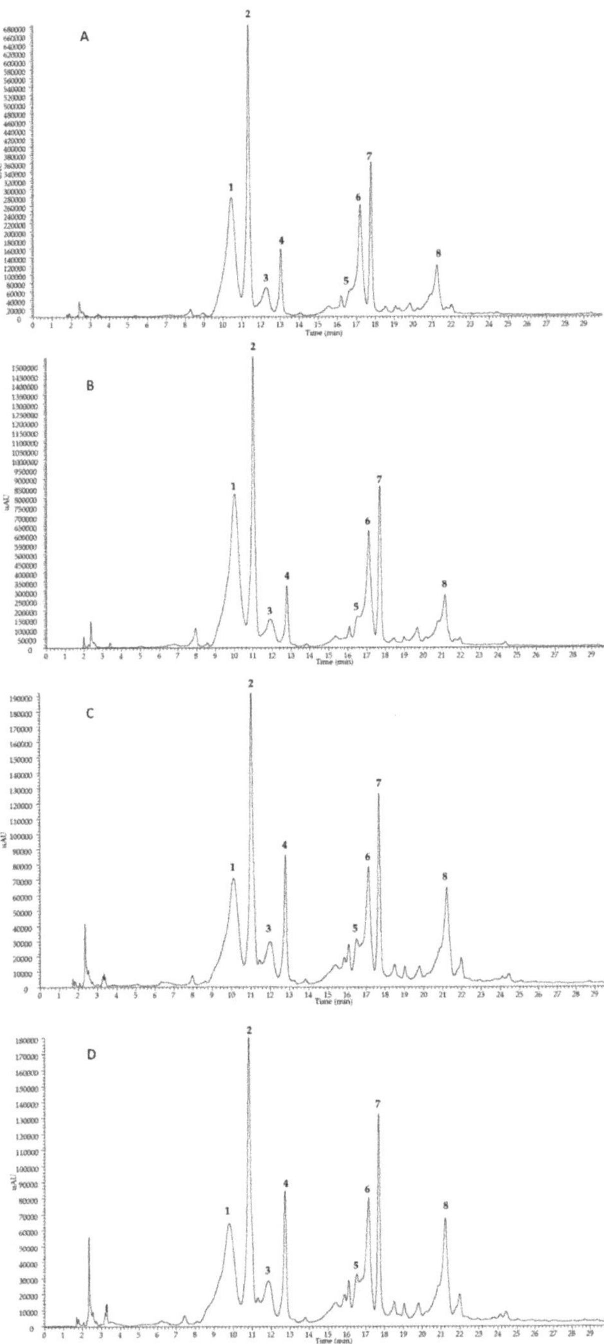

Figure 3. HPLC–PDA/UV profiles (detected at 280 nm) of secoisolariciresinol diglucoside (SDG) oligomers alkaline hydrolysates from the defatted seed meal extracts of flaxseed Sideral cultivated in Bologna (**A**) and Pisa (**B**), and Buenos Aires cultivated in Bologna (**C**) and Pisa (**D**). For peak data, see Table 3.

Table 3. Spectral (UV and ESI–MS/MS) and chromatographic data (t_R, retention time) of compounds **1–8**, detected in all defatted seed meal extracts of flaxseed after the alkaline hydrolysis of secoisolariciresinol diglucoside (SDG) oligomers.

Peak [a]	Compound	t_R (min)	$[M - H]^-$	$[M + CH_3COO]^-$	$[M + HCOO]^-$	MS/MS Ions (*m/z*) [b]	λ_{max} (nm)
	Phenolic acids						
1/2	*p*-Coumaric acid glucoside	10.3, 11.2	325	385	371	325, 163	228, 295
3/4	Ferulic acid glucoside	12.1, 13.0	355	415	401	355, 193	235, 290
8	Ferulic acid	21.2	193	253	239	178, 149, 134	237, 323
	Flavonoids						
5	Herbacetin diglucoside	16.2	625	–	–	463, 301	235, 284, 323
	Lignans						
6–7	Secoisolariciresinol diglucoside (SDG)	17.1, 17.6	685	745	731	685, 583	232, 280

[a] Compound numbers correspond to peak numbers in Figure 3. [b] For peaks **1–4**, **6**, and **7** MS/MS data were obtained by the fragmentation of the $[M + CH_3COO]^-$ precursor ions, whereas for peaks **5** and **8**, data were obtained by the fragmentation of the deprotonated molecule $[M - H]^-$.

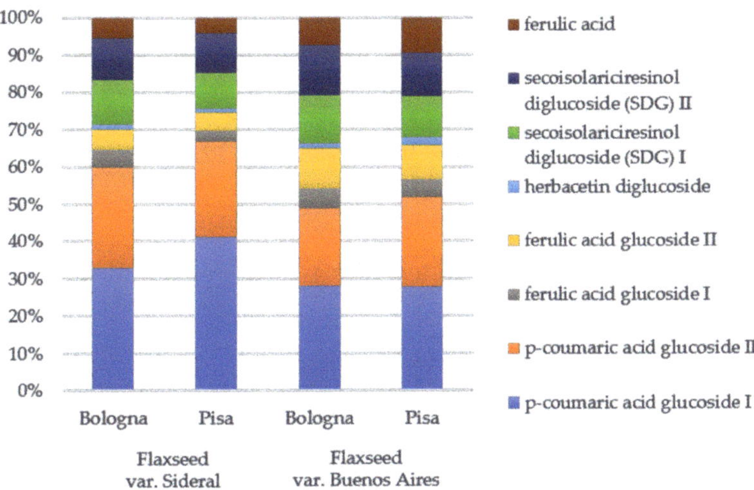

Figure 4. Percentage amount of identified phenols in secoisolariciresinol diglucoside (SDG) oligomers after the alkaline hydrolysis of flaxseed meal extracts from Sideral and Buenos Aires in two Italian cultivation sites, Pisa and Bologna.

2.3.2. Glucosinolate and Phenol Contents of Camelina Meal

The LC–ESI–MS analyses of camelina meal extracts (Figure 5), registered in negative ion mode, showed the presence of two major classes of compounds, represented by glucosinolates (peaks **9**, **10**, and **15**) and flavonoids (peaks **11–13**). The compounds were tentatively identified on the basis of spectral data. The MS/MS of all glucosinolates ([M − H]$^-$ at m/z 506, 520, and 534) showed diagnostic fragments due to the losses of a SO_2 molecule and were identified as glucoarabin (**9**), glucocamelinin (**10**), and 11-(methylsulfinyl)undecylglucosinolate (**15**), in agreement with previous studies [18,19]. The three detected flavonoids, characterized by two strong UV absorptions at 256–258 and 350–356 nm were all in the form of glycosides characterized by the presence of the same aglycone identified as quercetin due to the diagnostic fragment at m/z 301 in the MS/MS. Compound **11** ([M − H]$^-$ at m/z 741) was a triglycoside as deduced by its fragmentation pathway ([M − H − 132 − 146 − 162]$^-$) and was identified as quercetin 2″-*O*-apiosyl-3-*O*-rutinoside, previously isolated from camelina by Quéro et al. [18]. Both compounds **12** ([M − H]$^-$ at m/z 595) and **13** ([M − H]$^-$ at m/z 609) contain a disaccharide chain constituted by pentose–hexose and deoxyhexose–hexose, respectively. Thus, compound **13** was identified as quercetin 3-*O*-rutinoside (rutin) [19], while compound **12** is probably a quercetin apiosyl-glucoside in which the exact position of sugars cannot be determined only on the basis of MS/MS data; based on similar components previously found in camelina seeds, it can be assumed that compound **12** is quercetin 2″-*O*-apiosyl-3-*O*-glucoside. Finally, compound **14** ([M − H]$^-$ at m/z 623) remained not completely identified, but it could be a synapoil derivative as can be deduced by the presence of fragment [M − H − 206]$^-$ at m/z 417.

The LC–MS quantitative analyses (Table 5) showed that among glucosinolates glucocamelinin was the most representative in both camelina meal extracts, while rutin was the most abundant among flavonoids. Furthermore, meal extract from camelina cultivated in Bologna showed the highest content in term of glucosinolates [18.6 ± 0.3 vs. 14.5 ± 0.5 mg/g dry weight (DW)], whereas meal extract from camelina cultivated in Pisa showed the highest content in terms of flavonol glycosides (4.8 ± 0.08 vs. 4.5 ± 0.06 mg/g DW).

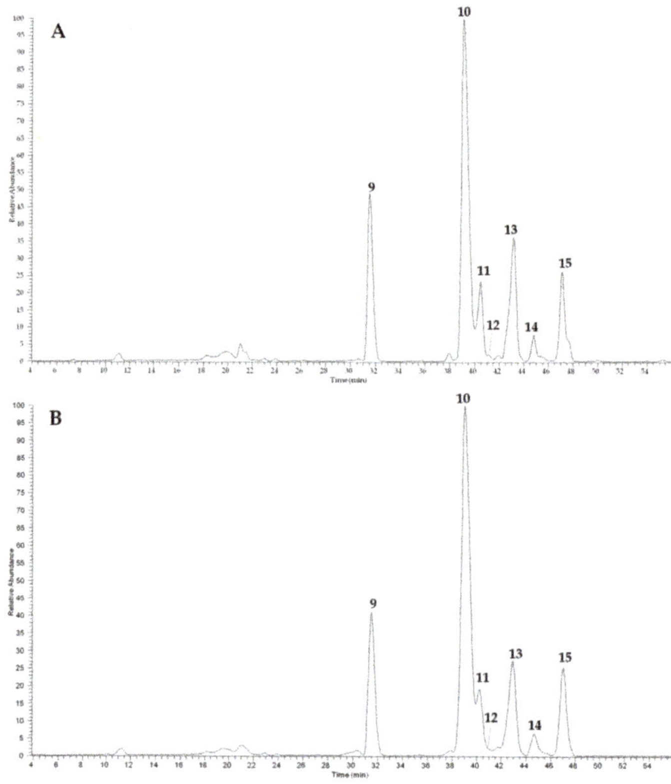

Figure 5. HPLC–ESI–MS profiles (registered in negative ion mode) of camelina meal extracts cultivated in Pisa (**A**) and Bologna (**B**). For peak data, see Table 4.

Table 4. Spectral (UV and ESI–MS/MS) and chromatographic data (t_R, retention time) of compounds **9–15**, detected in camelina meal extracts.

Peak [a]	Compound	t_R (min)	[M − H]⁻	MS/MS Ions (*m/z*)	λ_{max} (nm)
	Glucosinolates				
9	Glucoarabin (9-(methylsulfinyl)nonylglucosinolate)	31.4	506	491, 442, 248	240
10	Glucocamelinin (10-(methylsulfinyl)decylglucosinolate)	38.9	520	505, 456, 262	239
15	11-(methylsulfinyl)undecylglucosinolate	47.1	534	519, 470	256
	Flavonol glycosides				
11	Quercetin 2″-*O*-apiosyl-3-*O*-rutinoside	40.2	741	609, 300, 301	256, 354
12	Quercetin *O*-apiosyl-glucoside	40.9	595	463, 300, 301	258, 350
13	Quercetin 3-*O*-rutinoside (rutin)	42.9	609	463, 301	257, 356
	Other compound				
14	Synapoil derivative	44.6	623	417, 399, 209	249, 328

[a] Compound numbers correspond with peak numbers in Figure 5.

Table 5. Quantitative amount (mg/g $\pm$ SD DW) of constituents found in meal extracts from *Camelina sativa* Italia in two cultivation sites, Pisa and Bologna.

	Peak n. (in Figure 4)	Pisa	Bologna
Glucosinolates			
Glucoarabin	9	3.4 ± 0.04	3.9 ± 0.08
Glucocamelinin	10	9.3 ± 0.4	12.3 ± 0.2
11-(methylsulfinyl)undecylglucosinolate	15	1.8 ± 0.06	2.4 ± 0.04
Total		14.5 ± 0.5	18.6 ± 0.3
Flavonol glycosides			
Quercetin 2″-*O*-apiosyl-3-*O*-rutinoside	11	1.8 ± 0.03	1.7 ± 0.02
Quercetin apiosyl-glucoside	12	0.14 ± 0.004	0.21 ± 0.005
Quercetin 3-*O*-rutinoside	13	2.9 ± 0.05	2.6 ± 0.04
Total		4.8 ± 0.08	4.5 ± 0.06

DW: dry weight; SD: standard deviation.

3. Discussion

The obtained results showed that both flaxseed and camelina were able to reach satisfactory yields in both environments, even if clear effects of environment and variety were observed, with the highest productive performance reached by Sideral and Italia grown in Pisa. The obtained results are consistent with those reported in the literature for different environments and cultivars [4,20–22]. The analysis of the chemical composition of flax seeds showed that oil contents were not significantly different under the influence of variety or environmental conditions, with average values of 46%. This content was higher in comparison with those reported in previous studies [23,24] in which flaxseed oil content ranged between 34 and 45%, depending on the geographical area, genotype and environmental conditions. Flaxseeds were characterized by a very stable proportion of polyunsaturated fatty acids, in both varieties and environments, with α-linolenic acid as the most abundant fatty acid. It is known that this fatty acid is characterized by beneficial effects on the prevention of several diseases, such as cardiovascular diseases, hypercholesterolemia, chronic kidney diseases, atherosclerosis, and neurological disorders [25,26]. In the present study, the content of α-linolenic acid in both flaxseed varieties, ranging from 57 to 65%, consistent and sometimes higher than that reported in literature (from 50 to 59%, depending on genotype and environment) [27,28].

In camelina, oil content and fatty acid profiles were dependent on the environment, with higher oil and oleic acid contents in Pisa samples, compared to the Bologna ones. On the contrary, the seed oil of camelina grown in Bologna showed an increased content of eicosenoic acid and α-linolenic one. These differences could be due to the highest temperatures experienced during flowering and seed filling by the crop cultivated in Pisa, for which a spring sowing was performed. It is known that elevated temperatures during flowering and seed ripening determine a rise in oleic acid since temperature can interfere with the activity of the enzymes involved in the biosynthesis of fatty acids [29]. Interestingly, eicosenoic acid can be a valuable source of medium chain fatty acids for the bio-based industry, which nowadays are not produced in Europe as they are totally derived from palm and coconut oils [30]. Nonetheless, the oil content fell in the range typically reported for this oilseed crop [21,30]. Finally, as a general observation, the crude protein content of both types of oilseeds was negatively correlated with the oil content confirming previous findings [5,31].

The evaluation of the total phenols and flavonoids and their related antioxidant activities showed that both types of meal were characterized by interesting levels of these beneficial substances, regardless of the variety and growing environment, except for that given for total phenolic content and the anti-radical activity of camelina meals. In this case, in fact, higher phenols and stronger anti-radical activity, measured by DPPH, were observed for defatted meal deriving from camelina grown in Pisa, in comparison with camelina meal obtained from Bologna. In the literature, regarding total phenols and flavonoids in flaxseed and camelina cakes and/or meals, lower and higher values are reported, in comparison with our findings, depending on the variety, extraction method, and growing conditions [32–35]. For example, Teh and Birch [32] found, in defatted flaxseed cake, levels of total phenols and total flavonoids in the range of 474–807 mg GAE/100 g FW (FW = fresh weight) and 5.6–15.6 mg luteolin equivalents (LUE)/100 g FW, respectively, depending on the extraction method (ultrasonic and conventional method), solvent volume and extraction temperature. In the defatted camelina meal, Rahman et al. [34] observed a mean total phenolic content of 11.69 ± 0.44 mg GAE/g DW with a total flavonoid content equal to 6.81 ± 0.68 mg CAE/g DW.

Phenolic compounds are recognized as important food metabolites able to prevent several pathologies, such as cardiovascular and neurodegenerative diseases and cancer [36,37]. Polyphenols exhibit, in fact, interesting antioxidant activity thanks to their ability to transfer a hydrogen atom or an electron, acting as a reducing agent, as well as by the possible chelation of metal ions and the inhibition of the activity of oxidases [38]. The antioxidant activity of the hydroalcoholic extracts of the two kinds of meals tested in the present study was in line with the results on phenol and flavonoid concentration. In fact, a lower value of EC_{50} (for both DPPH and ABTS assays) were revealed for camelina meal, suggesting a positive correlation with the higher phenols and flavonoids detected in this kind of meal. The lower the EC_{50} value is, in fact, the higher the ability of the extract to scavenge radicals is. Previous reports underlined that camelina and flaxseed meals, after solvent oil extraction, contain good amounts not only of phytochemicals but also crude proteins (32–45%, with the presence of important essential amino acids, such as lysine, methionine and cysteine), insoluble fiber, carbohydrates and minerals [34,39–41], which make these meals good candidates for food and feed applications.

As expected, all flaxseed meal extracts were found to be a good source of SDG, herbacetin diglucoside, and hydroxycinnamic acid glucosides (ferulic and *p*-coumaric acid glucosides), herein characterized by HPLC–UV–MS analyses of alkaline hydrolysate, since they are accumulated in flaxseed seeds in form of oligomers [15]. Although the chemical compositions of the four extracts showed the same profiles in terms of constituents and their relative abundance, Buenos Aires flaxseed's meal cultivated in Bologna resulted, even if with small distances from others, the major source of SDG, a lignan whose potential health benefits are under investigation in many recent studies [8]. SDG, one of the most representative monomeric constituents of the lignan macromolecule, is reported to have many biological activities such as antioxidant and anti-inflammatory properties, playing a role in the prevention against chronic diseases, such as cardiovascular events and metabolic syndrome [8].

In agreement with previous investigations [18], three major glucosinolates were found in camelina meals, showing glucocamelinin as the most representative (about 64 and 66% of the total glucosinolate composition in the varieties growing in Pisa and Bologna, respectively). Long-chain glucosinolates predominate in camelina compared with short-chain glucosinolates that comprise the majority of glucosinolates in canola meal. Previous studies evidenced that glucosinolate content in camelina seeds is dependent on the geographic origin of seeds, as well as climatic factors and soil conditions [9]. A large variation was also observed between different cultivars [42]. In a recent work, Russo and Reggiani [43] reported an amount of total glucosinolates ranging from 19.6 to 40.3 mmol/kg DW in camelina meal from 47 accessions, with an average of 30.3 mmol/kg DW. Compared to these data, the amount of total glucosinolates in camelina meal obtained in the present work from the two Italian cultivation sites were moderately high, with camelina from Bologna richer (35.7 mmol/kg DW) than camelina from Pisa (27.9 mmol/kg DW), probably due to the different environmental conditions. In addition to glucosinolates, both camelina extracts were shown to contain phenolic constituents in quite a similar amount, with camelina from Pisa slightly above camelina from Bologna. Flavonoids, mainly kaempferol and quercetin derivatives, were previously reported in camelina seeds [19] and their profile compared to that of camelina meal [34,35]. The LC–MS analyses of camelina meal extracts from Pisa and Bologna showed that both samples contained three major quercetin glycosides, with quercetin 3-*O*-rutinoside the most representative, followed by quercetin 2″-*O*-apiosyl-3-*O*-rutinoside, herein more expressed than previous studies. The presence of glucosinolates and flavonoids is important for defining the nutraceutical value of camelina meals. Glucosinolates, secondary metabolites typical of the Brassicaceae family, have received great attention for their potential benefits in the prevention of carcinogenesis as well as cardiovascular and neurological diseases [44–46]. In a recent preliminary study, glucocamelinin and glucoarabin from defatted seed meal showed an ability to upregulate the phase II detoxification enzyme quinone reductase (NQO1) [18]. Similarly, rutin, a common flavonol glycoside found in a large number of plant species, is known for its antioxidant activity and its role in the treatment and prevention of various diseases [47].

4. Materials and Methods

4.1. Reagents and Standards

Methanol, formic acid and acetic acid for HPLC–MS analyses, and all analytical grade solvents and reagents were purchased from VWR (Milano, Italy). Water HPLC grade (18 mΩ) was prepared by Mill-Ω purification system (Millipore Co., Bedford, MA, USA). Folin–Ciocalteu reagent was purchased from Merck (Darmstadt, Germany). ABTS and DPPH were purchased from Sigma Aldrich (St. Louis, MO, USA). Rutin (purity ≥ 99%) and glucoraphanin (purity ≥ 98%) were purchased from Extrasynthese (Genay, France).

4.2. Experimental Conditions and Plant Material

Field plot experiments were carried out during the 2013–2014 growing season at the Centre for Agro-Environmental Research "Enrico Avanzi" of the University of Pisa located in San Piero a Grado (Pisa, central Italy, 43°40′ N; 10°19′ E, 1 m above sea level) and at the CREA (Council for Agricultural Research and Economics) experimental farm in Budrio (Bologna, northern Italy, 44°32′ N; 11°29′ E, 28 m above sea level), by adopting a randomized block design with four replicates (plots size of 6.5 m × 3.0 m) for each species and/or cultivars. Both sites were characterized by flat land with alluvial deep loam soils. In Pisa, the soil was a typic Xerofluvent, representative of the lower Arno river plain, characterized by a low level of organic matter (1.7%), and a medium content of available phosphorous (12.0 mg/kg) and total nitrogen (1.1 g/kg), with a moderately alkaline reaction (pH 8.2) and slightly calcareous (total $CaCO_3$ 3.1%). In Bologna, the soil was moderately alkaline (pH 8.1), characterized by good contents of organic matter (2.1%) and total nitrogen (1.4 g/kg), very good level of available P (33.3 mg/kg), and moderately calcareous (total $CaCO_3$ 10.3%). In both environments, the two flaxseed varieties used in the experiment were Sideral and Buenos Aires. Sideral is a variety registered in the EC (European Commission) common catalogue of varieties and commercially available (Semfor s.r.l., Verona, Italy), characterized by high resistance to cold and lodging and early ripening with blue-violet flowers and brown colored seeds [5]. Buenos Aires belongs to the germplasm collection of CREA-CI (Bologna, Italy), and it is characterized by a low cold resistance and an early ripening, with white flowers and yellow-colored seeds. Regarding camelina, the cultivar Italia, belonging to CREA-CI germplasm collection, was used [48]. The previous crop in both locations was durum wheat (*Triticum turgidum* L. subsp. *durum* (Desf.) Husn.), assuming a rotation with cereals. All flaxseed crops were sown during the fall, from mid- to the end of October, after the cereal crop harvest. Camelina was analyzed as a separate experiment comparing a winter crop in Bologna (sown in mid-October 2013) with a spring crop in Pisa (sown in mid-March 2014). Both crops have been harvested from the beginning to the end of June, at full seed maturity.

4.3. Agronomic Evaluations

At full seed maturity, four randomized sample areas of 2 m^2 were collected within each experimental plot for each crop and in each environment to assess harvestable crop yield. The plants were manually cut and gathered and then threshed by a fixed machine, using sieves suitable for small seeds, and evaluated for their moisture and seed yield.

4.4. Seed Processing and Analysis

Seed moisture was determined by oven-drying the seeds at 40 °C until constant weight for dry weight determination and the moisture content was calculated as the difference between the seed weight before and after the treatment. Oil was extracted by Buchi E-816 ECE (Soxhlet-like extractor) for 210 min, with hexane and *trans*-methylated in 2N KOH methanol solution [49]. Fatty acid profile was evaluated by a gas chromatography equipped with a flame ionization detector (Carlo Erba HRGC 5300 MEGA SERIES) and a capillary column Restek RT × 2330 (30 m × 0.25 mm × 0.2 µm), following the internal normalization method (ISO 12966–4:2015). The crude protein content was expressed as the percentage of dry matter and calculated from nitrogen using the conventional factor of 6.25. The obtained meals were dried at room temperature, vacuum-sealed and then stored away from light and heat, until the subsequent analysis.

4.5. Extraction of Bioactive Compounds

Flaxseed and camelina meals (0.25 g) were extracted with 5 mL of methanol–water (80% v/v) and sonicated for 30 min at room temperature. After centrifugation, the supernatant was filtered through a sterile 0.45 µm Minisart Syringe Filter and the resulting extracts were stored at −20 °C before use for up to a week.

4.6. Analysis of Total Phenols and Flavonoids

Total phenols were determined using the Folin–Ciocalteu method according to Singleton et al. [50]. The absorbance of the blue complex formation was determined at 765 nm by UV–vis spectrophotometer (Varian Cary 1E, Palo Alto, CA, USA). The results were expressed as mg gallic acid equivalent (GAE) per gram of seed meal on dry basis. Total flavonoids were determined using the method described by Jia et al. [51] measuring the absorbance of the pink complex at 510 nm using a UV–vis spectrophotometer (Varian Cary 1E, Palo Alto, CA, USA). The results were expressed as mg catechin equivalent (CAE) per gram of seed meal on dry basis. Measurements were replicated three times for each sample.

4.7. HPLC–PDA/UV–ESI–MS/MS Analyses of Camelina and Flaxseed Meal Extracts

4.7.1. Alkaline Hydrolysis of Flaxseed Oligomers

For the hydrolysis of the SDG oligomers, the four dried flaxseed extracts were dissolved in methanol (100 mg/mL) and mixed to an equal volume of 2M NaOH solution. The alkaline hydrolysis was carried out for 2 h at room temperature, then stopped by the addition of HCl 36% (1.2 M final concentration) [52]. The samples were successively centrifugated. Each supernatant was finally subjected to HPLC analysis at a concentration of 2.0 mg/mL.

4.7.2. HPLC–UV–MS Analyses

After alkaline hydrolysis, the chemical content of each flaxseed meal extract, together with camelina meal extracts, was analyzed by HPLC–PDA/UV–ESI–MS/MS technique. The LC–PDA/UV ESI–MS system was composed by a Surveyor LC pump, a Surveyor autosampler, coupled with a Surveyor PDA detector, and a LCQ Advantage ion trap mass spectrometer (ThermoFinnigan, San Jose, CA, USA) equipped with Xcalibur 3.1 software.

Analyses were performed using a 4.6 × 250 mm, 4 µm, Synergi Fusion-RP column (Phenomenex, Bologna, Italy). The eluent was a mixture of methanol (solvent A) and a 0.1% v/v aqueous solution of acetic acid (solvent B). For flaxseed meal extract analysis, a linear gradient of increasing 5 to 35% A was developed within 30 min, while for camelina meal extract, a linear gradient of increasing 5 to 60% A was used within 55 min. The column was successively washed with methanol an equilibrated with 5% A for 10 min.

Elutions were performed at a flow rate of 0.8 mL/min with a splitting system of 2:8 to MS detector (160 µL/min) and PDA detector (640 µL/min), respectively [53]. The volume of the injected methanol solutions was 20 µL. Analyses were performed with an ESI interface in the negative ion mode. The ionization conditions were optimized and the parameters used were as follows: capillary temperature, 270 °C; capillary voltage, −16.0 V; tube lens offset, −5 V; sheath gas flow rate, 60.00 arbitrary units; auxiliary gas flow rate, 3.00 arbitrary units; spray voltage, 4.50 kV; scan range of m/z 150–1500. N_2 was used as the sheath and auxiliary gas. PDA data were recorded within the 200–600 nm range, with the UV preferential channel as the detection wavelength of 280 nm.

For quantitative analyses of glucosinolates and flavonoids in camelina seed cake extracts, calibration curves were constructed by using glucoraphanin (concentration range 0.06–0.5 mg/mL) and rutin (concentration range 0.01–0.25 mg/mL), as external standards, respectively. Standard methanol solutions at different concentrations were prepared by serial dilution from stock solution (1 mg/mL), then analyzed by triplicate injections, and finally used with respect to the area obtained from the integration of the MS base peak [M − H]$^-$ of each standard to generate a calibration curve. The relations between variables were analyzed by using a linear simple correlation (R^2 = 0.9829 for rutin and 0.9837 for glucoraphanin). The phenol amounts were obtained by using a Microsoft$^{®}$ Office Excel (Redmond, WA, USA) and finally expressed as mg/g of dried cakes.

4.8. Free Radical-Scavenging Assay

The free radical-scavenging activity was evaluated by the DPPH (1,1-diphenyl-2-picryl-hydrazil radical) and ABTS free radical assay according to the method described by Brand-Williams et al. [54] and spectrophotometrically estimating the solution discoloration at 515 and 734 nm, respectively. The concentration required to obtain a 50% antioxidant effect (EC_{50}) was evaluated as the concentration of extract (mg of defatted seed meal $\times$ mL^{-1} of extraction solvent) causing the 50% inhibition of the initial color production. A series of dilutions in 80% methanol was prepared for each extract and standard. An Infinite M200 PRO microplate reader was used for spectrophotometric assays. Trolox and butylated hydroxyanisole (BHA) were used as the reference standards. Measurements were replicated three times for each sample.

4.9. Statistical Analyses

All the agronomic and phytochemical variables were subjected to the analysis of variance (ANOVA) using the statistical software CO-STAT Cohort, 2002 (CoHort Software, Monterey, CA, USA). A factorial design with variety (V) and cultivation site (S) as main treatments was used for flaxseed deriving data. Means were separated on the basis of least significance difference (LSD) post-hoc test only when the ANOVA F-test per treatment was significant at $\leq$ 0.05 probability level. For the camelina data, a Student's t-test analysis was performed in order to estimate the effect of a cultivation site.

5. Conclusions

In summary, this study pointed out the interesting functional properties of flaxseed and camelina meals, thanks to the abundant presence of antioxidants and phytochemicals, which represent high-value components of these important plant-derived products. In particular, all the examined flaxseed meals resulted as rich in phenol content, as deduced from LC–UV–MS analyses of each extract after alkaline hydrolysis of SDG oligomers. On the other hand, camelina defatted meals, showed comparable content of glucosinolates and flavonols glycosides quercetin derivatives, with glucocamelinin and rutin the most representative, respectively.

Thus, flaxseed and camelina meals investigated in the present work could be considered in further biological studies as potential dietary sources of health-beneficial compounds, able to play an important role in the reduction in the incidence of non-communicable diseases, including obesity, diabetes, cancer, and other chronic conditions. Thanks to their interesting chemical composition, their edible defatted meals can be used in human consumption as processed ingredients and/or as a source of antioxidants, and incorporated, for example, in bakery, infant products, and multipurpose supplements. At the same time, taking into account the increasing environmental issues, the use of these co-products and

Plants **2021**, 10, 156

---

the recovering/recapturing of valuable components can be a sustainable tool for reducing waste disposal and developing new environmental-friendly functional foods and/or ingredients.

Author Contributions: Conceptualization, L.L. and L.G.A.; methodology, S.T., M.D.L., A.B. and L.G.A.; validation, S.T. and M.D.L.; formal analysis, S.T., M.D.L. and R.M.; investigation, S.T., M.D.L. and R.M.; resources, L.L. and A.B.; data curation, S.T., M.D.L.; writing—original draft preparation, S.T. and M.D.L.; writing—review and editing, A.B. and L.G.A.; supervision, L.L., A.B. and L.G.A.; project administration, L.L.; funding acquisition, L.L. All authors have read and agreed to the published version of the manuscript.

Funding: This research was funded by the work was supported by MiPAAF, Ministry of Agricultural, Food and Forestry Policies (D.M. 2419, 20/02/08) and the trials were performed within the activities of the "SUSCACE *Sistema Integrato di Tecnologie per la valorizzazione dei sottoprodotti della filiera del Biodiesel*" Project.

Institutional Review Board Statement: Not applicable.

Informed Consent Statement: Not applicable.

Data Availability Statement: Data is contained within the article.

Acknowledgments: The authors wish to express their gratitude to Luisa Ugolini and Lorena Malagutti of CREA-CI for their useful help in executing some laboratory analyses.

Conflicts of Interest: The authors declare no conflict of interest. The funders had no role in the design of the study; in the collection, analyses, or interpretation of data; in the writing of the manuscript, or in the decision to publish the results.

References

1. Ancuța, P.; Sonia, A. Oil press-cakes and meals valorization through circular economy approaches: A review. *Appl. Sci.* **2020**, 10, 7432. [CrossRef]
2. Baiano, A. Recovery of biomolecules from food wastes-A review. *Molecules* **2014**, 19, 14821–14842. [CrossRef] [PubMed]
3. Tavarini, S.; Castagna, A.; Conte, G.; Foschi, L.; Sanmartin, C.; Incrocci, L.; Ranieri, A.; Serra, A.; Angelini, L.G. Evaluation of chemical composition of two linseed varieties as sources of health-beneficial substances. *Molecules* **2019**, 24, 3729. [CrossRef] [PubMed]
4. Berti, M.; Gesch, R.; Eynck, C.; Anderson, J.; Cermak, S. Camelina uses, genetics, genomics, production, and management. *Ind. Crop. Prod.* **2016**, 94, 690–710. [CrossRef]
5. Tavarini, S.; Angelini, L.G.; Casadei, N.; Spugnoli, P.; Lazzeri, L. Agronomical evaluation and chemical characterization of *Linum usitatissimum* L. as oilseed crop for bio-based products in two environments of Central and Northern Italy. *Ital. J. Agron.* **2016**, 11, 735. [CrossRef]
6. Angelini, L.G.; Abou Chehade, L.; Foschi, L.; Tavarini, S. Performance and potentiality of camelina (*Camelina sativa* L. Crantz) genotypes in response to sowing date under Mediterranean environment. *Agronomy* **2020**, 10, 1929. [CrossRef]
7. Sanmartin, C.; Taglieri, I.; Venturi, F.; Macaluso, M.; Zinnai, A.; Tavarini, S.; Botto, A.; Serra, A.; Conte, G.; Flamini, G.; et al. Flaxseed cake as a tool for the improvement of nutraceutical and sensorial features of sourdough bread. *Foods* **2020**, 9, 204. [CrossRef]
8. Adolphe, J.L.; Whiting, S.J.; Juurlink, B.H.J.; Thorpe, L.U.; Alcorn, J. Health effects with consumption of the flax lignan secoisolariciresinol diglucoside. *Br. J. Nutr.* **2010**, 103, 929–938. [CrossRef]
9. Zubr, J.; Matthäus, B. Effects of growth conditions on fatty acids and tocopherols in *Camelina sativa* oil. *Ind. Crop. Prod.* **2002**, 15, 155–162. [CrossRef]
10. Pagnotta, E.; Ugolini, L.; Matteo, R.; Lazzeri, L.; Foschi, L.; Angelini, L.G.; Tavarini, S. Exploring the *Camelina sativa* value chain: A new opportunity for bio-based products and overall crop sustainability. *Riv. Ital. Sost. Grasse* **2019**, XCVI, 259–268.
11. Simopoulos, A.P. The importance of the omega-6/omega-3 fatty acid ratio in cardiovascular disease and other chronic diseases. *Exp. Biol. Med.* **2008**, 233, 674–688. [CrossRef] [PubMed]
12. Matthäus, B.; Angelini, L.G. Anti-nutritive constituents in oilseed crops from Italy. *Ind. Crop. Prod.* **2005**, 21, 89–99. [CrossRef]
13. Gutiérrez, C.; Rubilar, M.; Jara, C.; Verdugo, M.; Sineiro, J.; Shene, C. Flaxseed and flaxseed cake as a source of compounds for food industry. *J. Soil Sci. Plant Nutr.* **2010**, 10, 454–463. [CrossRef]
14. Strandås, C.; Kamal-Eldin, A.; Andersson, R.; Åman, P. Composition and properties of flaxseed phenolic oligomers. *Food Chem.* **2008**, 110, 106–112. [CrossRef]
15. Li, X.; Yuan, J.P.; Xu, S.P.; Wang, J.H.; Liu, X. Separation and determination of secoisolariciresinol diglucoside oligomers and their hydrolysates in the flaxseed extract by high-performance liquid chromatography. *J. Chromatogr. A* **2008**, 1185, 223–232. [CrossRef]

16. Waszkowiak, K.; Barthet, V.J. Characterization of a partially purified extract from flax (*Linum usitatissimum* L.) seed. *J. Am. Oil Chem. Soc.* **2015**, *92*, 1183–1194. [CrossRef]

17. Struijs, K.; Vincken, J.P.; Verhoef, R.; van Oostveen-van Casteren, W.H.; Voragen, A.G.; Gruppen, H. The flavonoid herbacetin diglucoside as a constituent of the lignan macromolecule from flaxseed hulls. *Phytochemistry* **2007**, *68*, 1227–1235. [CrossRef]

18. Das, N.; Berhow, M.A.; Angelino, D.; Jeffery, E.H. *Camelina sativa* defatted seed meal contains both alkyl sulfinyl glucosinolates and quercetin that synergize bioactivity. *J. Agr. Food Chem.* **2014**, *62*, 8385–8391. [CrossRef]

19. Quéro, A.; Molinié, R.; Mathiron, D.; Thiombiano, B.; Fontaine, J.-X.; Brancourt, D.; Van Wuytswinkel, O.; Petit, E.; Demailly, H.; Mongelard, G.; et al. Metabolite profiling of developing *Camelina sativa* seeds. *Metabolomics* **2016**, *12*, 186. [CrossRef]

20. Lafond, G.P.; Irvine, B.; Johnston, A.M.; May, W.E.; McAndrew, D.W.; Shirtlie, S.J.; Stevenson, F.C. Impact of agronomic factors on seed yield formation and quality in flax. *Can. J. Plant Sci.* **2008**, *88*, 485–500. [CrossRef]

21. Berti, M.; Wilckens, R.; Fischer, S.; Solis, A.; Johnson, B. Seeding date influence on camelina seed yield, yield components, and oil content in Chile. *Ind. Crops Prod.* **2011**, *34*, 1358–1365. [CrossRef]

22. Fila, G.; Bagatta, M.; Maestrini, C.; Potenza, E.; Matteo, R. Linseed as a dual-purpose crop: Evaluation of cultivar suitability and analysis of yield determinants. *J. Agr. Sci.* **2018**, *156*, 162–176. [CrossRef]

23. Daun, J.; Barthet, V.; Chornick, T.; Duguid, S. Structure, composition, and variety development of flaxseed. In *Flaxseed in Human Nutrition*, 2nd ed.; Thompson, L., Cunanne, S., Eds.; AOCS Press: Champaign, IL, USA, 2003; pp. 1–40. [CrossRef]

24. Kaur, R.; Kaur, M.; Singh Gill, B. Phenolic acid composition of flaxseed cultivars by ultra-performance liquid chromatography (UPLC) and their antioxidant activities: Effect of sand roasting and microwave heating. *J. Food Process. Preserv.* **2017**, *41*, e13181. [CrossRef]

25. Oomah, B.D. Flaxseed as a functional food source. *J. Sci. Food Agric.* **2001**, *81*, 889–894. [CrossRef]

26. Klein, J.; Zikeli, S.; Claupein, W.; Gruber, S. Linseed (*Linum usitatissimum*) as an oil crop in organic farming: Abiotic impacts on seed ingredients and yield. *Org. Agr.* **2017**, *7*, 1–19. [CrossRef]

27. Silska, G.; Walkowiak, M. Comparative analysis of fatty acid composition in 84 accessions of flax (*Linum usitatissimum* L.). *J. Pre-Clin. Clin. Res.* **2019**, *13*, 118–129. [CrossRef]

28. Klimek-Kopyra, A.; Zając, T.; Micek, P.; Borowiec, F. Effect of mineral fertilization and sowing rate on chemical composition of two linseed cultivars. *J. Agric. Sci.* **2013**, *5*, 224–229. [CrossRef]

29. Vollmann, J.; Moritz, T.; Kargl, C.; Baumgartner, S.; Wagentristl, H. Agronomic evaluation of camelina genotypes selected for seed quality characteristics. *Ind. Crops Prod.* **2007**, *26*, 270–277. [CrossRef]

30. Righini, D.; Zanetti, F.; Monti, A. The bio-based economy can serve as the springboard for camelina and crambe to quit the limbo. *OCL* **2016**, *23*, D504. [CrossRef]

31. Zanetti, F.; Eynck, C.; Christou, M.; Krzyżaniak, M.; Righini, D.; Alexopoulou, E.; Stolarski, M.; Van Loo, E.; Puttick, D.; Monti, A. Agronomic performance and seed quality attributes of camelina (*Camelina sativa* L. Crantz) in multi-environment trials across Europe and Canada. *Ind. Crops Prod.* **2017**, *107*, 602–608. [CrossRef]

32. Teh, S.-S.; Birch, E.J. Effect of ultrasonic treatment on the polyphenol content and antioxidant capacity of extract from defatted hemp, flax and canola seed cakes. *Ultrason. Sonochem.* **2014**, *21*, 346–353. [CrossRef] [PubMed]

33. Sarkis, J.R.; Côrrea, A.P.F.; Michel, I.; Brandeli, A.; Tessaro, I.C.; Marczak, L.D.F. Evaluation of the phenolic content and antioxidant activity of different seed and nut cakes from the edible oil industry. *J. Am. Oil Chem. Soc.* **2014**, *91*, 1773–1782. [CrossRef]

34. Rahman, M.J.; Costa de Camargo, A.; Shahidi, F. Phenolic profiles and antioxidant activity of defatted camelina and Sophia seeds. *Food Chem.* **2018**, *240*, 917–925. [CrossRef] [PubMed]

35. Terpinc, P.; Polak, T.; Makuc, D.; Ulrih, N.P.; Abramovič, H. The occurrence and characterisation of phenolic compounds in *Camelina sativa* seed, cake and oil. *Food Chem.* **2012**, *131*, 580–589. [CrossRef]

36. Tangney, C.; Rasmussen, H.E. Polyphenols, inflammation, and cardiovascular disease. *Curr. Atheroscler. Rep.* **2013**, *15*, 324. [CrossRef]

37. Özçelik, B.; Kartal, M.; Orhan, I. Cytotoxicity, antiviral and antimicrobial activities of alkaloids, flavonoids, and phenolic acids. *Pharm. Biol.* **2011**, *49*, 396–402. [CrossRef]

38. Apak, R.; Özyürek, M.; Güçlü, K.; Çapanoğlu, E. Antioxidant activity/capacity measurement. 1. Classification, physicochemical principles, mechanisms, and electron transfer (ET)-based assays. *J. Agric. Food Chem.* **2016**, *64*, 997–1027. [CrossRef]

39. Aziza, A.E.; Quezada, N.; Cherian, G. Antioxidative effect of dietary camelina meal in fresh, stored, or cooked broiler chicken meat. *Poultry Sci.* **2010**, *89*, 2711–2718. [CrossRef]

40. Beltranena, E.; Oryschak, M. *Camelina sativa* Coproducts as Feedstuffs for Poultry. In Proceedings of the Western Poultry Conference, AB, Alberta, Canada, 29 February 2016.

41. Shim, Y.Y.; Gui, B.; Arnison, P.G.; Wang, Y.; Reaney, M.J.T. Flaxseed (*Linum usitatissimum* L.) bioactive compounds and peptide nomenclature: A review. *Trends Food Sci. Technol.* **2014**, *38*, 5–20. [CrossRef]

42. Berhow, M.A.; Polat, U.; Glinski, J.A.; Glensk, M.; Vaughn, S.F.; Isbell, T.; Ayala-Diaz, I.; Marek, L.; Gardner, C. Optimized analysis and quantification of glucosinolates from *Camelina sativa* seeds by reverse-phase liquid chromatography. *Ind. Crops Prod.* **2013**, *43*, 119–125. [CrossRef]

43. Russo, R.; Reggiani, R. Glucosinolates and sinapine in camelina meal. *Food Nutr. Sci.* **2017**, *8*, 1063–1073. [CrossRef]

44. Dinkova-Kostova, A.T.; Kostov, R.V. Glucosinolates and isothiocyanates in health and disease. *Trends Mol. Med.* **2012**, *18*, 337–347. [CrossRef] [PubMed]

45. Piragine, E.; Flori, L.; Di Cesare Mannelli, L.; Ghelardini, C.; Pagnotta, E.; Matteo, R.; Lazzeri, L.; Martelli, A.; Miragliotta, V.; Pirone, A.; et al. *Eruca sativa* Mill. seed extract promotes anti-obesity and hypoglycemic effects in mice fed with a high-fat diet. *Phytother. Res.* **2020**, 1–8. [CrossRef] [PubMed]

46. Lucarini, E.; Pagnotta, E.; Micheli, L.; Parisio, C.; Testai, L.; Martelli, A.; Calderone, V.; Matteo, R.; Lazzeri, L.; Di Cesare, M.L.; et al. Eruca sativa meal against diabetic neuropathic pain: An H2S-mediated effect of glucoerucin. *Molecules* **2019**, *24*, 3006. [CrossRef]

47. Gullón, B.; Lú-Chau, T.A.; Moreira, M.T.; Lema, J.M.; Eibes, G. Rutin: A review on extraction, identification and purification methods, biological activities and approaches to enhance its bioavailability. *Trends Food Sci. Technol.* **2017**, *67*, 220–235. [CrossRef]

48. Lazzeri, L.; Malaguti, L.; Cinti, S.; Ugolini, L.; De Nicola, G.R.; Bagatta, M.; Casadei, N.; D'Avino, L.; Matteo, R.; Patalano, G. The Brassicaceae biofumigation system for plant cultivation and defence. An Italian twenty-year experience of study and application. *Acta Hortic.* **2013**, *1005*, 375–382. [CrossRef]

49. Conte, L.S.; Leoni, O.; Palmieri, S.; Capella, P.; Lercker, G. Half-seed analysis: Rapid chromatographic determination of the main fatty acids of sunflower seed. *Plant Breed.* **1998**, *102*, 158–165. [CrossRef]

50. Singleton, V.L.; Orthofer, R.; Lamuela-Raventós, R.M. Analysis of total phenols and other oxidation substrates and antioxidants by means of Folin-Ciocalteu reagent. *Methods Enzymol.* **1999**, *299*, 152–178. [CrossRef]

51. Jia, Z.; Mengcheng, T.; Jianming, W. The determination of flavonoid contents in mulberry and their scavenging effects on superoxide radicals. *Food Chem.* **1999**, *64*, 555–559. [CrossRef]

52. Waszkowiak, K.; Gliszczyńska-Świgło, A.; Barthet, V.; Skręty, J. Effect of extraction method on the phenolic and cyanogenic glucoside profile of flaxseed extracts and their antioxidant capacity. *J. Am. Oil Chem. Soc.* **2015**, *92*, 1609–1619. [CrossRef]

53. Felice, F.; Fabiano, A.; De Leo, M.; Piras, A.M.; Beconcini, D.; Cesare, M.M.; Braca, A.; Zambito, Y.; Di Stefano, R. Antioxidant effect of cocoa by-product and cherry polyphenol extracts: A comparative study. *Antioxidants* **2020**, *9*, 132. [CrossRef] [PubMed]

54. Brand-Williams, W.; Cuvelier, M.; Berset, C. Use of a free radical method to evaluate antioxidant activity. *LWT* **1995**, *28*, 25–30. [CrossRef]

Article

Impact of Germination on the Microstructural and Physicochemical Properties of Different Legume Types

Denisa Atudorei, Silviu-Gabriel Stroe and **Georgiana Gabriela Codină** *

Faculty of Food Engineering, Stefan cel Mare University of Suceava, 720229 Suceava, Romania;
denisa.atudorei@outlook.com (D.A.); silvius@fia.usv.ro (S.-G.S.)
* Correspondence: codina@fia.usv.ro or codinageorgiana@yahoo.com; Tel.: +40-745460727

Abstract: The microstructural and physicochemical compositions of bean (*Phaseolus vulgaris*), lentil (*Lens culinaris Merr.*), soybean (*Glycine max* L.), chickpea (*Cicer aretinium* L.) and lupine (*Lupinus albus*) were investigated over 2 and 4 days of germination. Different changes were noticed during microscopic observations (Stereo Microscope, SEM) of the legume seeds subjected to germination, mostly related to the breakages of the seed structure. The germination caused the increase in protein content for bean, lentil, and chickpea and of ash content for lentil, soybean and chickpea. Germination increased the availability of sodium, magnesium, iron, zinc and also the acidity for all legume types. The content of fat decreased for lentil, chickpea, and lupine, whereas the content of carbohydrates and pH decreased for all legume types during the four-day germination period. Fourier transform infrared spectroscopic (FT-IR) spectra show that the compositions of germinated seeds were different from the control and varied depending on the type of legume. The multivariate analysis of the data shows close associations between chickpea, lentil, and bean and between lupine and soybean samples during the germination process. Significant negative correlations were obtained between carbohydrate contents and protein, fat and ash at the 0.01 level.

Keywords: legume; germination; lyophilization; microstructure; chemical compounds; FT-IR analysis; principal component analysis

Citation: Atudorei, D.; Stroe, S.-G.; Codină, G.G. Impact of Germination on the Microstructural and Physicochemical Properties of Different Legume Types. *Plants* **2021**, *10*, 592. https://doi.org/10.3390/plants10030592

Academic Editor: Ivo Vaz de Oliveira

Received: 24 February 2021
Accepted: 15 March 2021
Published: 22 March 2021

1. Introduction

Today, we are witnessing continuous progress at all levels. This is why it is necessary to synchronize the trends in the food industry with those of today, which should, of course, include the requirements and preferences of consumers. What is of increasing interest nowadays is the concept of innovation, which is required more and more by the food industry. In this field, the concept is related to the obtention of new food products that may satisfy and attract as many categories of consumers as possible. The possibility of using different legume types in various forms in order to create innovative foods such as pasta [1], yoghurt [2,3], bakery products [4,5], drinks [6], etc., is an increasing trend nowadays that may satisfy the consumer demand. Studying the consumer market and also the literature, it was observed that consumer interest in healthy diets has increased in recent years. This is a result of the fact that more and more people nowadays face various food deficiencies or various health conditions caused by inadequate nutrition [7–11]. By closely studying what people purchase from pharmacies, it was observed that more and more people buy food supplements based on vitamins and minerals [12]. However, these are produced in laboratories by chemical methods. However, nutritionists recommend that the population adopt a balanced diet that fully meets the vitamin and mineral needs of consumers of all ages. Unfortunately, even if the population understands this, it is not always possible to adopt a healthy lifestyle because some foods, even basic ones, are poor in nutrients and rich in high-calorie compounds [13–16].

Considering the fact that legumes have a balanced nutritional composition, being an ideal source of minerals, vitamins, proteins, etc. [17,18], but also the fact that they help prevent and treat various diseases [19,20], their use for consumption would be desirable. However, they are also known to contain a number of antinutritive compounds [21–23], which prevent the absorption of nutrients into the body. A suitable solution to decrease or even to eliminate the antinutritive compounds from legumes would be to use the germination process to balance them. Studies in the field have repeatedly highlighted, through concrete data, the advantages of the germination process over the nutritional profile of legumes [24,25]. If we also think about the fact that germination is a relatively simple process that does not require special uses and techniques and is also friendly to the environment, then the preference for this process increases. However, studies conducted so far show that the number of advantages depends on how the germination process is conducted. Therefore, in the first phase, it would be necessary to carefully study the influence of the germination process on legumes in order to establish an optimum process.

The importance of this study is supported by the idea that germination is of particular interest in the food industry. Therefore, there are a lot of studies in the literature that have highlighted the possibility of using different legumes in germinated form or flours from them, emphasizing the influence of this addition, both in terms of optimizing the nutritional profile of the food products and on their quality characteristics. Thus, the authors of various articles have outlined the fact that legumes in a germinated form can be used as an addition (as such or in powder form) in a wide range of foods, such as drinks [26], sweet products [27], various salty foods, bakery products, dairy products [28], pasta [29], crackers [30], etc. Depending on the food products, the addition of legumes in a germinated form has different effects. Therefore, studies are needed to highlight all these effects, but these studies should first start by highlighting all the transformations that take place in the grain during germination to better understand how the addition of it in a germination form would influence the food product quality.

Germination is an advantageous process that can be used successfully to improve the nutritional profile of grains subjected to this process, as evidenced by studies in the field [31,32]. This is of interest process because grains in a germinated form can be incorporated as such or in powder form in various recipes of many categories of foods in order to improve their sensory and nutritional quality, but also to reduce the number of chemical additives used. For example, they can be used successfully to replace the addition of enzymes in the bread-making process because during germination the enzymes in the grain are activated, having an essential role in producing bakery products of a high-quality. This is an example of the possibility of using these grains in a germinated form, but the examples can continue with other food products.

This study highlights the influence of the germination process on the internal profile of the different types of legumes (bean, lentil, soybean, chickpea and lupine) and on their nutritional profile (minerals, lipids, proteins, ash) before the germination process begins, at two days and four days, respectively, of the germination period. The germination process effects on different types of legume seeds were analyzed using different modern instruments such as a stereo microscope, scanning electron microscope (SEM), atomic absorption spectrometry, Fourier transform infrared spectroscopic (FT-IR) spectrometer, etc., on nongerminated and germinated seeds for 2 and 4 days. All this must be highlighted in the literature because, before the legumes in the germinated form are used as an addition in the recipes for the manufacture of various foods, it is necessary to understand the germination mechanism and the physical and physiological transformations in the grain, so that the germination form to be used with success, depending on the desired goal. The study is of particular interest due to the fact that, to our knowledge, in the literature there are not many publications related to the germination of different legumes, especially since each legume undergoes specific changes during the germination process.

2. Results

2.1. *Appearance of Legume Seeds during Germination Period*

Figure 1 shows the images captured with the stereo microscope device before subjecting the legume seeds (bean, lentil, soybean, chickpea and lupine) to the germination process and during the second and the fourth days of the germination period.

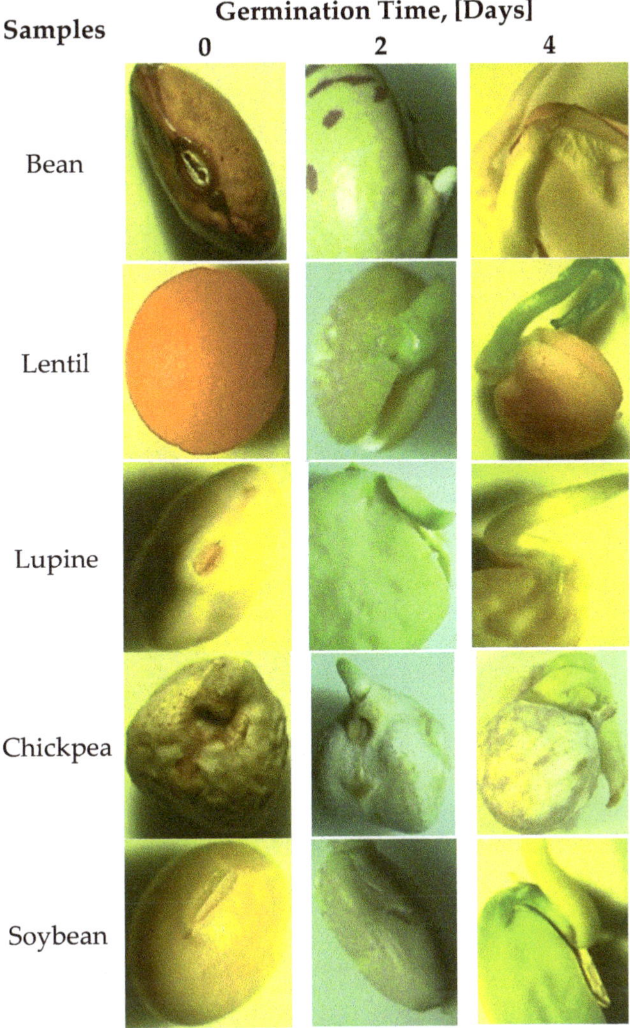

Figure 1. Stereomicroscope images of legumes during germination.

2.2. *SEM Analysis*

In Figure 2a, the microscopy images (SEM) of raw bean, lentil, soybean, chickpea, and lupine seeds can be seen. These were similar to those published in the literature [33–37].

Beans

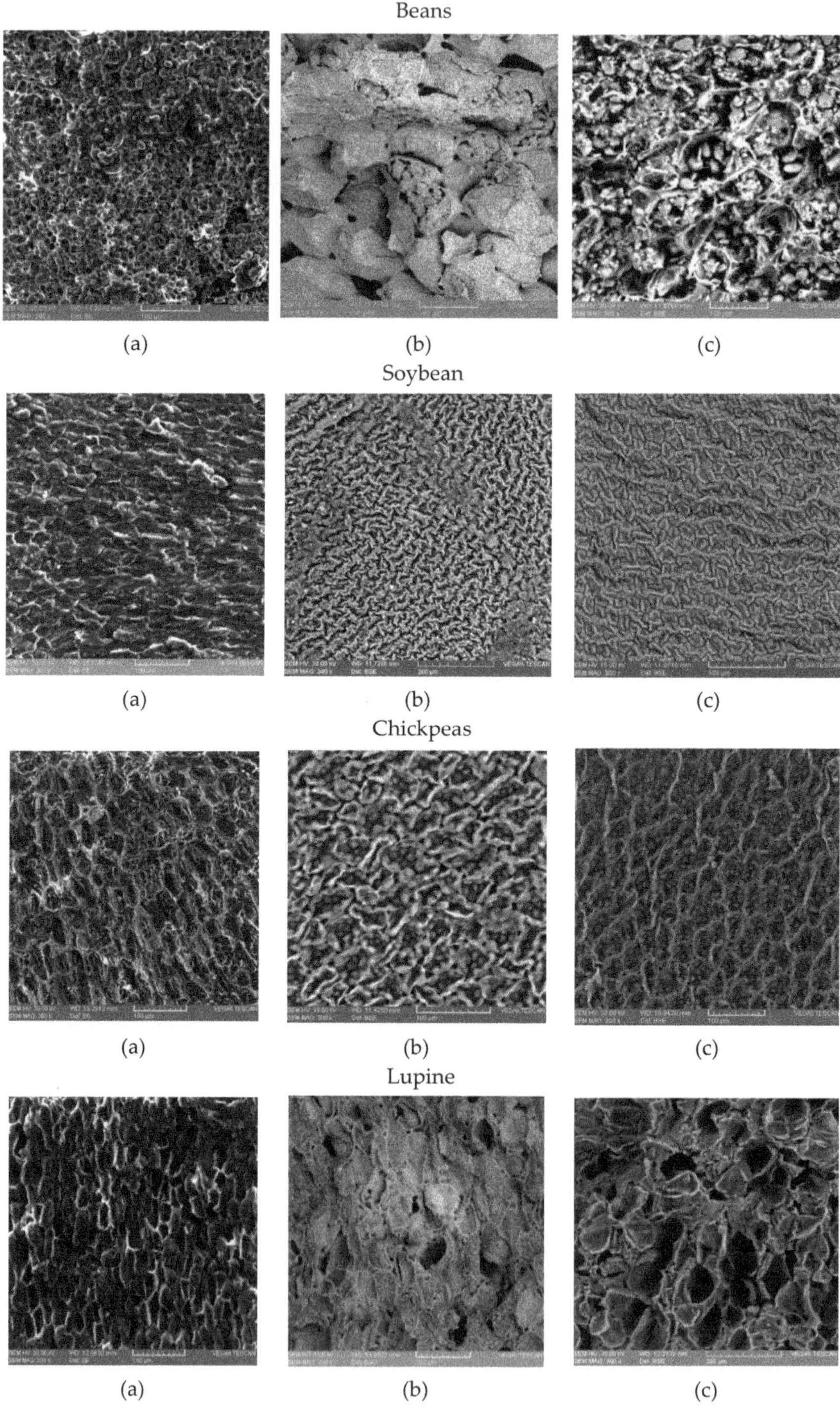

Figure 2. *Cont.*

Lentil

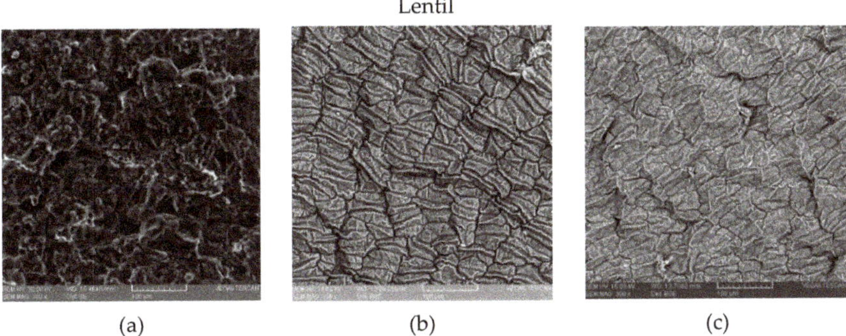

(a) (b) (c)

Figure 2. Scanning electron microscope (SEM) images showing the microstructures of legumes during germination: (**a**) nongerminated; (**b**) germinated 2 days; (**c**) germinated 4 days.

2.3. Physical-Chemical Characterization of Legume Seeds during the Germination Period

The results from Table 1 show that proteins in legume seed flours varied between 19.4 and 40.3% (dry weight), this amount being high due to the fact that the legume seeds are among the richest sources of proteins for animal and human nutrition [38]. Soybean and lupine flours are richer in protein compared to the bean, lentil and chickpea flours, which are around the 20–30% value. Similar data for the protein content of these legume types has also been reported by different researchers [39,40]. Bean, chickpea and lentil flours in germinated form presented an increase in total protein content compared to nongerminated seeds flour. As may be seen from the data obtained, germination for up to four days increased the total protein content by 15.04% in beans, by 8.76% in chickpea and by 3.14% in lentil. Similar results were also reported for germinated bean [41–43], chickpea [44–46] and lentil seeds [45]. However, for lupine flour the total protein content value decreased by 1.25%, whereas for soybean flour any significant variation ($p < 0.05$) during the four-day germination period was not noticed.

Table 1. Physical-chemical properties of legume seeds during germination period.

Legume Type	Germination Period, Days	Protein (%)	Fat (%)	Ash (%)	Moisture (%)	Carbohydrates (%)	pH	Acidity (°)
Bean	0	22.6 ± 0.28 cC	1.6 ± 0.07 aD	3.6 ± 0.07 aB	10.9 ± 0.07 aA	61.3 ± 0.14 aB	6.55 ± 0.007 aB	7.15 ± 0.007 cB
	2	24.1 ± 0.14 bO	1.2 ± 0.14 aO	3.4 ± 0.07 abM	10.5 ± 0.14 abL	60.6 ± 0.14 bL	6.50 ± 0.021 aL	8.12 ± 0.021 bP
	4	26.0 ± 0.28 aY	1.4 ± 0.07 aY	3.0 ± 0.14 bV	10.1 ± 0.14 bUV	59.7 ± 0.14 cV	6.41 ± 0.021 bV	15.07 ± 0.056 aY
Lentil	0	28.6 ± 0.28 bB	1.2 ± 0.07 aD	2.6 ± 0.07 bD	7.6 ± 0.07 aC	60.0 ± 0.07 aC	6.63 ± 0.014 aA	5.16 ± 0.035 cD
	2	30.0 ± 0.14 aN	1.1 ± 0.00 aP	2.8 ± 0.07 bN	8.9 ± 0.14 aN	58.9 ± 0.07 bM	6.51 ± 0.021 bL	10.66 ± 0.021 bM
	4	29.5 ± 0.14 abX	1.0 ± 0.07 aY	3.1 ± 0.14 aV	8.8 ± 0.28 aX	59.0 ± 0.14 bX	6.48 ± 0.007 bU	16.28 ± 0.021 aX
Soybean	0	40.3 ± 0.07 aA	16.6 ± 0.14 bA	4.5 ± 0.07 aA	9.8 ± 0.28 bC	28.8 ± 0.14 aE	6.62 ± 0.028 aA	5.68 ± 0. 007 cC
	2	40.3 ± 0.07 aL	17.6 ± 0.07 aL	4.7 ± 0.07 aL	10.4 ± 0.07 aL	27.0 ± 0.07 bO	6.50 ± 0.007 bL	9.83 ± 0.042 bO
	4	40.2 ± 0.07 aU	17.9 ± 0.07 aU	5.1 ± 0.28 aU	10.5 ± 0.14 aU	26.3 ± 0.14 cZ	6.32 ± 0.021 cX	16.29 ± 0.028 aX
Chickpea	0	19.4 ± 0.07 bD	5.9 ± 0.14 aC	3.1 ± 0.07 bC	10.3 ± 0.28 aB	61.3 ± 0.14 aA	6.42 ± 0.007 aC	4.2 ± 0.141 cE
	2	20.7 ± 0.28 aP	5.4 ± 0.00 bN	3.5 ± 0.07 aM	9.7 ± 0.07 abM	60.7 ± 0.07 bL	6.22 ± 0.028 bM	10.24 ± 0.021 bN
	4	21.1 ± 0.00 aZ	5.2 ± 0.07 bX	3.6 ± 0.07 aV	9.4 ± 0.07 bVX	60.7 ± 0.07 bU	6.11 ± 0.014 cY	20.45 ± 0.629 aV
Lupine	0	39.9 ± 0.14 aA	9.3 ± 0.21 aB	3.3 ± 0.07 aB	7.7 ± 0.14 bD	39.8 ± 0.14 bD	5.62 ± 0.007 aD	14.53 ± 0.014 cA
	2	39.3 ± 0.07 bM	7.5 ± 0.00 bM	3.3 ± 0.07 aM	10.1 ± 0.07 aL	39.8 ± 0.07 bN	5.55 ± 0.007 bN	20.87 ± 0.035 bL
	4	39.4 ± 0.07 bV	6.9 ± 0.07 cV	3.4 ± 0.07 aV	10.3 ± 0.14 aU	40.0 ± 0.07 aY	5.53 ± 0.007 bZ	31.9 ± 0. 007 aU

Means followed by the different letter (^{a,b,c}) within the same column, for each legume type, are significantly different ($p < 0.05$). Means followed by different letter within the same column, for each germination period (^{A,B,C,D,E} for 0 day; ^{L,M,N,O,P} for 2 days; ^{U,V,X,Y,Z} for 4 days), are significantly different ($p < 0.05$).

For soybean, the lipid content increased in a significant way ($p < 0.05$), especially after 2 days of germination period; these data are in agreement with those reported by [47]. Considering all legume types analyzed, soybean presented the highest fat contents. It is a species of legume that it is classified as an oilseed due to its high fat content. It is well known that oil crop fat accumulation takes place during maturation and seed development. Regarding the variation of lipid content from the bean flour, it may be seen that its content did not significantly change ($p < 0.05$) during the germination process.

Ash content of nongerminated seeds ranged from 2.6% in lentil flour to 4.5% in soybean flour. On germination, there was a significant increase ($p < 0.05$) in ash content for lentil, soybean and chickpea flours. However, for bean flour the ash content decreased, whereas for lupine flour the ash content did not present any significant variations ($p < 0.05$) during the germination process.

The moisture content in nongerminated flour samples ranged from 7.6 to 10.9%. After germination and lyophilization, the moisture content of the germinated seed flour did not exceed the 10.5% value, a fact that allowed us to store them for a longer period of time in order to be used as ingredients in different food products.

Carbohydrate contents after germination decreased for all the analyzed samples, except lupine, whose value did not vary in a significant way ($p < 0.05$). However, seed types whose protein levels decreased during the germination process, such as lupine, had slightly increased carbohydrate contents.

The pH decreased and acidity increased in a significant way ($p < 0.05$) for all seed samples during the germination period.

Sodium, magnesium, iron and zinc contents increased for all the seed samples during the germination period, as may be seen in Table 2. Germination led to an increase in sodium and magnesium for all types of legumes during the germination period. This increase was significant ($p < 0.05$) for all seeds in the four-day germination compared to the control samples. The beneficial influence on iron and zinc contents was similar to that of sodium and magnesium for all germinated seed samples. However, this increase was not significant ($p > 0.05$) one for legume seeds in the case of zinc and for bean in the case of iron during the germination period.

Table 2. Minerals of legumes during germination period.

Legume Type	Germination Period, (Days)	Na (mg 100 g^{-1})	Mg (mg 100 g^{-1})	Fe (mg 100 g^{-1})	Zn (mg 100 g^{-1})
	0	38.15 ± 2.61 [bB]	141.65 ± 0.17 [cB]	7.57 ± 0.75 [aA]	3.22 ± 0.20 [aA]
Bean	2	62.62 ± 2.18 [aL]	148.50 ± 0.11 [bM]	7.83 ± 0.75 [aL]	3.23 ± 0.20 [aL]
	4	64.42 ± 0.71 [aV]	152.15 ± 0.01 [aX]	7.89 ± 0.80 [aU]	3.25 ± 0.20 [aU]
	0	26.08 ± 0.61 [cD]	122.35 ± 0.06 [cD]	2.62 ± 0.18 [aC]	3.07 ± 0.20 [aA]
Lentil	2	48.89 ± 0.94 [bM]	126.82 ± 0.09 [bO]	2.66 ± 0.18 [aN]	3.10 ± 0.20 [aL]
	4	53.52 ± 1.57 [aX]	146.63 ± 0.12 [aY]	2.75 ± 0.20 [aX]	3.12 ± 0.22 [aU]
	0	47.50 ± 1.06 [cA]	90.30 ± 0.10 [cE]	5.31 ± 0.14 [aB]	1.91 ± 0.15 [aB]
Soybean	2	60.38 ± 1.59 [bL]	92.41 ± 0.27 [bP]	5.32 ± 0.20 [aM]	1.93 ± 0.10 [aM]
	4	70.19 ± 2.03 [aUV]	106.88 ± 0.08 [aZ]	5.38 ± 0.21 [aV]	1.94 ± 0.15 [aV]
	0	40.11 ± 1.03 [cB]	168.31 ± 0.05 [cA]	5.27 ± 0.30 [aB]	3.25 ± 0.20 [aA]
Chickpea	2	58.66 ± 1.31 [bL]	172.56 ± 0.16 [bL]	5.29 ± 0.29 [aM]	3.26 ± 0.18 [aL]
	4	74.81 ± 3.50 [aU]	174.06 ± 0.03 [aU]	5.36 ± 0.30 [aV]	3.28 ± 0.20 [aU]
	0	31.94 ± 0.66 [cC]	128.98 ± 0.02 [cC]	4.58 ± 0.26 [aB]	3.22 ± 0.20 [aA]
Lupine	2	45.20 ± 1.25 [bM]	137.75 ± 0.11 [bN]	4.61 ± 0.20 [aM]	3.23 ± 0.20 [aL]
	4	69.43 ± 2.08 [aUV]	166.63 ± 0.12 [aV]	4.73 ± 0.35 [aV]	3.25 ± 0.20 [aU]

Means followed by different letter ([a,b,c]) within the same column, for each legume type, are significantly different ($p < 0.05$). Means followed by different letter within the same column, for each germination period ([A,B,C,D,E] for 0 day; [L,M,N,O,P] for 2 days; [U,V,X,Y,Z,] for 4 days), are significantly different ($p < 0.05$).

2.4. FT-IR Analysis

Analyzing the spectra shown in Figure 3, which were obtained using Fourier transform infrared spectroscopy, it can be seen that the compositions in terms of chemical compounds for nongerminated legume seeds and seeds subjected to germination for 2 days and 4 days, respectively, were different. From the analysis of Figure 3, it can be concluded that in the studied spectral range (4000–800 cm^{-1}), there were several peaks that correspond to the different molecular bonds of chemical compounds (proteins, lipids, carbohydrates) that interact with infrared radiation. In the literature, it was highlighted that the spectral range between 3000 and 2800 cm^{-1} corresponds to lipid compounds, due to vibrations produced by carbonyl groups of triglycerides (C-H) [48–50].

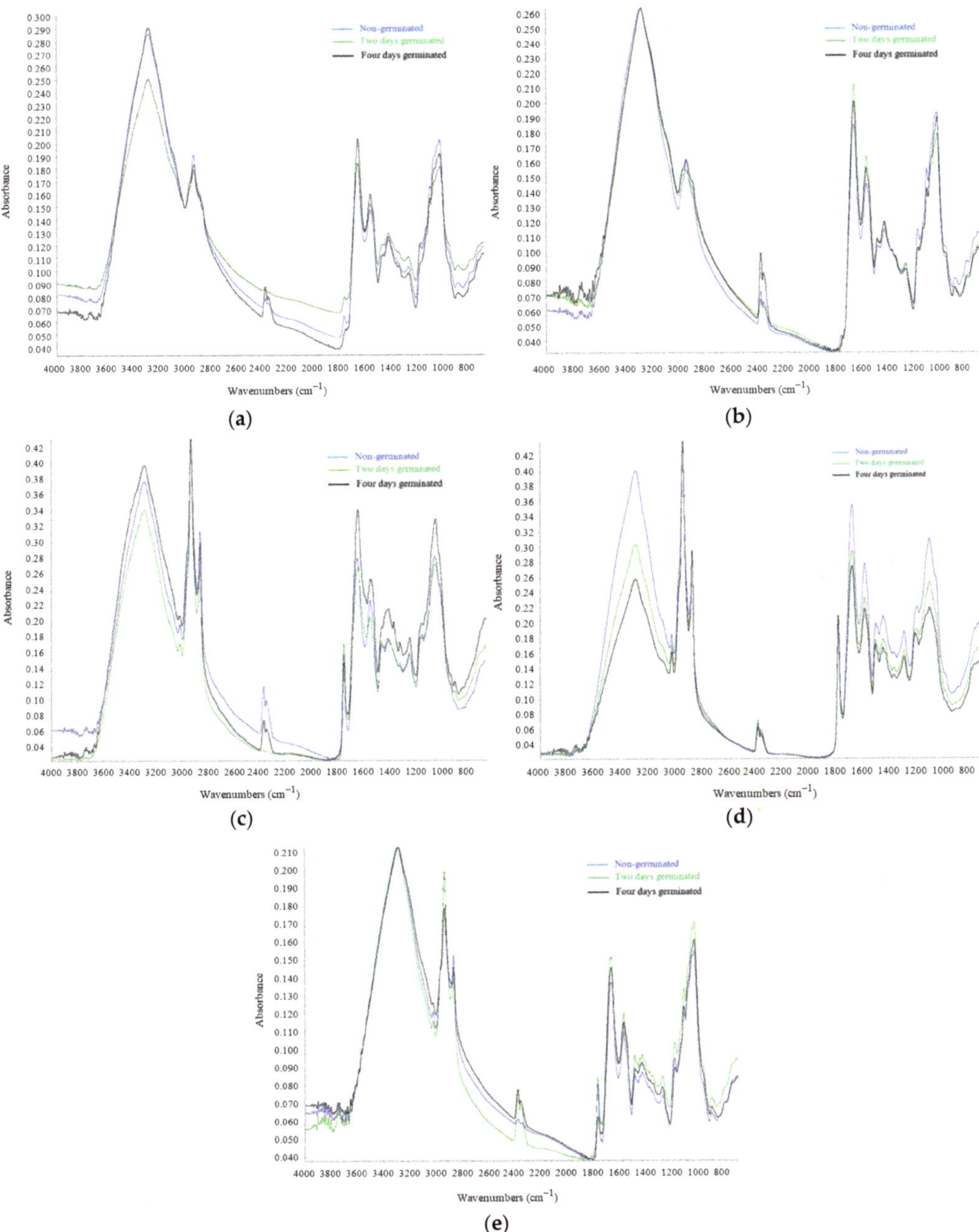

Figure 3. Fourier transform infrared spectroscopic (FT-IR) spectra of legumes during germination period: (**a**) bean; (**b**) lentil; (**c**) lupine; (**d**) soybean; (**e**) chickpea.

2.5. Relationships between Physico-Chemical Values of Legume Seeds during the Germination Period

In order to underline the correlations between the physico-chemical values of legume seeds during the germination period, PCA was used. As can be seen in Figure 4, the first two principal components (PCs) explain 66.25% (PC1—44.26% and PC2—21.99%) of the total variation. The plot of PC1 vs. PC2 loadings shows a close association between soybean and lupine samples, as well as and between chickpea, lentil and bean samples during the germination period. This may be explainable since soybean and lupine contain less carbohydrates and more proteins than the others legume samples analyzed. The carbohydrate parameter, which is predominant in the contents of chickpea, lentil and bean samples, is the closest placed to them and in opposition to the soybean and lupine samples of which protein are closely situated. Additionally, a close association was noticed between nongerminated lentil, bean and chickpea seeds, probably due to the fact that they present more similar compositions than those of nongerminated soybean and lupine seeds.

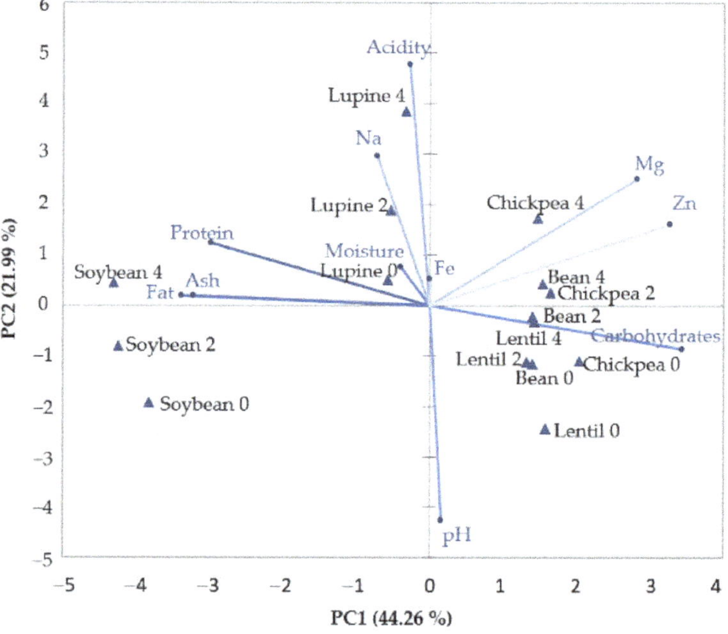

Figure 4. Principal component analysis for physico-chemical values of legume seeds during the germination period.

3. Discussion

3.1. Appearance of Legume Seeds during Germination Period

In the case of beans, an increase in the volume of the grain due to the fact that it absorbed water can be clearly seen. On the second day, the radicle developed up to 2.5 mm, which increased by five times at the fourth day of the germination period. In the four day of the germination period, it may also be seen that the root of the future plant developed. All these physical and physiological changes of the bean seeds during the germination period are in accordance with those described in the international literature by different researchers [51,52].

The influence of germination on the lentil development was also in agreement with different studies reported in the international literature [53]. On the second day of germination, it was clearly seen that the lentil increased in volume and began to develop the radicle, which had a size of 2.5 mm. On day 4 of germination, the radicle was high, the plumule had developed, and the dimensions of the two components of the germ were 6 mm for the radicle and 5 mm for the plumule.

The development of lupine seeds during the germination process has also been previously reported by other authors [54]. As can be seen, on the second day of germination, in the case of lupine, the radicle was 3 mm in size. Starting on the fourth day, the germs began to synthesize chlorophyll, the radicle size being 11 mm.

In the case of chickpea germination, on the second day of germination period seed radicle development was clearly seen, of which the size was of 3 mm. In the fourth day of the germination period, the radicle size increased five times more, its length being of 16 mm. Additionally, on the fourth day of germination period how the leaves of the future plant developed was clearly seen.

For soybean, it can be said that radicle development began on the second day of germination and had the size of 3 mm on this day. The image from the second day of soybean germination clearly shows the degradation of the outer shell of the grain in order to allow the development of the radicle. In the image showing the fourth day of germination, it can be seen how the leaf developed inside the bean. The size of the radicle reached 24 mm in length on this day of germination. Similar changes during soybean germination have also been reported by different researchers [55,56].

3.2. SEM Analysis

Analyzing the microscopy images (SEM) of raw bean, lentil, soybean, chickpea and lupine seeds, middle lamellas separating seed cells and cell walls were clearly visible in all images of nongerminated seeds. In the case of bean, chickpea and lentil, cotyledon cells composed of protein materials which cover elongated starch granules were seen. In the case of soybean and lupine seeds, which contain less starch, more dense compactness of the protein agglomerates was visible.

In germinated seeds, significant structural changes may be seen in cotyledon cells for all analyzed samples. Middle lamellas separating seeds cells and cell walls were visible. In the case of chickpea and bean, a high number of depicted starch granules, spherical ones with smooth surfaces covered in some parts by amorphous protein structures, could clearly be seen. As the germination period went on, cell structures from the germinated seeds continued to change. It may be seen that starch granules were still separated by proteins and presented smooth surfaces but with higher disruption of the cellular order, probably due to the increased activity of the hydrolytic enzymes. More tiny filaments between starch granules and their neighboring proteins appeared. Furthermore, some precipitate was seen on starch granules, probably due to the increase in the proteolytic activity with the increase in the germination period [36]. With the germination process, the changes were more pronounced in the cotyledon cells. Some breakages of the surface were clearly distinguishable. Lupine seeds, as well as soybean and also lentil seeds, presented a different microstructure during germination than those of bean and chickpea. These were probably due to their higher content of protein and lower amount in starch in the case of soybean and lupine seeds. Spherical to oval protein bodies were visible in the continuous cell matrix which appeared to be denser materials. After germination, the cells were more loosely packed with large intercellular spaces between them. These intercellular spaces varied from a few in nongerminated seeds to many in seeds where the germination period was longer. A shorter germination period led to a higher compactness of the protein structure, whereas a longer germination period led to a more open structure of protein agglomerates. The protein structure appeared to be granular with multiple cracking caused by seed germination, which was only slightly visible in the case of nongerminated seeds. The middle lamella, which is a pectin layer, absorbs water during germination and expands, being more visible to the germination seeds [37]. It can be clearly seen that, after germination, ruptures appear around the cells, which makes visible the middle lamella and the cell walls. Secondly, the cracks presented on the seed coat surface may be related to the water uptake during the germination process. Crack distribution may also coincide with the sites of initial water uptake [34]. After lyophilization, the water from the germination process was removed, leaving many cracks in the seeds' structures, which were clearly visible in all the germinated seed samples.

3.3. Physical-Chemical Characterization of Legume Seeds during the Germination Period

The increase in the amount of protein after germination may be due to the fact that, during this process, hydrolysis of some proteins, which are inaccessible within the nongerminated seeds structure, may take place [56]. In legume seeds, the proteins are mainly localized to the seed cotyledon tissues but also to the embryonic axis and testas, which contribute less to the total protein content, mainly due to the fact that these components represent small amounts of the seed mass [57]. During seed development, stored proteins are hydrolyzed to provide carbon skeletons, nitrogen and energy required for growth and protein synthesis [58]. The proteins stored in the vegetative cells are used for seed formation, whereas the embryonic proteins stored are used for seed germination [56]. Protein release from the cell structure during the germination process varies between seed type. However, in the case of lupine and soy, the influence on the amount of protein can be attributed to a stronger hydrolysis effect on storage proteins which causes their degradation and loss for seed development [40]. It must be mentioned that the small protein variation in the case of soy may be better highlighted by the water-soluble protein through FTIR analysis [59]. Studies indicate that during the germination process the solubility of proteins increases for enzymatic activity and the development of the new tissues (the components of germs). Therefore, the amount of water-soluble proteins increases due to the germination process [56]. In this regard, FTIR analysis (presented in Section 2.4) highlighted in a better way how the amount of water-soluble protein varies depending on the increase in germination time.

In general, the germination of the legume seeds decreased the lipid content during the germination period. These data are in agreement with other studies which have also reported a decreased lipid content of some legume seeds during the germination period [42,43]. This decrease may be due to the fact that the lipid was used as an important source of energy during the sprouting process necessary for the development of seed embryos during the germination process [42]. The reduction in the lipid content of germinated flours may be an advantage for their storage due to the fact that may improve its stability [43]. In germinating soybean seeds, also characterized by high protein contents, the main metabolites in the first stage of germination may be amino acids which are important respiratory substrates and may be used in many anabolic processes, and may cause a lower breakdown of storage lipid content [39]. Additionally, even if an increase in free fatty acids and a decrease in triacylglycerol take place during the germination process, some studies [60] have reported that some complex lipids, such as phospholipids, can be synthesized during germination. These facts may lead to a constant or a slight increase in the total lipid content, especially in the first stage of the germination period. In the case of bean, a nonsignificant change of lipid content during the germination process has also been reported by Ferreira (2019) [46] for chickpea germination over a 48 h period. According to this study, this behavior may be due to the short germination period reported for this seed type, which presented the highest size from all the analyzed legume types. Lipids are polymeric structures used for storage, mainly in the form of triacylglycerol. In order to be used for energy generation, the lipases must act on triglycerides to release free fatty acids. In the case of beans, the use of lipids may not have been so high during the 4 days of germination.

The significant increase ($p < 0.05$) in ash content for lentil, soybean and chickpea flour could be due to the increase in phytase activity during germination, which hydrolyzed the bond between the proteins, enzymes and minerals, to release the minerals [43,61]. The decrease in the amount of ash in the case of beans may be due to a shorter germination period reported to the seed type. A reduction in the ash content during the germination period has also been reported by Ferreira (2019) [46] for chickpea germination during the

48 h period. It seems that the ash variation during the germination process depends on the grain type and the germination period [42,62]. Different studies reported that the ash value presented the highest value after 72 h of germination period, the time used by us for the legume germination [63,64]. Additionally, some studies reported that the increase in ash content occurs due to the reduction in total soluble solids [65]. This is probably due to hydrolytic enzymes that have not favored the accumulation of total soluble solids, which led to their increased levels [66].

The reduction in carbohydrate contents after germination may be attributed to amylases activity which increases during germination, leading to a starch degradation process [67]. Additionally, this decrease may be due to the carbohydrates use as a substrate during the germination period leading to an increase in the protein levels [46]. However, the amount of carbohydrate variation during the germination process depends on the legume type probably due to a stronger or weaker hydrolyses of them. During the germination period, the carbohydrates were solubilized in order to support the development of the components of the germs [56].

The pH decrease and acidity increase could be due to the activity of the lipase, which acts on triacylglycerols converting them into free fatty acids necessary for energy generation [46].

The increase in the availability of the analyzed minerals was a general effect of the germination process, which is related to the phytase action, phytate content, extent of binding of minerals within the matrix, or interaction of these factors. Legumes are rich in α-amylase inhibitors, polyphenolic compounds, protease inhibitors, tannins, lectins and phytic acid that cause poor absorption and digestibility of minerals and nutrients. More, legumes contain phytase enzymes that are activated by germination to destroy phytate [68]. Phytate is found in legumes in the form of phytic acid and is considered to be a chelating agent because it forms phytates with minerals. During germination, there is a decrease in the amount of phytate due to its hydrolysis by the enzyme phytase [69]. Phytase is a phytate-specific phosphatase. This hydrolyzes phytate to inositol and free orthophosphate and releases minerals [70]. This may be a reason why the amount of minerals increases after germination. Studies have also shown that with the increase in the amount of minerals after germination, there is an improvement in their digestive availability. Therefore, in the longer germination period, the legumes' mineral contents increase. These data agree with those obtained by different researchers or various types of vegetables during the germination process [71,72]. The literature does not indicate particular mechanisms for each mineral substance that may elucidate the increase in the amount of this following the germination process. Each increase is attributed to the three processes—namely, the decrease in the amount of phytic acid, the activation of phytase and the release of minerals found in legumes in bound form (in the constitution of different complex substances) [65,68,70,73,74]. These transformations occur due to the fact that germination is a catabolic process by which, in the first stage, with the absorption of water, the reserve substances from seeds are hydrolyzed in order to provide nutrients important for the development of germs. A decrease in phytic acid content lad to an increase in some mineral contents from the seed subjected to germination [75]. However, all these variations are different for each grain.

3.4. FT-IR Analysis

By analyzing the spectra obtained for bean in both raw and germinated forms, it can be seen that the highest absorbance at the wavelength range specified above was recorded in the case of nongerminated beans and the lowest absorbance value was recorded in the case of beans that germinated for two days. The peaks recorded in this spectral range are also due to the vibrations of the -C-H(CH$_2$) and -C-H(CH$_3$) groups in the composition of fatty acids, as explained in the literature [50]. Additionally, the literature highlights the fact that the peaks recorded around the wavelength 1750 cm^{-1} are determined by the C=O bonds of the ester or carboxylic acid groups, in the FT-IR spectra this peak is in the range of 1800–1700 cm^{-1}. The spectral range between 1700 and 1500 cm^{-1} corresponds to the protein content of the samples. Analyzing the spectra obtained in the case of beans, it can be concluded that the absorbance recorded a higher value in the case of beans subjected to germination for 4 days and the lowest value for nongerminated beans. The two peaks that are very clearly visible are suitable for amide I (C=O, C-N), with a wavelength of about 1650 cm^{-1}, and for amide II (N-H, C-N), with a wavelength of about 1550 cm^{-1}, as shown by the interpretation of the bean spectra. Additionally, the second peak recorded is due to the peptide (CO-NH) bond [76,77]. Thus, it can be concluded that the peaks in the range of 1700–1500 cm^{-1} are due to the vibrations produced by the peptide bonds and thus they can be an indicator for the protein content of the samples. To characterize the carbohydrates in the samples using FT-IR determination, different researchers have showed that the wavelengths in the range 1200–900 cm^{-1} correspond to carbohydrates [78]. By analyzing the obtained spectra, it can be concluded that the absorbance recorded a higher value in the case of nongerminated beans and the lowest value for two days of bean germination. The highest intensity of the peak was recorded for the wavelength of about 1050 cm^{-1}, which corresponds to the OH group of carbohydrates. It was also observed that the peaks differ more in intensity values, not in the shapes of the peaks.

Analyzing the spectra obtained for nongerminated lentil and for the lentil germinated (two and four days), it can be concluded that the wavelength between 1700 and 15,000 cm^{-1} corresponds to the amount of protein. The maximum height of the peak was obtained for the wavelength of 1700 cm^{-1}. The recording of the peak at this wavelength is due to the vibration made by the OH and C=O groups in the amide I region [79,80], which provides indications of the protein content of the sample. The second peak in this spectral region is due to the vibration caused by the N-H and C-N groups associated with amide II [81]. The wide peak corresponding to the maximum wavelength of 1050 cm^{-1} corresponds to the vibration of the C-O groups from the carbohydrate content [80,82].

From the analysis of the obtained spectra for the nongerminated and germinated lupine grains (at two days and at four days of germination, respectively), it can be concluded that the spectra for nongerminated lupine and those for 2 days and 4 days of germination, respectively, are similar—only the value of absorbance differs. For example, in the spectral range of 1200–800 cm^{-1}, there were peaks with high absorbance values in the case of germinated chickpeas for 4 days. In this spectral range, peaks provide indications of carbohydrate concentration [83]. In this case, the values recorded correspond to those indicated by us through physicochemical data.

The spectral range between wavelengths 1700 and 1300 cm^{-1} provides indications of the amount of protein. The maximum value of absorbance was recorded in the case of nongerminated soybeans, with a wavelength value of approximately 1650 cm^{-1}. The lowest peak height in this spectral range was recorded for soybeans germinated for 4 days. Regarding carbohydrate compounds that correspond to the wavelength between 1200 and 1900 cm^{-1}, and according to the spectra obtained, it can be concluded that the peak had a higher value in the case of nongerminated soybeans and a lower value with the increase in the germination period. Thus, it can be concluded that the amount of carbohydrates decreased with an increasing germination period. The recording of peaks in this spectral range was achieved due to the vibrations of the C-OH and C-O-C groups [84]. The interpretation of the data obtained by using FT-IR analysis corresponds to the physicochemical data which also indicated that in the case of germinated soybeans the amount of protein and carbohydrates was lower when the germination period increased.

In the case of chickpeas, from the spectra obtained from the FT-IR analysis, it may be clearly seen that in the wavelength range between 3000–2850 cm^{-1}, corresponding to lipid compounds [85], the highest value of the absorbance was recorded for nongerminated chickpea, and the value decreased with a longer germination period. Previous studies indicate similar data, namely that the peaks from the spectral range 3050–2800 cm^{-1} are due to the CH$_3$ and CH$_2$ groups which are in the composition of hydrocarbons from the lipid constitution. The wavelength around 3000 cm^{-1} corresponds to the vibration of the groups =C-H from the composition of fatty acids [86]. Comparing the data obtained from the FT-IR analysis and the physicochemical ones, it can be concluded that there is a correlation between the data obtained, in the sense that by physicochemical determinations it was also shown that the amount of lipids decreased with increasing germination time in the case of chickpea seeds.

3.5. Relationships between Physico-Chemical Values of Legume Seeds during the Germination Period

The first main component (PC1) opposes the pH and acidity values, obtaining a significant reverse correlation (r = −0.776) for a level of 0.01. This fact is explainable since it is well known that the higher the acidity, the lower the pH. Additionally, the carbohydrate contents, which are in predominant in all legume types analyzed, are in opposition with all chemical data of protein, fat and ash, which present high significant negative correlations of r = −0.930, r = −0.922 and r = −0.809 for a level of 0.01. It is known that nitrogenous substances accumulate in the grains before the starch. The more starch that accumulates in the grain during the maturation process, the lower its contents of protein and other substances are. Of the carbohydrates content, starch is predominant, and therefore the negative correlation between carbohydrates and protein, fat and ash is somehow predictable. Closeness of moisture and iron (Fe) to the center of PC shows that these variables are not useful when describing the differences between legumes analyzed. As for the second component (PC2), the lupine and soybean legume types were placed in the left part of the graph, while chickpea, bean and lentil were placed in the right part of the graph, with very good correlations shown between these samples. From the mineral association point of view with the analyzed legumes, it seems that magnesium (Mg) and zinc (Zn) minerals are more correlated with chickpea samples, which, according to the data obtained, are in the highest amount, whereas sodium (Na) is the most associated with lupine samples.

4. Materials and Methods

4.1. Materials

Bean (*Phaseolus vulgaris*), lentil (*Lens culinaris Merr.*), soybean (*Glycine max* L.), chickpea (*Cicer aretinium* L.) and lupine (*Lupinus albus*) from the 2019 crop year were cultivated in Romania and were not genetically modified according to the manufacturer declaration.

4.2. Germination and Lyophilization of Legume Seeds

Before the germination, the legumes were soaked for 6 h (small grains: lentils, soybeans) or for 12 h (larger grains: beans, lupine, chickpeas), in pure water, with a temperature of 20 °C in order to activate the germination process. The germination of the grains was carried out using filter paper. The germination temperature was 25 °C. The humidity of 80% was kept constant throughout the entire germination period. The germination process was carried out exclusively in dark conditions. The maximum germination time was 4 days. At the end of the germination period, this process was stopped. For this, the freeze-drying process was used because, compared to other moisture removal alternatives, this process has the least influence on the nutritional profile of germinated seed samples [87,88]. For this purpose, an Alpha 1–4 LSC plus lyophilizer was used for lyophilization. Lyophilization was performed at −50 °C, for 72 h, at a pressure of 4.2 Pa.

4.3. Appearance of Legume Seeds Analysis

In order to highlight the physical and physiological changes that occur in the legume profiles during the germination period, the Motic SMZ-140 stereo microscope device was used, which allowed the capturing of images with legumes in the dorsal, frontal, and ventral positions and in the section of them. For this, the stereo microscope was equipped with a Moticam 10× camera. The use of the stereo microscope made it possible to highlight the changes in legume profiles: increase in volume due to water absorption, degradation of the outer layer, the appearance of the radicle, root and plumule depending on legume type. The determinations were made taking into account the procedure described in the literature [89,90]. All these transformations were correlated with the determination of the size of the constituent parts of the germs (radicle and plumule), using a Modelcraft Vernier Caliper of 125 mm. Additionally, the caliper was used taking into account the procedure described in the literature [91].

4.4. SEM Analysis

The study of the microstructures of the vegetable samples was performed using a scanning electron microscope (SEM) Vega II LMU-Tescan (Brno-Kohoutovice, Czech Republic) equipped with a SE detector, which works only in a high vacuum environment, and a BSE detector, a scintillation detector that works in high and low vacuum environments. The Vega II LMU-Tescan microscope, used in this research, allows the study of samples in a pressure range between 10–2 Pa and 2 kPa. The vegetable samples were sectioned longitudinally in order to obtain a flat surface, were placed on an aluminum sample holder using a double-sided adhesive tape and were studied at an acceleration potential of 30 kV.

4.5. Physico-Chemical Composition Analysis

Protein, fat, ash and moisture contents were determined using AACC International Approved Methods 46-12.01, 30-25.01, 08-01.01 and 44-15.02, respectively. The pH was measured with a HQ30d portable pH Meter (HACK, Germany) on a slurry prepared with 10 g legume flours in 40 mL of boiled, deionized water according to official AOAC procedures (AOAC, 02-52.01). The total acidity value was determined according to Romanian standard method SR 90:2007. Carbohydrate contents of the legume samples were calculated according to the following formulas [92]: carbohydrates, % = 100 − (protein, % + fat, % + ash, % + moisture, % content of legumes).

The Na, Mg, Fe and Zn contents of the vegetable samples were analyzed by flame atomic absorption spectrometry (FAAS) (AA-6300 Shimadzu, Kyoto, Japan) equipped with air-acetylene flame. Hollow cathode lamps of Na, Mg, Fe and Zn were used. In total, 10 g with an accuracy of 10 mg from each sample was used for calcination. The calcination temperature was increased with a maximum speed of 50 °C/h up to 450 °C. The calcination time was 8 h. Ash digestion was performed using 10 mL 0.1 mol/L nitric acid (HNO_3) (Sigma-Aldrich/Merck, Darmstadt, Germany) on a hot plate. After digestion of the ash samples, up to 50 mL was filled with bidistilled and deionized water. Standard solutions of Na, Mg, Fe and Zn (Sigma-Aldrich/Merck, Darmstadt, Germany) were used and diluted as necessary to obtain working standards. In order to eliminate the risk of contamination, all glassware was washed after each use with HNO_3 solution and rinsed with bidistilled and deionized water. The instrumental conditions for determining the mineral content of vegetable samples by FAAS method are shown in Table 3.

Table 3. Instrumental conditions for mineral analysis to flame atomic absorption spectrometry (FAAS).

Element	Wavelength (nm)	Slit Width (nm)	Fuel Gas Flow Rate (L/min)	Support Gas Flow Rate (L/min)	Flame Type	Pre-Spray Time (s)	Integration Time (s)	Response Time (s)
Na	589.0	0.2	1.8	15.0	Air-C_2H_2	10	5	1
Mg	285.2	0.7	1.8	15.0	Air-C_2H_2	10	5	1
Fe	248.3	0.2	2.2	15.0	Air-C_2H_2	10	5	1
Zn	213.9	0.7	2.0	15.0	Air-C_2H_2	10	5	1

The number of injections of the solutions in the flame was: standard solution—5; blank solution—3; sample solution—3. In order to eliminate analytical errors, each sample was analyzed three times.

4.6. FT-IR Analysis

In order to highlight the changes in the composition of grains subjected to germination process, FT-IR analysis was also used. This allows the realization of a correlation between the spectra and the physicochemical data obtained previously. To obtain the spectra, a FT-IR spectrometer (Thermo Scientific, Karlsruhe, Dieselstraße, Germany) was used, equipped with the ATR IX option, which allowed the obtention of accurate data, using a detector at 4 cm^{-1}. Therefore, an attenuated total reflectance-Fourier transform infrared spectroscopy (ATR-FTIR) was used to obtain the spectra. This technique is based on tracking the interaction between infrared radiation and the material under analysis. Fourier transform infrared spectroscopy (FT-IR) contains a source of light with infrared radiation which passes through the sample and this absorbs luminous energy. At the same time, occur vibrational movements due to the chemical bonds inside the molecules. These vibrational movements provide information about the chemical structure of the sample and these are provided as an FT-IR spectrum. FT-IR spectra were recorded in the spectral range situated between 800 and 4000 cm^{-1}. This spectral range was suitable for characterizing the chemical compounds present in the seed samples: proteins, carbohydrates, lipids, etc.

4.7. Statistical Analysis

All of the data were made in duplicate and are expressed as the means of the measurements ± standard deviation. The one-way analysis of variance (ANOVA) with Tukey's test were made to test the differences between means at a 5% significance level by using the XLSTAT statistical package (2021 version, Addinsoft, Inc., Brooklyn, NY, USA).

5. Conclusions

The germination period significantly influenced the growth of the seeds, leading gradually to development of the radicle and first leaves. The seed development was captured by a stereo microscope device which showed that a four-day germination may be optimum for legumes used in food consumption. The microstructures of the seeds during germination changed, ascribed in the case of bean, chickpea and lentil mainly to the starch, and in the case of lupine and soybean to the protein. Germination of seeds resulted in increased protein and ash contents for lentil and chickpea, whereas for the rest of the legume seeds their contents showed different variations. Additionally, during the germination period the fat content varied for bean and soybean, whereas it decreased for lentil, chickpea, and lupine. Mineral contents (sodium, magnesium, zinc, iron) of the germinated legume seeds increased during germination in all legume seeds, showing the beneficial influences of germination on the nutritional profile of legumes. An increase in acidity values and a decrease in pH and carbohydrate contents have also been recorded for all legume types during the germination period. Fourier transform infrared spectroscopy (FT-IR) highlighted variation in chemical compounds of legume seeds during the germination period. According to the wavelength and peak height of the FT-IR spectra, we clearly showed the fact that different compounds such as protein, carbohydrates and lipids varied depending on germination period and each legume type. Principal component analysis was performed on the combined physico-chemical and minerals data, clustering the legume types obtained during the germination process in the PC space. PCA highlighted an association between lentil, bean and chickpea samples which were placed in the right part of the graph and between soybean and lupine samples which were placed in the left part of the graph, indicating similar compositions of these samples. From the physical-chemical data point of view, significant negative correlations were obtained between carbohydrate contents and protein, fat and ash variables at a level of 0.01.

Author Contributions: D.A., S.-G.S. and G.G.C. contributed equally to the study design, collection of data, development of the sampling, analyses, interpretation of results, and preparation of the paper. All authors have read and agreed to the published version of the manuscript.

Funding: This research received no external funding.

Institutional Review Board Statement: Not applicable.

Informed Consent Statement: Not applicable.

Data Availability Statement: Not applicable.

Acknowledgments: This work was supported by a grant of the Romanian Ministry of Education and Research, CNCS—UEFISCDI, project number PN-III-P1-1.1-TE-2019-0892, within PNCDI III.

Conflicts of Interest: The authors declare no conflict of interest.

References

1. Laleg, K.; Cassan, D.; Abecassis, J.; Micard, V. Procédé de Fabrication de Pâte Destinée a L'alimentation Humaine et/ou Animale Comprenant au Moins 35% de Legumineuse. Patent No. WO 2016097328 A1, 23 June 2016.
2. Eker, M.E.; Karakaza, S. Influence of the Addition of Chia Seeds and Germinated Seeds and Sprouts on the Nutritional and Beneficial Properties of Yogurt. *Int. J. Gastron. Food Sci.* **2020**, *22*, 100276. [CrossRef]
3. Benmeziane, F.; Raigar, R.K.; Ayat, N.E.H.; Aoufi, D.; Djermoune-Arkoub, L.; Chala, A. Lentil (*Lens culinaris*) Flour Addition to Yogurt: Impact on Physicochemical, Microbiological and Sensory Attributes during Refrigeration Storage and Microstructure changes. *LWT* **2021**, *140*, 110793. [CrossRef]
4. Martins, Z.E.; Pinho, O.; Ferreira, I.M.P.L.V.O. Food Industry By-products Used as Functional Ingredients of Bakery Products. *Trends Food Sci. Technol.* **2017**, *67*, 106–128. [CrossRef]
5. Mitelut, A.C.; Popa, E.E.; Popescu, P.A.; Popa, M.A. Chapter 7–Trends of Innovation in Bread and Bakery Production. In *Trends in Wheat and Bread Making*; Academic Press: Cambridge, MA, USA, 2021; pp. 199–226.
6. Mridula, D.; Sharma, M. Development of Non-dairy Probiotic Drink Utilizing Sprouted Cereals, Legume and Soymilk. *LWT* **2015**, *62*, 482–487. [CrossRef]
7. Ben-Porat, T.; Weiss, R.; Sherf-Dagan, S.; Nabulsi, N.; Maayani, A.; Khalaileh, A.; Abed, S.; Brodie, R.; Harari, R.; Mintz, Y.; et al. Nutritional Deficiencies in Patients with Severe Obesity before Bariatric Surgery: What Should Be the Focus During the Preoperative Assessment? *J. Acad. Nutr. Diet* **2020**, *120*, 874–884. [CrossRef]
8. McLoughlin, R.J.; McKie, K.; Hirsh, M.P.; Cleary, M.A.; Aidlen, J.T. Impact of Nutritional Deficiencies on Children and Young Adults with Crohn's Disease Undergoing Intraabdominal Surgery. *J. Pediatr. Surg.* **2020**, *55*, 1556–1561. [CrossRef] [PubMed]
9. Zule, S.; Wanik, J.; Holm, E.M.; Bruder, M.B.; Shanley, E.; Sherman, C.Q.; Fitterman, M.; Lerner, J.; Marcello, M.; Parenchuck, N.; et al. Nutritional Deficiency Disease Secondary to ARFID Symptoms Associated with Autism and the Broad Autism Phenotype: A Qualitative Systematic Review of Case Reports and Case Series. *J. Acad. Nutr. Diet.* **2021**, *121*, 467–492.
10. Carubbi, F.; Barbato, A.; Burlina, A.B.; Francini, F.; Mignani, R.; Pegoraro, E.; Landini, L.; De Danieli, G.; Bruni, S.; Strayyullo, P. Nutrition in Adult Patients with Selected Lysosomal Storage Diseases. *Nutr. Metab. Cardiovasc. Dis.* **2020**, *31*, 733–744. [CrossRef]
11. Wang, H.; Liu, F.; Ma, H.; Zin, H.; Wang, P.; Bai, B.; Guo, L.; Geng, Q. Associations between Depression, Nutrition, and Outcomes among Individuals with Coronary Artery Disease. *Nutrition* **2021**, *86*, 111157. [CrossRef]
12. Rajakumari, R.; Oluwafemi, O.S.; Thomas, S.; Kalarikkal, N. Dietary Supplements Containing Vitamins and Minerals: Formulation, Optimization and Evaluation. *Powder Technol.* **2018**, *336*, 481–492. [CrossRef]
13. Scaife, J.C.; Godier, L.R.; Reinecke, A.; Harmer, C.J.; Park, R.J. Differential Activation of the Frontal Pole to High vs. Low Calorie Foods: The Neural Basis of Food Preference in Anorexia Nervosa? *Psychiatry Res. Neuroimaging* **2016**, *258*, 44–53. [CrossRef]
14. de Vries, R.; de Vet, E.; de Graaf, K.; Boesveldt, S. Foraging Minds in Modern Environments: High-calorie and Savory-taste Biases in Human Food Spatial Memory. *Appetite* **2020**, *152*, 104718. [CrossRef]
15. Lee, M.; Lee, J.H. Automatic Attentional Bias Toward High-calorie Food Cues and Body Shape Concerns in Individuals with a High Level of Weight Suppression: Preliminary Findings. *Eat. Behav.* **2021**, *40*, 101471. [CrossRef]
16. Racine, S.; Suissa-Rocheleau, L.; Martin, S.J.; Benning, S.D. Implicit and Explicit Motivational Responses to High- and Low-calorie Food in Women with Disordered Eating. *Int. J. Psychophysiol.* **2021**, *159*, 37–46. [CrossRef]
17. Kan, L.; Nie, S.; Hu, J.; Wang, S.; Bai, Z.; Wang, J.; Zhou, Y.; Jiang, J.; Zeng, Q.; Song, K. Comparative Study on the Chemical Composition, Anthocyanins, Tocopherols and Carotenoids of Selected Legumes. *Food Chem.* **2018**, *260*, 317–326. [CrossRef]
18. Kumar, S.; Pandey, G. Biofortification of Pulses and Legumes to Enhance Nutrition. *Heliyon* **2020**, *6*, e03682. [CrossRef] [PubMed]
19. Oreno-Valdespino, C.A.; Luna-Vital, D.; Camacho-Ruiz, R.M.; Mojica, L. Bioactive Proteins and Phytochemicals from Legumes: Mechanisms of Action Preventing Obesity and Type-2 diabetes. *Food Res. Int.* **2020**, *130*, 108905. [CrossRef]
20. Liu, W.; Dehghan, M.; Mente, A.; Wang, C.; Yan, R.; Rangarajan, S.; Tse, L.A.; Yusuf, S.; Liu, X.; Wang, Y.; et al. Fruit, Vegetable, and Legume Intake and the Risk of All-cause, Cardiovascular, and Cancer Mortality: A Prospective Study. *Clin. Nutr.* **2021**. [CrossRef] [PubMed]
21. Hefnawy, T.H. Effect of Processing Methods on Nutritional Composition and Anti-nutritional Factors in Lentils (*Lens culinaris*). *Ann. Agric. Sci.* **2011**, *56*, 57–61. [CrossRef]
22. Rui, S.; Hua, W.; Rui, G.; Qin, L.; Lei, P.; Jianan, L.; Zhihui, H.; Chanyou, C. The Diversity of Four Anti-nutritional Factors in Common Bean. *Hortic. Plant J.* **2016**, *2*, 97–104.
23. Rahate, K.A.; Madhumita, M.; Pradhakar, P.K. Nutritional Composition, Anti-nutritional Factors, Pretreatments-cum-processing Impact and Food Formulation Potential of Faba Bean (*Vicia faba* L.): A Comprehensive Review. *LWT* **2021**, *138*, 110796. [CrossRef]
24. Swieca, M.; Gawlik-Dziki, U.; Jakubczyk, A.; Bochnak, J.; Sikora, M.; Suliburska, J. Nutritional Quality of Fresh and Stored Legumes Sprouts—Effect of *Lactobacillus plantarum* 299v Enrichment. *Food Chem.* **2019**, *288*, 325–332. [CrossRef]
25. Ohanenye, I.C.; Tsopmo, A.; Ejike, C.E.C.C.; Udenigwe, C.C. Germination as a Bioprocess for Enhancing the Quality and Nutritional Prospects of Legume Proteins. *Trends Food Sci. Technol.* **2020**, *101*, 213–222. [CrossRef]

26. Chavan, M.; Gat, Y.; Harmalkar, M.; Waghmare, R. Development of Non-dairy Fermented Probiotic Drink Based on Germinated and Ungerminated Cereals and Legume. *LWT* **2018**, *91*, 339–344. [CrossRef]

27. Kaczmarska, K.T.; Chandra-Hioe, M.V.; Frank, D.; Arcot, J. Enhancing Wheat Muffin Aroma through Addition of Germinated and Fermented Australian Sweet Lupin (*Lupinus angustifolius* L.) and Soybean (*Glycine max* L.) Flour. *LWT* **2018**, *96*, 205–214. [CrossRef]

28. Ujiroghene, O.J.; Liu, L.; Zhang, S.; Lu, J.; Zhang, C.; Lv, J.; Pang, X.; Zhang, M. Antioxidant Capacity of Germinated Quinoa-based Yoghurt and Concomitant Effect of Sprouting on Its Functional Properties. *LWT* **2019**, *116*, 108592. [CrossRef]

29. Torres, A.; Frias, J.; Granito, M.; Vidal-Valverde, C. Germinated *Cajanus cajan* Seeds as Ingredients in Pasta Products: Chemical, Biological and Sensory Evaluation. *Food Chem.* **2007**, *101*, 202–211. [CrossRef]

30. Polat, H.; Capar, T.D.; Inanir, C.; Ekici, L.; Yalcin, H. Formulation of Functional Crackers Enriched with Germinated Lentil extract: A Response Surface Methodology Box-Behnken Design. *LWT* **2020**, *123*, 109065. [CrossRef]

31. López-Martínez, L.X.; Leyva-López, N.; Gutiérrez-Grijalva, E.P.; Heredia, J.B. Effect of Cooking and Germination on Bioactive Compounds in Pulses and Their Health Benefits. *J. Funct. Foods* **2017**, *38*, 624–634. [CrossRef]

32. Xu, M.; Jin, Z.; Peckrul, A.; Chen, B. Pulse Seed Germination Improves Antioxidative Activity of Phenolic Compounds in Stripped Soybean Oil-in-water Emulsions. *Food Chem.* **2018**, *250*, 140–147. [CrossRef]

33. Miano, A.C.; Garcia, J.A.; Augusto, P.E.D. Correlation between Morphology, Hydration Kinetics and Mathematical Models on Andean Lupin (*Lupinusmutabilis* Sweet) Grains. *LWT Food Sci. Techol.* **2015**, 290–298. [CrossRef]

34. Ma, F.; Cholewa, E.; Mohamed, T.; Peterson, C.A.; Gijzen, M. Cracks in the Palisade Cuticle of Soybean Seed Coats Correlate with Their Permeability to Water. *Ann. Bot.* **2004**, *94*, 213–228. [CrossRef]

35. Marconi, E.; Ruggeri, S.; Cappelloni, M.; Leonardi, D.; Carnovale, E. Physicochemical, Nutritional, and Microstructural Characteristics of Chickpeas (*Cicerarietinum* L.) and Common Beans (*Phaseolus vulgaris* L.) Following Microwave Cooking. *J. Agric. Food Chem.* **2000**, *48*, 598–694. [CrossRef]

36. Biaszczak, W.; Doblado, R.; Frias, J.; Vidal-Valverde, C.; Sadowska, J.; Fornal, J. Microstructural and Biochemical Changes in Raw and Germinatedcowpea Seeds upon High-pressure Treatment. *Food Res. Int.* **2007**, *40*, 415–423. [CrossRef]

37. Swanson, B.; Hughes, J.S.; Rasmussen, H.P. Seed Microstructure: Review of Water Imbition in Legumes. *Food Struct.* **1985**, *4*, 115–124.

38. Atudorei, D.; Stroe, S.G.; Codină, G.G. Physical, Physiological and Minerals Changes of Different Legumes Types during the Germination Process. *Food Technol.* **2020**, *9*, 844–863.

39. Borek, S.; Ratajczack, W.; Ratajczack, L. Regulation of Storage Lipid Metabolism in Developing and Germinating Lupin (*Lupinus* spp.) Seeds. *Acta Physiol. Plant* **2015**, *37*, 119. [CrossRef]

40. Atudorei, D.; Codină, G.G. Perspectives on the Use of Germinated Legumes in the Bread Making Process, A Review. *Appl. Sci.* **2020**, *10*, 6244. [CrossRef]

41. Kassegn, H.H.; Atsbha, T.W.; Weldeabezgi, L.T.; Yildiz, F. Effect of Germination Process on Nutrients and Phytochemicals Contents of Faba Bean (*Viciafaba* L.) for Weaning Food Preparation. *Cogent Food Agric.* **2018**, *4*. [CrossRef]

42. Huang, P.V.; Yen, N.T.H.; Phi, N.T.L.; Tien, N.P.H.; Trung, N.T.T. Nutritional Composition, Enzyme Activities and Bioactive Compounds Ofmung Bean (*Vignaradiata* L.) Germinated under Dark and Light Conditions. *LWT Food Sci. Techol.* **2020**, *133*, 110100. [CrossRef]

43. Chinma, C.E.; Adedeji, O.E.; Etim, I.I.; Aniaka, G.I.; Mathew, E.O.; Ekeh, U.B.; Anumba, N.L. Physicochemical, Nutritional, and Sensory Properties of Chips Produced from Germinated African Yam Bean (*Sphenostylisstenocarpa*). *LWT Food Sci. Techol.* **2021**, *136*, 110330. [CrossRef]

44. Domínguez-Arispuro, D.M.; Cuevas-Rodríguez, E.O.; Milán-Carrillo, J.; León-López, L.; Gutiérrez-Dorado, R.; Reyes-Moreno, C. Optimal Germination Condition Impacts on the Antioxidant Activity and Phenolic Acids Profile in Pigmented Desi Chickpea (*Cicerarietinum* L.) Seeds. *J. Food. Sci. Technol.* **2018**, *55*, 638–647. [CrossRef]

45. Xu, M.; Jin, Z.; Simsek, S.; Hall, C.; Rao, J.; Chen, B. Effect of Germination on the Chemical Composition, Thermal, Pasting, and Moisture Sorption Properties of Flours from Chickpea, Lentil, and Yellow Pea. *Food Chem.* **2019**, *295*, 579–587. [CrossRef] [PubMed]

46. Ferreira, C.D.; Bubolz, V.K.; da Silva, J.; Dittgen, C.L.; Ziegler, V.; de Oliveira Raphaelli, C.; de Oliveira, M. Canges in the Chemical Composition and Bioactive Compounds of Chickpea (*Cicer arietinum* L.) Fortified by Germination. *LWT* **2019**, *111*, 363–369. [CrossRef]

47. Kayembe, N.C.; Rensburg, J. Germination as a Processing Technique for Soybeans in Small-scale Farming. *S. Afr. J. Anim. Sci.* **2013**, *42*, 167–173. [CrossRef]

48. Upadhyay, N.; Jaiswal, P.; Jha, S.N. Application of Attenuated Total Reflectance Fourier Transform Infrared Spectroscopy (ATR–FTIR) in MIR Range Coupled with Chemometrics for Detection of Pig Body Fat in Pure *Ghee* (Heat Clarified Milk Fat). *J. Mol. Struct.* **2018**, *1153*, 275–281. [CrossRef]

49. Jaiswal, P.; Jha, S.N.; Kaur, J.; Borah, A.; Ramya, H.G. Detection of Aflatoxin M1 in Milk Using Spectroscopy and Multivariate Analyses. *Food Chem.* **2018**, *238*, 209–214. [CrossRef] [PubMed]

50. Andrade, J.; Pereira, C.G.; de Almeida Junior, J.C.; Viana, C.C.R.; de Oliveira Neves, L.N.; da Silva, P.H.F.; Bell, M.J.V.; de Calvalho dos Anjos, V. FTIR-ATR Determination of Protein Content to Evaluate Whey Protein Concentrate Adulteration. *LWT* **2019**, *99*, 166–172. [CrossRef]

51. Fazio, E.; Gualandi, G.; Palleschi, S.; Footitt, S.; Silvestroni, L. Optically Functionalized Biomorphism of Bean Seeds. *J. Lumin.* **2017**, *182*, 189–195. [CrossRef]

52. Cheng, Y.D.; Bai, Y.X.; Jia, M.; Chen, Y.; Wang, D.; Wu, T.; Wang, G.; Yang, H.W. Potential Risks of Nicotine on the Germination, Growth, and Nutritional Properties of Broad Bean. *Ecotoxicology* **2021**, *209*, 111797.

53. Elezz, A.A.; Ahmed, T. The Efficacy Data of Two Household Cleaning and Disinfecting Agents on *Lens Culinaris Medik* and *Vicia Faba* Seed Germination. *Data Brief* **2021**, *35*, 106811. [CrossRef] [PubMed]

54. Cruz, F.; Batista-Santos, P.; Monteiro, S.; Neves-Martins, J.; Ferreira, R.B. Maximizing Blad-containing Oligomer Fungicidal Activity in Sweet Cultivars of Lupinus Albus Seeds. *Ind. Crops Prod.* **2021**, *162*, 113242. [CrossRef]

55. Kim, S.L.; Lee, J.E.; Kwon, Y.U.; Kim, W.H.; Jung, G.H.; Kim, D.W.; Lee, C.K.; Lee, Y.Y.; Kim, M.J.; Kim, Y.H.; et al. The Germination of Soybeans Increases the Water-soluble Components and Could Generate Innovations in Soy-based Foods. *Food Chem.* **2013**, *136*, 491–500. [CrossRef] [PubMed]

56. Bueno, D.B.; da Silva Júnior, S.I.; Chiarotto, A.B.S.; Cardoso, T.M.; Neto, J.A.; dos Reis, G.C.; Glória, M.B.A.; Tavano, O.L. The Germination of Soybeans Increases the Water-soluble Components and Could Generate Innovations in Soy-based Foods. *LWT* **2020**, *117*, 108599. [CrossRef]

57. Baudoin, J.P.; Maquet, A. Improvement of Protein and Amino Acids Content in Seeds of Food Legumes. A Case Study in Phaseolus. *Biotechnol. Agron. Soc. Environ.* **1999**, *3*, 220–224.

58. Zhao, M.; Zhang, H.; Yan, H.; Qiu, L.; Baskin, C.C. Mobilization and Role of Starch, Protein, and Fat Reserves During Seed Germination of Six Wild Grassland Species. *Front. Plant Sci.* **2018**, *9*, 234. [CrossRef]

59. Ulrichs, T.; Drotleff, A.M.; Ternes, W. Determination of Heat-induced Changes in the Protein Secondary Structure of Reconstituted Livetins (Water-soluble Proteins from Hen's Egg Yolk) by FTIR. *Food Chem.* **2015**, *172*, 909–920. [CrossRef]

60. Harwood, J.L. Lipid Synthesis by Germinating Soya Bean. *Phytochemistry* **1975**, *14*, 1985–1990. [CrossRef]

61. Inyang, C.; Zakari, U.M. Effect of Germination and Fermentation of Pearl Millet on Proximate, Chemical and Sensory Properties of Instant "Fura"-A Nigerian Cereal Food. *Pak. J. Nutr.* **2008**, *7*, 9–12. [CrossRef]

62. Otutu, O.L.; Ikuomola, D.S.; Oloruntoba, R.O. Effect of Sprouting Days on the Chemical and Physicochemical Properties of Maize Starch. *J. Food Nutr. Res.* **2014**, *2*, 131–149.

63. Chinma, C.E.; Adewuyi, O.; Abu, J.O. Effect of Germination on the Chemical, Functional and Pasting Properties of Flour from Brown and Yellow Varieties of Tigernut (*Cyperus esculentus*). *Food Res. Int.* **2009**, *42*, 1004–1009. [CrossRef]

64. Echendu, C.A.; Obizoba, I.C.; Aniyka, J.U. Effects of Germination on Chemical Composition of Ground Bean (*Kerstingiellageocarpa harm*) Seeds. *Pak. J. Nutr.* **2009**, *8*, 1849–1854. [CrossRef]

65. Cornejo, F.; Novillo, G.; Villacrés, E.; Rosell, C.M. Evaluation of the Physicochemical and Nutritional Changes in Two Amaranth Species (*Amaranthus quitensis* and *Amaranthus caudatus*) after Germination. *Food Res. Int.* **2019**, *121*, 933–939. [CrossRef] [PubMed]

66. Chinma, C.E.; Anuonze, J.C.; Simon, O.C.; Ahiare, R.O.; Danbaba, N. Effect of Germination on the Physicochemical and Antioxidant Characteristics of Rice Flour from Three Rice Varieties from Nigeria. *Fodd. Chem.* **2015**, *185*, 454–458. [CrossRef] [PubMed]

67. Maetens, E.; Hettiarachchy, N.; Dewettinck, K.; Horax, R.; Meons, K.; Moseley, D.O. Physicochemical and Nutritional Properties of a Healthy Snack Chip Developed from Germinated Soybeans. *LWT* **2017**, *84*, 505–510. [CrossRef]

68. Nkhata, S.G.; Ayua, E.; Kamau, E.H.; Shingiro, J.B. Fermentation and Germination Improve Nutritional Value of Cereals and Legumes through Activation of Endogenous Enzymes. *Food Sci. Nutr.* **2018**, *6*, 2446–2458. [CrossRef]

69. Sokrab, A.M.; Ahmed, I.A.M.; Babiker, E.E. Effect of Germination on Antinutritional Factors, Total, and Extractable Minerals of High and Low Phytate Corn (*Zea mays* L.) Genotypes. *J. Saudi Soc. Agric. Sci.* **2012**, *11*, 123–128. [CrossRef]

70. Luo, Y.; Xie, W.; Jin, X.; Wang, Q.; He, Y. Effects of Germination on Iron, Zinc, Calcium, Manganese, and Copper Availability from Cereals and Legumes. *CYTA J. Food* **2014**, *12*, 22–26. [CrossRef]

71. El-Adawy, T.A.; Rahma, E.H.; El-Bedawey, A.A.; El-Beltagy, A.E. Nutritional Potential and Functional Properties of Germinated Mung Bean, Pea and Lentil Seeds. *Plants. Food Hum. Nutr.* **2003**, *58*, 1–13. [CrossRef]

72. Yasmin, A.; Zeb, A.; Khalil, A.W.; Paracha, G.M. Effect of Processing on Anti-nutritional Factors of Red Kidney Bean (*Phaseolus vulgaris*) Grains. *Food Bioproc. Tech.* **2008**, *1*, 415–419. [CrossRef]

73. Ogbonna, A.C.; Abuajah, C.I.; Udofia, U.S. Effect of Malting Conditions on the Nutritional and Anti-nutritional Factors of Sorghum Grist. *Ann. Univ. Dunarea Jos Galati* **2012**, *36*, 64–72.

74. Chinma, C.E.; Abu, J.O.; Asikwe, B.N.; Sunday, T.; Adebo, O.A. Effect of Germination on the Physicochemical, Nutritional, Functional, Thermal Properties and in vitro Digestibility of Bambara Groundnut Flours. *LWT* **2021**, *140*, 110749. [CrossRef]

75. Kajla, P.; Sharma, A.; Sood, D.R. Effect of Germination on Proximate Principles, Minerals and Antinutrients of Flaxseeds. *J. Dairy Res.* **2017**, *36*, 52–57. [CrossRef]

76. Andrade, J.; Pereira, C.G.; Ranquine, T.; Azarias, C.A.; Bell, M.J.V.; de Carvalho dos Anjos, V. Long-Term Ripening Evaluation of Ewes' Cheeses by Fourier-Transformed Infrared Spectroscopy under Real Industrial Conditions. *J. Spectrosc.* **2018**, 1381864. [CrossRef]

77. Wang, X.; Esquerre, C.; Downey, G.; Henihan, L.; O'Callaghan, D.; O'Donnell, C. Feasibility of Discriminating Dried Dairy Ingredients and Preheat Treatments Using Mid-Infrared and Raman Spectroscopy. *Food Anal. Methods* **2017**, *11*, 1380–1389. [CrossRef]

78. El Darra, N.; Rajha, H.N.; Saleh, F.; Al-Oweini, R.; Maroun, R.G.; Louka, N. Food Fraud Detection in Commercial Pomegranate Molasses Syrups by UV–VIS Spectroscopy, ATR-FTIR Spectroscopy and HPLC Methods. *Food Control* **2017**, *78*, 132–137. [CrossRef]

79. Oroian, M.; Ursachi, F.; Dranca, F. Ultrasound-Assisted Extraction of Polyphenols from Crude Pollen. *Antioxidants* **2020**, *9*, 322. [CrossRef]

80. Pini, R.; Furlanetto, G.; Castellano, L.; Saliu, F.; Rizzi, A.; Tramelli, A. Effects of Stepped-combustion on Fresh Pollen Grains: Morphoscopic, Thermogravimetric, and Chemical Proxies for the Interpretation of Archeological Charred Assemblages. *Rev. Palaeobot. Palynol.* **2018**, *259*, 142–158. [CrossRef]

81. Kasprzyk, I.; Depciuch, J.; Grabek-Lejko, D.; Parlinska-Wojtan, M. FTIR-ATR Spectroscopy of Pollen and Honey as a Tool for Unifloral Honey Authentication. The Case Study of Rape honey. *Food Control* **2018**, *84*, 33–40. [CrossRef]

82. Castiglioni, S.; Astolfi, P.; Conti, C.; Monaci, E.; Stefano, M.; Carloni, P. Morphological, Physicochemical and FTIR Spectroscopic Properties of Bee Pollen Loads from Different Botanical Origin. *Molecules* **2019**, *24*, 3974. [CrossRef] [PubMed]

83. Elbakush, A.E.; Güven, D. Evaluation of Ethanol Tolerance in Relation to Intracellular Storage Compounds of Saccharomyces cerevisiae Using FT-IR Spectroscopy. *Process Biochem.* **2021**, *101*, 266–273. [CrossRef]

84. Feng, G.D.; Zhang, F.; Cheng, L.H.; Xu, X.H.; Zhang, L.; Chen, H.L. Evaluation of FT-IR and Nile Red Methods for Microalgal Lipid Characterization and Biomass Composition Determination. *Bioresour. Technol.* **2013**, *18*, 107–112. [CrossRef] [PubMed]

85. Klaassen, M.; De Vries, E.G.; Masen, M.A. Interpersonal Differences in the Friction Response of Skin Relate to FTIR Measures for Skin Lipids and Hydration. *Colloids. Surf. B Biointerfaces* **2020**, *189*, 110883. [CrossRef] [PubMed]

86. Grace, C.E.E.; Laksmi, P.K.; Meenakshi, S.; Vaidyanathan, S.; Srisudha, S.; Mary, M.B. Biomolecular Transitions and Lipid Accumulation in Green Microalgae Monitored by FTIR and Raman Analysis. *Spectrochim. Acta A Mol. Biomol. Spectrosc.* **2020**, *224*, 117382. [CrossRef] [PubMed]

87. Mujumdar, A.S.; Law, C.L.; Woo, M.W. Freeze Drying: Effects on Sensory and Nutritional Properties. In *Encyclopedia of Food and Health*; Caballero, B., Finglas, P.M., Toldrá, F., Eds.; Academic Press: Cambridge, MA, USA, 2016; pp. 99–103. ISBN 9780123849533.

88. Waghmare, R.B.; Perumal, A.B.; Moses, J.A.; Anandharamakrishnan, C. Recent developments in freeze drying of foods. In *Innovative Food Processing Technologies*; Elsevier: Amsterdam, The Netherlands, 2021; pp. 82–99.

89. Franceschi, C.R.B.; Smidt, E.C.; Vieira, L.N.; Ribas, L.L.F. Storage and in vitro Germination of Orchids (Orchidaceae) Seeds from Atlantic Forest–Brazil. *An. Acad. Bras. Ciênc.* **2019**, *91*, e20180439. [CrossRef] [PubMed]

90. Koene, F.M.; Amano, É.; Ribas, L.L.F. Asymbiotic Seed Germination and in vitro Seedling Development of *Aciantheraprolifera* (Orchidaceae). *S. Afr. J. Bot.* **2019**, *121*, 83–91. [CrossRef]

91. Athapattu, M.; Saveh, A.H.; Kazemi, S.M.; Wang, B.; Chizari, M. Measurement of the Femoral Head Diameter at Hemiarthroplasty of the Hip. *Proc. Technol.* **2014**, *17*, 217–222. [CrossRef]

92. Ertas, N.; Aslan, M. Antioxidant and Physicochemical Properties of Cookies Containing Raw and Roasted Hemp Flour. *Acta Sci. Pol. Technol. Aliment.* **2020**, *19*, 177–184.

Article

Nutritional Evaluation of Beetroots (*Beta vulgaris* L.) and Its Potential Application in a Functional Beverage

Eman Abdo [1], Sobhy El-Sohaimy [2,3,*], Omayma Shaltout [1], Ahmed Abdalla [1] and Ahmed Zeitoun [1]

[1] Faculty of Agriculture (Saba Basha), Alexandria University, Alexandria 21531, Egypt; eman-abdo@alexu.edu.eg (E.A.); prof.dr.Omimaelsaidshaltout@alexu.edu.eg (O.S.); prof.dr.Ahmedelsaidmohmedabdallah@alexu.edu.eg (A.A.); zaytoun19@alexu.edu.eg (A.Z.)

[2] Department of Technology and Organization of Public Catering, Institute of Sport Tourism and Service, South Ural State University, 454080 Chelyabinsk, Russia

[3] Department of Food Technology, Arid Lands Cultivation Research Institute, City of Scientific Research and Technological Applications, New Borg El Arab 21934, Alexandria, Egypt

* Correspondence: alsukhaimisa@susu.ru

Received: 27 October 2020; Accepted: 8 December 2020; Published: 10 December 2020

Abstract: Beetroot is a good source of minerals, fibers, and bioactive components. The present research work was conducted to evaluate the nutritional quality of beetroots (juice, peels, leaves and pomace) enhancing the extracted bioactive components, and developing a functional probiotic beverage. Chemical composition and minerals content of beetroot parts were estimated. The bioactive components were extracted by instant extraction method (IEM) and overnight extraction method (at −20 °C) (OEM) to determine total phenolics, flavonoids, and DPPH inhibition ratio. The extracted beetroot juice was mixed with milk for valorization of the beverage nutritional value and fermented with LA-5 and ABT-5 cultures to create a novel functional beverage. Chemical composition, minerals content, and bioactive components of beverages were estimated. The leaves exhibited the highest calcium content (1200 mg/100 g). Juice showed the highest amount of all minerals except for calcium and magnesium. Overnight extraction method (OEM) increased the antioxidant activity in peels and stems. Natural juice exhibited the highest activity compared to extracts. Fermentation of beet-milk beverage with LA-5 and ABT-5 cultures enhanced the beverage taste, flavor, and antioxidant capacity. Beetroot wastes and juice comprise a valuable nutritional source. Fermentation improved the nutritional value of beetroot and the acceptability of the product.

Keywords: Beetroot (*Beta vulgaris* L.); juice; bioactive components; fermented beverage; probiotics

---

1. Introduction

During the latest decades, the humans' awareness regarding the importance of vegetable consumptions elevated with believe that vegetables and fruits are a rich source of bioactive components which confirmed their participation in health improvement rather than the use of supplements [1,2]. Accordingly, the production of vegetables increased worldwide significantly from 682.43 million tons in 2000 to 1088.9 million tons in 2018 [1–3]. Consequently, root and tuber vegetable production raised from 8.99 million tons in 2008 to 10.53 million tons in 2018 worldwide, where Egypt ranked as the first producer country in North Africa by nearly 5221 tons [4]. Beetroot (*Beta vulgaris* L.) is an herbaceous biennial plant classified as one of the *Chenopodiaceae* family. The taproot found either in yellow pulp color or red [5–7] where the red root utilized in salad, juice, food coloring, and as a medicine [6,8] that emerged along the Mediterranean coast. Beets are considered as one of the most effective vegetables, they are a source of betalain pigment in addition to phenolic acids such as gallic, syringic, and caffeic acids and flavonoids. It has anti-inflammatory, and antioxidant effects, which scavenge free radical from the cells promoting cancer prevention by inhibiting the tumor cells proliferation, reducing the risk of cardiovascular diseases, and expelling kidney stones [5,6]. Studies also revealed that it reduced the low-density lipoprotein (LDL) oxidation by 50% [9], decline the blood glucose after beetroot consumption by 40% [10]. Beetroot also considered as a good source of minerals such as iron, calcium, phosphorus, potassium, sodium, and zinc, in addition to vitamins like biotin, niacin, folate [11]. Exposure of beetroot to thermal processing increases the loss and the degradation of vitamins, minerals, and bioactive components, resulting in a significant reduction in the concerning health benefits. Fermentation of beetroot by lactic acid bacteria aids in the prevention of bioactive components degradation, where fermented beetroot juice retained its antimutagenic effect for 30 days under refrigerating [12]. Juice is a good source of carbohydrates, which is a suitable medium for homo-fermentation by probiotic strains [13] such as *Lactobacillus acidophilus*, *Lactobacillus plantarum*, *Lactobacillus casei*, and *Bifidobacterium bifidum* [14]. During fermentation, lactic acid bacteria produce vitamins that have the ability to enhance the nutritional value of the products [13], and lactic acid that is responsible for lowering the pH of the juice thus reducing the growth of the spoilage microorganisms [15], thus prolonging the shelf life [13]. The globally widespread of fermented beverages could be due to its health properties; as lactic acid bacteria decrease the B-glucuronidase activity which resulted in the prevention of cancer, particularly colon cancer besides, the reduction of the pathogens, and it is also proper for the lactose-intolerance people [14]. Additionally, studies revealed that probiotics could enhance immunity and participate in memory impairment [16]. Incorporating beetroot juice in the production of fermented product could affect the human' health positively, as the phytochemicals can bind the carcinogens reducing their transferability into the cell, thus preventing the cellular DNA mutation [12]. Extraction of the juice from the beetroot plant resulted in extra peels and pomace, in addition to the removed stems and leaves wastes. The residues from processing of the beetroots like peels, seeds, stems, and pomaces reached up to 1.3 billion tons yearly as estimated by FAO which represented one-third of food industry production [17]. Those wastes were discarded for a long time by manufacturers or used as animal feed, or fertilizers until recent studies showed that they are a valuable source of bioactive components which could be used as food additives or formulating novel functional foods [17,18]. Furthermore, those by-products exhibited a significant anti-microbial effect compared to the influence of synthetic antibiotics [19,20]. Beet peels showed a high antioxidant activity because it contains the highest betalain content compared to the other parts [9,19,20]. Like peels, the beet pulp revealed an extraordinary antioxidant effect due to its significant amount of betalain, as well as the presence of other phenolic compounds like "ferulic, vanillic, p-hydroxybenzoic, caffeic, and catechuic acid [19,20]. Leaves are an impressive nutritive source as they contain a tremendous concentration of polyunsaturated fatty acids particularly, alpha-linolenic acid. However, it is cut off from pulps and discarded, mostly because of the dietary habits and the low information about its health benefits [8]. In Egypt, limited studies were carried out to determine the nutritional quality of beetroot peels and pomaces resulting from the juice extraction, stems, and leaves. Therefore, the current study was conducted to evaluate the nutritional quality of the beetroots to ascertain its potential use in food technology. Furthermore, explore the effect of using the probiotic strains [LA5 strain (*Lactobacillus acidophilus*), and ABT5 strain which consist of (*Lactobacillus acidophilus*, *Bifidobacterium bifidum*, and *Streptococcus thermophilus*)] on the nutritional quality and sensory attributes of beetroot juice mixed with 40% milk.

2. Results and Discussion

2.1. Chemical Composition

The proximate chemical composition of beetroot plant parts is shown in (Table 1). There are no significant differences ($p > 0.05$) detected between peels and pomace regarding all the chemical parameters, this might be due to the difficulty of removing the thin peel from the pulp. Consequently, the juice also showed no significant differences ($p > 0.05$) with peels and pomace concerning total lipids, total sugars, and ash content. Interestingly, the leaves had the highest amount of protein (5.64%), which comprised a promising good source of protein. The obtained results were higher than that previously reported by Biondo et al. [8] where protein content was 3.81%. Furthermore, there was a discrepancy between the protein content of pomace (1.13%) in the present study and the other previous studies; It was much less compared to 45.53% by Shyamala and Jamuna [21], and slightly fewer than 1.6% as noted by Neha et al. [22]. These differences reasonably attributed to the different nitrogen content that resulted from the variations in nitrogen fertilization, the properties of the soil and other environmental conditions. It is also worth mentioning that leaves and stems exhibited the highest total lipids content (0.43 and 0.41%), respectively. These findings disagreed with Biondo et al. [8] who reported 0.78%, with a high concentration of the essential fatty acid (linolenic acid). Whereas, pomace contain fewer lipids portion (0.15%) which was in agreement with the USDA value (0.17%) described by Neha et al. [22], but it was lower than 0.31% as evaluated by Shyamala and Jamuna [21]. The sugar content in pomace and peel was 8.79% and 8.4%, respectively, which was slightly less than 9.56% in beetroot plant as reported by Neha et al. [22]. The sugar content in leaves was 0.44% which was nearly twenty times less than the sugar content in the pomace. These results were much less than 3.98% evaluated by Biondo et al. [8]. On the other hand, the juice sugar content was 4.8% which was higher than Kazimierczak et al. [23], who reported that the sugar content was 3.33% in the juice. The crude fiber was 2.6, 1.97, 2, and 2.15 in peels, pomaces, stems, and leaves, respectively. The value in pomace was less than the expected value (2.8%) as in USDA nutritional data reported by Neha et al. [22], and 35.53% as reported by Shyamala and Jamuna [21]. It might be the filtration of pomace during preparation decreased its fiber content. The obtained results revealed that the beetroots plant is a promising nutritional source for macronutrients which make it a good source for supporting several kinds of food products.

Table 1. Proximate chemical composition and minerals content of beetroot plant parts.

Sample Type	Peel	Pomace	Stems	Leaves	Juice
Proximate analysis [1]					
Moisture %	86.3 ± 0.98 [b]	86.8 ± 1.98 [b]	91 ± 0.99 [a]	90.7 ± 0.42 [a]	92.9 ± 0.1 [a]
Total lipids %	0.2 ± 0.04 [b]	0.15 ± 0.04 [b]	0.41 ± 0.02 [a]	0.43 ±0.04 [a]	0.16 ± 0.03 [b]
Protein %	1.02 ± 0.1 [c]	1.13 ± 0.18 [c]	2.45 ± 0.2 [b]	5.64 ± 0.28 [a]	1.21 ± 0.14 [c]
Crude fibers %	2.6 ± 0.12 [a]	1.97 ± 0.23 [a]	2 ± 1.41 [a]	2.15 ± 0.42 [a]	ND *
Ash %	1.48 ± 0.21 [a]	1.16 ± 0.3 [a,b]	1.27 ± 0.15 [a]	0.64 ± 0.22 [b]	0.93 ± 0.01 [a,b]
Total sugars % **	8.4 ± 0.76 [a]	8.79 ± 2.28 [a]	2.87 ± 2.22 [b]	0.44 ± 1.31 [b]	4.8 ± 1.11 [a,b]
Minerals (mg/100 g) [2]					
P	32.43 ± 0.01 [b]	41.02 ± 0.74 [b]	36.24 ± 7.42 [b]	40.23 ± 0.83 [b]	256 ± 12.01 [a]
K	635 ± 134.6 [c]	1971.6 ± 129.53 [b]	2831.3 ± 242.26 [a]	2196.1 ± 146.94 [b]	3053.7 ± 75.97 [a]
Ca	235.36 ± 89.24 [c,d]	154.92 ± 20.46 [d]	495.7 ± 71.74 [b]	1200 ± 127.1 [a]	412.52 ± 0.18 [b,c]
Mg	311.44 ± 65.37 [a]	116.4 ± 19.81 [c]	48.22 ± 3.47 [c]	58.02 ± 3.11 [c]	217.6 ± 39.84 [b]
Fe	121.19 ± 13.62 [b]	99.19 ± 19.01 [b]	1.29 ± 0.45 [c]	13.71 ± 3.36 [c]	911.65 ± 20.11 [a]
Cu	1.95 ± 0.1 [b]	1.32 ± 0.18 [b,c]	1.65 ± 0.13 [b]	0.17 ± 0.06 [c]	6.32 ± 1.01 [a]
Mn	5.19 ± 0.74 [b]	4.73 ± 1.51 [b]	0.10 ± 0.32 [b]	0.89 ± 0.44 [b]	27.30 ± 3.68 [a]
Zn	3.81 ± 0.68 [b]	1.77 ± 0.6 [b]	2.03 ± 0.35 [b]	1.98 ± 0.91 [b]	319.03 ± 26.78 [a]

Mean values in a raw having different superscript are significantly different at ($p \leq 0.05$); ND * (Not detected); ** Total sugars calculated by difference; ([1]) proximate analysis parameters expressed in (g/100 g); ([2]) minerals content expressed in (mg/100 g).

2.2. Minerals Content

The analysis of minerals content revealed that, juice contained the highest level of all detected minerals, except for calcium (412.52 mg/100 g) and magnesium (217.6 mg/100 g) (Table 1). At the same time, it is worth mentioning that all samples in the present study were high in potassium content ranged from 635 mg/100 g to 3053.7 mg/100 g; which is known to manage the blood pressure and cardiovascular system on the long-term usage [24,25]. In addition, the juice showed the highest value of iron (911.65 mg/100 g) which plays a role in anemia prevention [22]. On the other hand, Phosphorus in pomace was 41.02 mg/100 g which is in a variation range of (32.43–256 mg/100 g), and close to 40 mg as reported by Neha et al. [22], but less than 293.81 mg/100 g detected by Shyamala and Jamuna [21]. Results also revealed that potassium in leaves was 2196.1 mg/100 g which agreed with Biondo et al. [8] being 2078.4 mg/100 g. Potassium in pomace were 1971.6 mg/100 g which is six times higher than the USDA result reported by Neha et al. [22]. The highest amount of calcium found in leaves (1200 mg/100 g), which was significantly higher than that reported by Biondo et al. [8] being 186.46 mg/100 g. Although, pomaces had the lowest calcium content (154.92 mg/100 g), it was higher than that reported by Neha et al. [22] being 16 mg/100 g. Iron is considered one of the most crucial minerals, as it has a vital role in anemia treatment. The highest amount of Fe was in juice (911.65 mg/100 g), followed by peels (121.19 mg/100 g) and pomaces (99.19 mg/100 g). This value in pomaces was much higher than that reported by Neha et al. [22] being 0.8 mg, and Shyamala and Jamuna [21] who reported 11.61 mg/100 g of Fe. But the resulted value in leaves (13.71 mg/100 g) was less than that reported by Biondo et al. [8] being 25.63 mg/100 g. The obtained results in the current study emphasized the considerable content of crucial minerals that are necessary for human health and confirmed the importance of the beetroots as a good source of micro-and macro-elements.

2.3. Extraction Yield

Generally, phytochemicals extraction varied according to the solvent polarity and the nature of the extracted molecules, in addition to temperature [26]. Accordingly, using different solvents, temperatures, and extraction time resulted in various amounts of bioactive components (Figure 1). Ethanol (II) (OEM) exhibited the highest phenolic extraction capacity from juice (21.44%) compared to ethanol (I) (IEM) (6.70%). However, methanol (II) revealed the highest yield from stems and leaves, (18.70 and 17.10%) respectively, while methanol (I) extracted 12 and 14.70% from stems and leaves, respectively. On the other hand, methanol (I) extracted the highest bioactive compounds from peel and pomace (22.70 and 17.30%) respectively. The obtained yield from peel and pomace by methanol (I) was close to the amount obtained by ethanol (II) (19.80 and 15.34%) respectively (Figure 1). The extraction efficiency of phenolic compounds from different plant materials and different parts depends on the nature of the phenolic substances, polarity of the extraction solvents, time and temperature of the extraction.

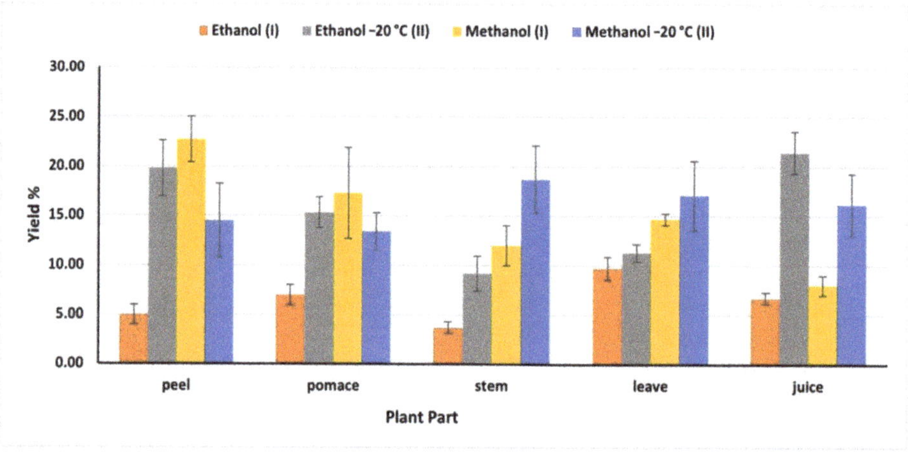

Figure 1. Yield of extraction of different solvents expressed in % of the dry matter; (I): IEM; (II): OEM at −20 °C.

2.4. Total Phenolic Content

The natural juice had the highest phenolic content value (11.58 mg/g) compared to the phenolics obtained in the juice extracts (Figure 2), being nine times higher than the obtained amount by Kazimierczak et al. [23] (1.29 mg). It is also worth to be mentioning that (OEM) revealed a high phenolic extraction for stems and peels, while no significant differences were detected in the total phenolics obtained by IEM and OEM in pomace and leaves. As mentioned previously, OEM, particularly ethanol (II), exhibited high extraction efficiency with stems recording 14.58 mg/g, compared to 9.35 mg/g obtained by ethanol (I). Whereas in leaves, ethanol (I) extracted 8.54 mg/g, which is close amount as that extracted by methanol (II) (8.12 mg/g). The obtained results were less than the results reported by Biondo et al. [8]. Phenolic compounds extracted from pomace by methanol (II) were 6.66 mg/g which was significantly higher than that reported by Shyamala and Jamuna [21] (2.2 mg/g), while less than that obtained by Čanadanović-Brunet et al. [27] (376.4 mg/g). However, methanol II extracted 9.96 mg/g from peels, which was less than that value reported by Kujala et al. [28] (15.5 mg/g). This discrepancy in the extraction capacity of different methods in different parts and plant origins might have resulted to the environmental biotic and abiotic stresses, which influence the presence and distribution of the phenolic compounds in the plant [26] and also might be resulted to the essence and nature of the phenolics compounds and their concentration in the plant materials.

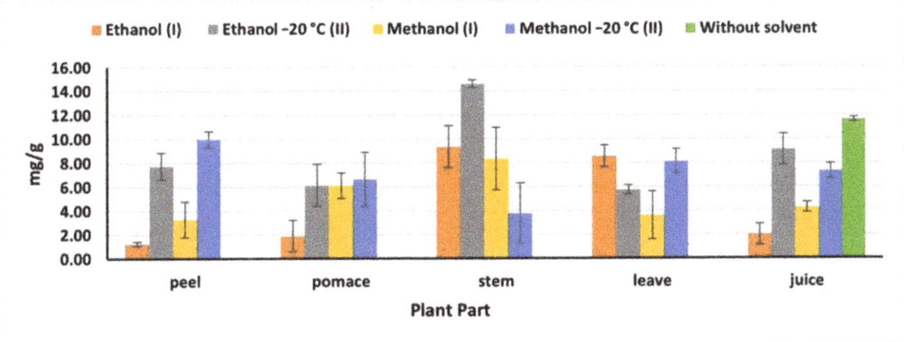

Figure 2. Total phenolics in beetroot extracts (mg gallic acid/g); (I): IEM; (II): OEM at −20 °C.

2.5. Total Flavonoids

Juice showed a high flavonoids content (10.73 mg/g) compared to the flavonoids content determined in the other beet parts which declined significantly by 81% in the juice methanolic (I) (IEM) extract (Figure 3). However, this value was higher than the amount obtained by Kazimierczak et al. [23] (0.2 mg/g). While peels and stems ethanolic (II) (OEM) extracts exhibited the highest values of flavonoids being 4.78 and 4.84 mg/g, respectively, compared to 0.93 mg/g and 1.19 mg/g obtained by ethanol (I) peel and stems extract, respectively. Methanol (I) is more effective in the extraction of flavonoids in pomace (1.80 mg/g), which agreed with results of El-Beltagi et al. [29] being 1.54 mg/g, however, it was less than Čanadanović-Brunet et al. [27] being 253.5 mg/g. On the other hand, ethanol (I) extract exhibited the highest extraction ability of flavonoids from leaves (4.85 mg/g). Based on our findings, the beetroot is considered a good source of phenolic compounds and flavonoids which raises its nutritional value and benefits in food processing.

2.6. Betalain Content

Betalains are composed of red-violet betacyanins and yellow-orange betaxanthins [6]. The distribution of betalain pigment differs not only according to the beetroot parts, but also the extraction method (Figure 4). Generally, OEM extracted the highest betalain content compared to IEM in all parts. On the other hand, beetroot peels, pulps, and juice exhibited the highest betalain content compared to of leaves and stems, regardless of the used solvent. Methanol (II) OEM extracted the highest betalain content from peel (0.81 mg/g) compared to extracted by methanol (I) IEM (0.39 mg/g) (Figure 4). While ethanol (II) extracted the highest amount of betalain from pulp (0.81 mg/g), leaves (0.48 mg/g), and juice (0.79 mg/g) compared to 0.40, 0.18, and 0.40 mg/g obtained by ethanol (I) from pomace, leaves, and juice, respectively. The obtained betalain from beetroot peel, pulp, and juice was higher than 274.4 mg/kg detected in ethanolic extract (50%) of beetroot puree treated at 120 °C for 60 min [30]. Methanol (II) extracted 0.5 mg of betacyanin from peel and pulp, while 0.31 and 0.29 mg/g of betacyanin were detected in peel and pulp, respectively. The obtained result of betacyanin and betaxanthin was higher than 37.22 mg/100 g of betanin and 0.71 mg/100 g of vulgaxanthin I detected by Vulic et al. [20] in fresh beetroot pomace ethanolic extract. However, the obtained betalain from peel and pulp (0.81 mg/g) agreed with the betalain content of Redval and Forono varieties from Australia 853 and 826 mg/L, respectively [31]. On the other hand, Wruss et al. [31] detected another betalain concentrations ranged from 767 to 1309 mg/L in the different Austrian beetroot varieties. The findings in the current invistigation showed that the different parts of Egyptian beetroots are rich betalain sources particularly peels, pomace and juice. At the same time methanol is the best solvent for the extraction of betalain from different parts of beetroots.

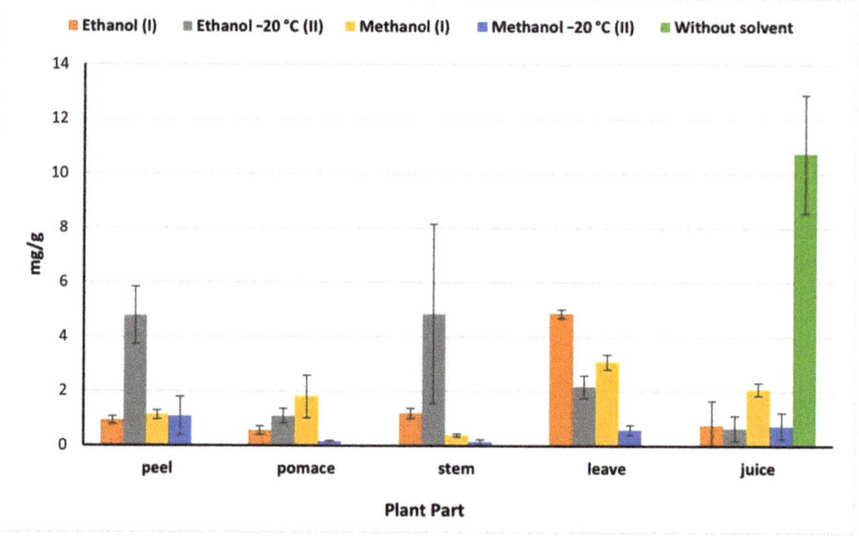

Figure 3. Total flavonoids of beetroot extract (mg catechin/g) based on the dry weight; (I): IEM; (II): OEM at −20 °C.

2.7. HPLC Analysis for Total Phenolics

The HPLC chromatogram classified the phenolic acids and flavonoids in natural juice, and ethanolic (OEM) extracts of peels, leaves, and stems (Table 2). Chlorogenic (32.96 µg/g), gallic (25.19 µg/g), syringic (2.74 µg/g), and cinnamic acid (0.35 µg/g) were recognized in juice as phenolic acids in addition to catechin (93.56 µg/g). This result indicated not only the composition of juice but also the pomace. More phenolic acids and flavonoids were observed in peel extract, where catechin and gallic acid recorded the highest content than other parts being 184.50 and 137.23 µg/g, respectively. The obtained resulted compounds in peels were disagreed with the phenolics identified by Koubaier et al. [32] and El-Beltagi et al. [29]. The obtained results indicated that, the stems contained abundant phenolic compounds than the other parts, where rutin was the highest content followed by catechin being 241.58 and 149.9 µg/g, respectively. However, gallic and chlorogenic acids were not recognized in stem extract, which is disagreed with Koubaier et al. [32]. These differences in the level of phenolic compounds in different parts may refer to changes in growth conditions, especially availability which plays a crucial role in the accumulation level of phenolic compounds in different plant parts [33]. The higher of level of the phenolic compounds is responsible for the higher DPPH scavenging activity of all parts of the plant.

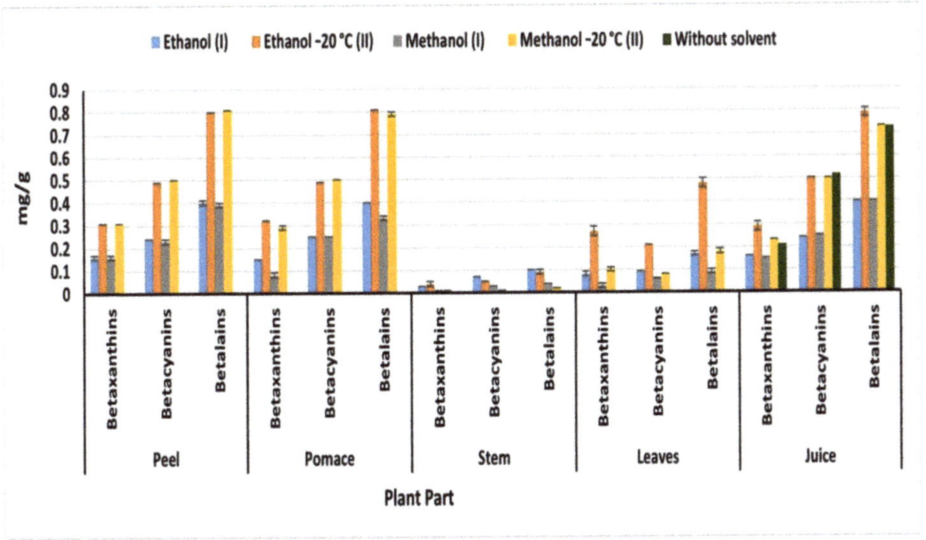

Figure 4. Betaxanthin, betacyanin, and betalain content of beetroot parts (mg/g); (I): IEM; (II): OEM at −20 °C.

Table 2. Phenolic acids and flavonoids identification of beetroot plant parts.

Sample/Compound	Juice	Peel	Leaves	Stem
Phenolics (µg/g)				
Gallic acid	25.19	137.23	5.54	ND *
Chlorogenic acid	32.96	7.52	7.27	ND *
Cinnamic acid	0.35	ND *	0.14	9.94
Ferulic acid	ND *	1.81	1.21	59.18
Caffeine	ND *	ND *	1.47	39.66
Coffeic acid	ND *	ND *	ND *	61.36
Syringic acid	2.74	20.73	ND *	50.85
Ellagic acid	ND *	ND *	ND *	87.56
Coumaric acid	ND *	4.11	28.97	59.90
Vanillin	ND *	ND *	ND *	70.44
Flavonoids (µg/g)				
Rutin	ND *	ND *	ND *	241.58
Naringenin	ND *	2.75	12.83	61.43
Propyl Gallate	ND *	ND *	0.67	18.18
4′,7-DihydroxyisoFlavone	ND *	ND *	0.79	20.33
Querectin	ND *	0.88	ND *	87.68
Catechin	93.56	184.50	22.31	149.91

ND * Not detected.

2.8. DPPH Scavenging Activity

DPPH$^\bullet$ is a stable free radical that reduced in the presence of antioxidants, resulting in color changing from purple to yellow [34]. Methanol (OEM) extracts exploited high DPPH$^\bullet$ inhibition ratio compared to the other extracts, except for stem and juice (Figure 5). Where the natural juice revealed the highest antioxidant, activity compared to the extracts being 205.32%. Similarly, the ethanolic OEM extract of stems revealed the highest inhibition ratio (375.35%) at 91.8 mg/mL compared to 109.86% obtained from stem ethanol IEM extract, which agreed with the phenolic content detected in this extract. In leaves, methanol II extract revealed an antioxidant activity higher than that detected in the other extracts being 185.21% at 170 mg/mL. The inhibition ratios in peel and pomace methanolic OEM extracts were179.58 and 254.23%, respectively, which agreed with the phenolic content of this extract. The obtained results in the present investigation were higher than that reported by Vodnar et al. [19], who reported 45% in one gram of dried beetroot waste. In all cases, the beetroots extracts exhibited a high antioxidant power, which is greatly beneficial for human health and encourage us to fortify our food products with the beetroots extract.

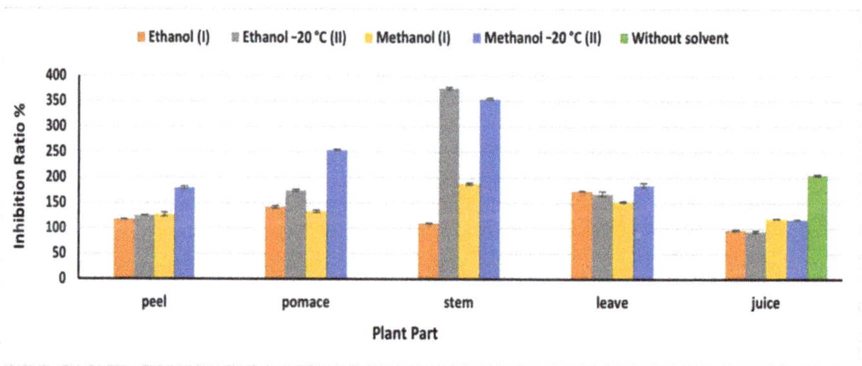

Figure 5. DPPH inhibition ratio% of beetroot extract based on the dry weight; (I): Instant Extraction method (IEM); (II): Overnight extraction method at −20 °C.

2.9. Fermented Beet-Milk Beverage

2.9.1. Total Viable Bacterial Count (TVBC) and Acidity

Probiotic products have proved to have at least 10^8 colony forming unit (cfu) of prebiotic strains/mL juice [15]. Probiotic LA-5 and ABT-5 strains were used in this study to produce a probiotic beetroot-milk fermented beverage. The total viable cells of LA-5 and ABT-5 in addition to the beverage's pH and acidity % during the fermentation were illustrated in Table 3. The milk was added to the beet juice to enhance the bacterial growth, as the LA-5 and ABT-5 numbers declined significantly after 6, 24, and 48 h of the fermentation of beet juice according to the preliminary studies. As noticed from the obtained results, pH value was reduced significantly at the end of the fermentation process from 6.6 in both LA-5 and ABT-5 fermented juices to 5.14 and 5.16, respectively ($p < 0.05$). The reduction of the pH was due to the production of lactic acid [35], which was confirmed by the % acidity of the juices. The acidity significantly increased from 0.41% of both samples at (zero time) to 1.84% and 1.94% at 6 h of fermentation for LA-5 and ABT-5 beverages, respectively ($p < 0.05$). *L. acidophilus* was increased significantly in LA-5 fermented beet-milk beverage after six hours of fermentation from 1.75×10^7 to 1.25×10^{10} ($p < 0.05$). Similarly, the ABT-5 probiotic strain raised significantly from 2.37×10^7 to 3.7×10^9 after six hours of fermentation ($p < 0.05$). The obtained viable bacterial count in the present study was higher than 27.8×10^8 resulted in beetroot juice after 72 h of fermentation with LA-5 strain as mentioned by Yoon et al. [35], and 7×10^8 resulted from the fermentation of beet juice by LA-5 for 8 h [13]. Thus, adding 40% full-fat milk to the beetroot enhanced the bacterial growth and reduced the fermentation period.

Table 3. PH, acidity, and viable cell count of LA-5 and ABT-5 beet-milk beverage.

Time (h)	PH	Acidity *	CFU/mL **
	LA-5 Beet-milk Fermented Juice		
0	6.6 ± 0.01 [a]	0.41 ± 0.01 [c]	$1.75 \times 10^7 \pm 1.1$ [b]
3	6.17 ± 0.02 [b]	0.59 ± 0.01 [b]	$5.45 \times 10^7 \pm 0.28$ [b]
6	5.14 ± 0.03 [c]	1.84 ± 0.05 [a]	$1.25 \times 10^{10} \pm 25.07$ [a]
	ABT-5 Beet-milk Fermented Juice		
0	6.6 ± 0.01 [a]	0.41 ± 0.01 [c]	$2.37 \times 10^7 \pm 0.65$ [b]
3	6.16 ± 0.01 [b]	0.61 ± 0.00 [b]	$6.5 \times 10^7 \pm 0.28$ [b]
6	5.16 ± 0.05 [c]	1.94 ± 0.06 [a]	$3.7 \times 10^9 \pm 3.54$ [a]

Different superscripts are significantly different at ($p \leq 0.05$); * Acidity expressed as % of lactic acid; ** CFU/mL: colony forming unit of probiotic culture/mL of beet-milk beverage.

2.9.2. Chemical Analysis

Table 4 illustrates the chemical composition of control and fermented beet-milk beverages. Generally, no significant differences were detected in moisture, ash, lipids, and protein content of LA-5, ABT-5 fermented beverage, and the control sample ($p > 0.05$). Where the moisture content was ranged from 84.85 to 85.83 g/100 mL, total lipids were in a range of 3.03 to 3.06 g/100 mL which constituted about 21.5% of the beverage' total dry matter. While the protein represented nearly 11.31% of the beverages dry matter, ranged from 1.48 to 1.68 g/100 mL. Ash content was about 1 g/100 mL. This result might be attributed to the short fermentation period, as the bacterial strains did not affect the nutrients in the beverage. On the other hand, carbohydrates constituted the highest part of all beverages, it was 8.61 g/100 mL for the control. However, it was reduced in the fermented beverages by nearly 30% as a result of the presence of the bacteria that converted the lactose into lactate to reach a value of 6.52 g/100 mL in LA-5 beverage, and 5.89 g/100 mL in ABT-5 beverage. As noticed in Table 4, no significant differences were detected between the fermented beet-milk beverages and the control beet-milk in magnesium, manganese, iron, copper, potassium, and sodium content ($p > 0.05$). Whereas, the calcium content was reduced after fermentation by 40% to reach 12.71 mg/100 mL in LA-5 fermented beverage, and 12.64 mg/100 mL in ABT-5 beverage compared to control (21.11 mg/100 mL). As noticed by Tang et al. [36], the calcium content was reduced after 12 h of fermentation of soymilk with *lactobacillus acidophilus* strains before being increased after 24 h of fermentation. Phosphorous content similarly was reduced after fermentation of the beverages by nearly 12%, being 53.71 mg/100 mL and 54.64 mg/100 mL in LA-5 and ABT-5 fermented beverages, respectively. The reduction of calcium and phosphorous could be due to the utilization of the probiotic culture of those elements in their growth [37]. On the other hand, fermentation recorded a slight increase in the zinc content of LA-5 and ABT-5 fermented beverages by 16% (0.27 mg/100 mL and 0.29 mg/100 g, respectively) compared to control, which might be resulted from releasing the metal from chelated complex compounds by bacterial activity and increasing its bioavailability [37].

2.9.3. Phenolics Content and Antioxidant Activity

Table 5 shows total phenolics, flavonoids content, and % DPPH free radical scavenging activity of control and fermented beet-milk beverages. Fermentation of beet-milk beverage with LA-5 and ABT-5 probiotic strains raised the total phenolic content significantly compared to the control beverage. As total phenolics in LA-5 fermented beverage increased by 40% to reach a value of 17.51 mg/mL (Table 5). While a 31% increase in the total phenolics content was reported in ABT-5 fermented beverage to reach 16.38 mg/mL ($p < 0.05$). The increase in the total phenolics might be associated with the probiotic activity during the fermentation. The total phenolic content of fermented beverages could demonstrate the rise in the DPPH free radicals scavenging activity of LA-5 and ABT-5 beverages compared to the control, being 98.11, 97.89, 95.2%, respectively. On the other hand, no significant differences ($p > 0.05$) were determined concerning the flavonoid content of the LA-5 fermented beverages (9.91 mg/mL), ABT-5 fermented beverage (10.06 mg/mL), and the control sample (9.66 mg/mL).

2.9.4. Sensory Evaluation

The data of the current investigation showed that, the fermentation process enhanced the taste and the flavor and consequently, the overall acceptance of LA-5 and ABT-5 fermented beverages compared to control beet-milk beverage. These results might be attributed to the formed lactic acid by probiotic strains [13]. On the other hand, no significant differences were detected in the acceptability of the fermented beverages and the control concerning color and consistency (Figure 6). So, the formulated fermented beverage-based-beetroots did not affect the sensory properties of the formulated functional beverage.

Table 4. Proximate chemical composition of control beet-milk and fermented beet-milk.

Parameters	Beet-Milk (Control)	Beet-Milk (LAB-5)	Beet-Milk (ABT-5)
Chemical Compositional Analysis *			
Moisture %	85.83 ± 2.78 [a]	84.85 ± 1.31 [a]	85.67 ± 0.93 [a]
Total lipids %	3.06 ± 0.06 [a]	3.03 ± 0.04 [a]	3.04 ± 0.08 [a]
Total protein %	1.48 ± 0.11 [a]	1.65 ± 0.03 [a]	1.68 ± 0.04 [a]
Total carbohydrates %	8.61 ± 2.9 [a]	6.52 ± 1.43 [b]	5.89 ± 0.67 [b]
Ash %	1.00 ± 0.03 [a]	0.98 ± 0.13 [a]	1.1 ± 0.14 [a]
Minerals (mg/100 g) **			
Ca	21.11 ± 1.29 [a]	12.71 ± 1.99 [b]	12.64 ± 3.31 [b]
P	61.87 ± 1.23 [a]	53.71 ± 0.7 [b]	54.64 ± 1.58 [b]
mg	14.6 ± 1.56 [a]	14.02 ± 1.81 [a]	13.32 ± 0.17 [a]
K	121 ± 1.4 [a]	120 ± 0.71 [a]	119.3 ± 1.1 [a]
Na	37.5 ± 5.79 [a]	48.25 ± 4.17 [a]	50.5 ± 0.7 [a]
Fe	2.95 ± 0.31 [a]	2.78 ± 0.47 [a]	2.63 ± 0.49 [a]
Mn	0.07 ± 0.03 [a]	0.13 ± 0.01 [a]	0.12 ± 0.02 [a]
Cu	0.09 ± 0.1 [a]	0.07 ± 0.02 [a]	0.44 ± 0.55 [a]
Zn	0.24 ± 0.31 [b]	0.27 ± 0.02 [a,b]	0.29 ± 0.0a [a]

Different superscripts are significantly different at ($p \leq 0.05$); * Calculated as g/100 gm wet matter; ** Expressed in mg/100 gm dry weight basis.

Table 5. Total phenolics, total flavonoids, and DPPH% free radicals scavenging activity of control and fermented beet-milk.

Beverage	TP *	TF **	DPPH % ***
Control Beet-Milk	12.48 ± 1.00 [b]	9.66 ± 0.52 [a]	95.2 ± 0.88 [b]
LA-5 Beet-Milk	17.51 ± 0.16 [a]	9.91 ± 0.37 [a]	98.11 ± 0.93 [a]
ABT-5 Beet-Milk	16.38 ± 1.17 [a]	10.06 ± 1.00 [a]	97.89 ± 1.00 [a]

Different superscript is significantly different at ($p \leq 0.05$); TP * (total phenolics), TF ** (total flavonoids) are expressed in mg/mL wet weight basis; DPPH % ***: % of free radical's inhibition ratio.

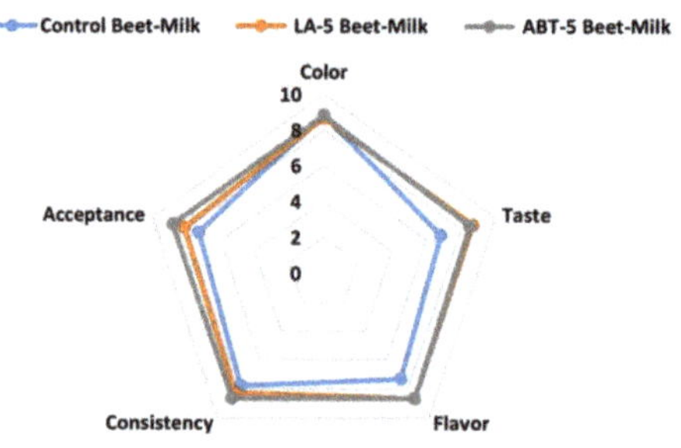

Figure 6. Sensory evaluation of control and fermented beet-milk beverages with LA-5 and ABT-5 probiotic cultures.

3. Materials and Methods

3.1. Materials and Reagents

Twenty-five kilograms of beetroot collected from Alexandria's local market, Egypt. Fresh full cream milk was purchased from a local market in Alexandria (protein 2.6%, fat 3%, carbohydrates 4.6% calcium 97 mg/100 mL and phosphorus 79.61 mg/100 mL). Two freeze-dried lactic probiotic cultures; ABT-5 probiotic consists of (*Lactobacillus acidophilus LA-5*, *Bifidobacterium bifidum BB-12*, and *Streptococcus thermophilus*), and (*Lactobacillus acidophilus LA-5*) were obtained from Christian Hansen's, Denmark. Absolute ethanol and methanol, chloroform, sodium hydroxide, sulphuric acid, boric acid, monopotassium phosphate, sodium carbonate, aluminium chloride and other commercial chemicals were supplied from Aljumhoria Company for chemicals, Alexandria, Egypt, Fine chemicals such as Folin-Ciocalteu, DPPH and phenolics standards purchased from Merk, Germany.

3.2. Preparation of the Beetroot Samples

Stems, and leaves were separated and washed, while the roots washed thoroughly to get rid of any soil residues. After peeling, the juice was extracted from the pulp. Peel, pomace, juice, leaves, and stems were used for further analysis. The plant parts were dried at 50 ± 2 °C for three days in (Wt-binder) drying oven; then each was milled until getting a fine powder by grinder (Kenwood FP691, UK). Afterward, the powder was packed into polyethylene bags as well as the extracted juice and stored at −20 ± 9 °C till used [21].

3.3. A Proximate Chemical Analysis

Moisture content was carried out using (Wt-binder) oven at 105 ± 2 °C for 24 h. till getting a constant weight of the samples. **The ash content** was determined by using a muffle furnace at 500 °C. **Crude fibers** was determined according to AOAC [38] by boiling the samples for 30 min with 1.25% of H_2SO_4, then NaOH after filtration and washing with hot water. The samples were dried in an oven, weighed (W1) and re-dried in a muffle till gray ash was formed and re-weighed (W2), then the crude fiber was calculated as g/100 g by the following Equation (1):

$$\text{Crude fibers} = \frac{W1 - W2}{\text{Sample weight}} \times 100 \tag{1}$$

Total fat extracted with chloroform-methanol solvent according to Folch method [39]. **The protein content** was determined by the micro-Kjeldahl method according to Peach and Tracy [40]. **Total sugar** was calculated by difference according to the following Equation (2):

$$\text{Total sugar content} = 100 - (\text{moisture} + \text{lipids} + \text{protein} + \text{fibers} + \text{ash}) \tag{2}$$

3.4. Minerals

Dried samples (1 g) of each wet-digested using conc. H_2SO_4-H_2O_2 mixture as described by Lowther [41]. Phosphorus content estimated by Vanadomolybdophosphoric yellow color method and the absorbance of the sample was measured at 405 nm. The concentration of phosphorus was determined by using monopotassium phosphate standard curve [42], Whereas the potassium content measured by the Backman flame photometer as described by Jackson [42]. Calcium, Magnesium, manganese, copper, iron, and zinc estimated by using the Inductively Coupled Argon Plasma (ICAP 6500 Duo, Thermo Scientific, Gloucester, UK), which standardized by 1000 mg/L multi-element certified standard solution, Merck, Germany.

3.5. Extraction Methods

The phytochemical compounds in the samples were extracted by the following two methods:

Instant extraction method (IEM): the extraction performed using ethanol (I) 70%, methanol (I) 80%, and water (I) as Vasconcellos et al. [43] with a slight modification, where each sample was vortexed with each solvent in a ratio of 1:10 (w/v) for 1 min, centrifuged for 10 min at 6000× g, and filtered. The pellet of each sample was re-extracted with the same solvent twice, and the filtrates combined before evaporating the solvent at 40 °C. The yield calculated as g/100 g, then the lyophilized samples were stored at −20 ± 9 °C till used.

Overnight extraction method (at −20 °C) (OEM): Each sample was vortexed with methanol (II) 80%, and ethanol (II) 70% solvents in the same prior ratios and stored overnight at −20 ± 9 °C before extracting to increase the time of exposure of the plant part to the solvent for enhancing the extraction without being affected by the high temperature followed by the extraction of bioactive component as mentioned previously.

3.6. Total Phenolic Content (TP)

Total phenolic content was determined according to the Folin-Ciocalteu method as described by Vondar et al. and Raupp et al. [19,44]. 200 µL of the extract was mixed with 1 mL of 0.2N Folin-Ciocalteu reagent, and 800 µL of Na_2CO_3 (7.5%). The mixture incubated for 2 h in the dark at the room temperature, before reading the absorbance at 760 nm by using (Jenway 6405UV/VIS) spectrophotometer. The total phenolics expressed in mg/g as gallic acid equivalent based on the dry weight using a standard curve of gallic acid.

3.7. Total Flavonoids Content (TF)

Total flavonoids content was determined by aluminum chloride method as described by Čanadanović-Brunet et al. and Baba and Maik [27,45]. 1 mL of each extract was mixed with 4 mL dH_2O and 0.3 mL 5% $NaNO_2$ before incubating the mixture for 5 min. Afterward, 0.3 mL of 10% $AlCl_3$ was added, and the samples were incubated for 6 min. Later, 2 mL of NaOH (1 mol/L) was added, and the volume completed with dH_2O up to 10 mL. The mixture then incubated for 15 min before measuring the absorbance at 510 nm. The total flavonoid value expressed as mg/g of catechin on dry matter basis.

3.8. Betalain Content

Betalain content of beetroot parts extracts were estimated spectrophotometrically as described by Anand et al. and Castellanos-Santiago and Yahia [46,47]. Betacyanin and betaxanthin content were measured in the extracts at 535 and 483 nm, respectively by using (Jenway 6405UV/VIS) spectrophotometer. Betalain content was calculated as mg/g by the following Equations (3) and (4):

$$\text{Betacyanin / Betaxanthin (mg/g)} = \frac{A \times DF \times MW \times V}{\varepsilon LW} \tag{3}$$

where: A is the maximum recorded absorption for betacyanins and betaxanthins, respectively, DF is the dilution factor, V is the extract volume (mL), W is the dried sample weight (g), and L is the path-length (1 cm) of the cuvette. The molecular weight and molar extinction coefficient (ε) of betacyanin are 550 g/mol, 60,000 L/(mol cm) in water, and of betaxanthin are 308 g/mol, 48,000 L/(mol cm) in water, respectively.

$$\textit{Betalain content mg/g = betacyanin (mg/g) + betaxanthin (mg/g)} \tag{4}$$

3.9. HPLC Profile of Phenolic Compounds

Phenolic compounds profile identified in natural juice, ethanolic (II) extract of peels, stem, and leaves by using an Agilent 1260 series HPLC with a multi-wavelength detector monitored at 280 nm. The injected 10 µL of each extract separated on a C18 column (4.6 mm × 250 mm i.d., 5 µm) at 35 °C, and a mobile phase consisting of water (A) and acetonitrile (B) at a flow rate 1 mL/min. Where the mobile phase was programmed consecutively in a linear gradient as follows: 0 min (80% A); 0–5 min (80% A); 5–8 min (40% A); 8–12 min (50% A); 12–14 min (80% A), and 14–16 min (80% A). The unknown compounds characterized by matching its retention time to the standard's retention time.

3.10. DPPH Free Radical Scavenging Assay

DPPH assay was carried out according to Do et al. [34]. 0.5 mL of freshly prepared 0.3 mM methanolic DPPH mixed with 0.5 mL of the extract, then the mixture incubated at room temperature for 20 min. Afterward, the absorbance of the control (DPPH methanolic solution), and the samples measured at 517 nm. The inhibition ratio % was calculated by the following equation:

% Inhibition = (absorbance of control − absorbance of sample/absorbance of control) × 100

3.11. Preparation of the Fermented Juice

Beetroot juice was mixed with 40% full fat milk before being homogenized and heated at 80 °C for 20 min to mimic pasteurization step [16]. After heating, the mixture was cooled to 40 °C and aseptically inoculated with 0.2 gm/L of ABT-5 and 0.33 gm/L of LA-5 probiotic direct vat cultures for the preparation of ABT-5 juice and LA-5 juice, respectively. Afterwards, the inoculated juices were homogenized incubated at 37 °C for six h [48].

3.12. Total Viable Bacterial Count (TVBC)

The total viable bacterial count was determined in ABT-5 and LA-5 juices at the zero time, 3 and 6 h of fermentation. The viable cell count (CFU/mL) in the ABT-5 and LA-5 juice samples was counted in an MRS medium by using ten-fold serial dilution in peptone-water and a MRS standard plate count, the results were expressed as log of (cfu/mL juice) [35,48].

3.13. pH and Acidity Analysis

pH of the functional beverage samples was measured by a calibrated PH meter (Adwa, AD1030). The acidity was determined according to Kazimierczak et al. and Yoon et al. [23,35] by titrating the juices with 0.2N NaOH till the pH reached 8.2. The total acidity was expressed as % of lactic acid by using the following Equation (5):

$$\% \text{ Acidity} = (V \times N \times 0.09/W \times 100) \tag{5}$$

where: (V) is mL of NaOH, (N) normality of NaOH, and (W) weight of the sample.

3.14. A Proximate Chemical Analysis of Fermented Beverage

Moisture and ash content of LA-5 and ABT-5 fermented beet-milk beverages and beet-milk control were determined according to AOAC [38]. Total fat of beverages was estimated by Folch method [39], protein was determined by the micro-Kjeldahl method according to Peach and Tracy [40]. While the total carbohydrates were estimated by Phenol-Sulfuric acid method [49]. Mineral content was determined as described previously at **Section** 3.4.

3.15. Bioactive Components and Antioxidant Power of Fermented Beverage

Total phenolics, total flavonoids, and DPPH free radicals scavenging were determined as mentioned by Vodnar et al. [19], Baba and Malik [45], and Do et al. [34], respectively.

3.16. Sensory Evaluation

Taste, flavor, consistency, and the overall acceptance were estimated in the ABT-5 and LA-5 juice compared to the control juice by ten panelists (6 females and 4 males) in ages average between 25 to 60.

The control juice was served to the panelists, followed by randomized, coded fermented beetroot-milk juices to assess the different sensory characteristics by recording a score of nine based on (1–9) Hedonic scale [50].

3.17. Statistical Analysis

All experiments were conducted in triplicates. SPSS program version 16 was used to determine the difference between means by one-way ANOVA test and univariate test for bioactive component yield, total phenolics, flavonoids, and DPPH assay. Means were compared by using Duncan's test at ($p < 0.05$).

4. Conclusions

The current study was conducted to evaluate the nutritional quality of beetroots to ascertain its potential use for formulation of novel functional foods. Furthermore, using beetroot juice in the development of a probiotic beetroot beverage using LA-5 and ABT-5 probiotic strains and evaluate the effect of the fermentation on the nutritional value of the beverage. The discarded beetroot wastes are rich in phenolics and antioxidants rather than crude fibers and minerals. Where the leaves had significant amounts of fibers, and lipids. Additionally, consuming 100 g of dried leaves powder meets the daily calcium recommendations as described by FDA (1000 mg/day). Beet juice, peels, and pomace exceed the daily iron recommendations (18 mg/day), which is considered an excellent source for iron and function as anti-malnutrition and anemia. Stems shared high antioxidant activity due to the high presence of phenolic compounds. Adding 40% milk to beet juice enhanced the growth of the probiotic strains on fermented beet beverage. The fermentation process resulted in a high lactate formation, which in turns resulted in a better taste, flavor, and thickening the consistency. Moreover, the fermentation increased zinc content in addition to antioxidant capacity. Hence, the finding of the current study emphasized the health benefits of beetroot' leaves and stems -which represent beetroots by-products-due to their high content of antioxidants and minerals. That is encourage utilizing beetroots by-products as a good source of novel food additives, food supplements and formulation of some novel functional foods.

Author Contributions: Conceptualization, S.E.-S. and E.A.; Formal analysis, E.A.; Investigation, E.A.; Methodology, S.E.-S. and A.Z.; Resources, S.E.-S. and A.Z., Supervision, O.S. and S.E.-S.; Validation: A.A.; Visualization, A.A.; Writing—Original draft, E.A.; Writing—Review & Editing: O.S. and S.E.-S. All authors have read and agreed to the published version of the manuscript.

Funding: This research received no external funding.

Acknowledgments: Authors acknowledge the department of Food science and Technology, Faculty of Agriculture (Saba Basha), Alexandria University and Food Technology Department, Arid Lands Cultivation Research Institute, City of Scientific Research and Technological Applications (SRTA-City), and A. Kh. A. Khalil, Specialist of Water Resources and Desert Soil, Desert Research Center, Cairo, Egypt for supporting this study.

Conflicts of Interest: The authors declare no conflict of interest.

References

1. Zandstra, J.; Hovius, C.; Weaver, L.; Marowa-Wilkerson, T. *Nutritional and Health Benefits of Fresh Vegetables-Past, Present and Future: A Literature Review*; Fresh Vegetable Growers of Ontario: Ridgetown, ON, Canada, 2007.
2. Dias, J.S. Nutritional quality and health benefits of vegetables: A review. *Food Nutr. Sci.* **2012**, *3*, 1354. [CrossRef]
3. Statista. Global Production Volume of Vegetables from 2000 to 2018. Available online: https://www.statista.com/statistics/264059/production-volume-of-vegetables-and-melons-worldwide-since-1990/ (accessed on 16 November 2020).
4. FAO. FAOSTAT, Food and Agriculture Organization of United Nation. Available online: http://www.fao.org/faostat/en/#data/QC (accessed on 16 November 2020).
5. Kale, R.; Sawate, A.; Kshirsagar, R.; Patil, B.; Mane, R. Studies on evaluation of physical and chemical composition of beetroot (*Beta vulgaris* L.). *Int. J. Chem. Stud.* **2018**, *6*, 2977–2979.

6. Singh, B.; Hathan, B.S. Chemical composition, functional properties and processing of beetroot—A review. *Int. J. Sci. Eng. Res.* **2014**, *5*, 679–684.

7. Srivastava, S.; Singh, K. Physical, Sensory and nutritional evaluation of biscuits prepared by using beetroot (*Beta vulgaris*) Powder. *Int. J. Innov. Res. Adv. Stud.* **2016**, *3*, 281–283.

8. Biondo, P.B.F.; Boeing, J.S.; Barizão, É.O.; Souza, N.E.D.; Matsushita, M.; Oliveira, C.C.D.; Boroski, M.; Visentainer, J.V. Evaluation of beetroot (*Beta vulgaris* L.) leaves during its developmental stages: A chemical composition study. *Food Sci. Technol.* **2014**, *34*, 94–101. [CrossRef]

9. Baião, D.D.S.; Da Silva, D.V.T.; Aguila, E.M.D.; Paschoalin, V.M.F. Nutritional, bioactive and physicochemical characteristics of different beetroot formulations. In *Food Additives*; InTech: Surrey, BC, Canada, 2017; pp. 22–43.

10. Ninfali, P.; Angelino, D. Nutritional and functional potential of *Beta vulgaris* cicla and rubra. *Fitoterapia* **2013**, *89*, 188–199. [CrossRef]

11. Vanajakshi, V.; Vijayendra, S.; Varadaraj, M.; Venkateswaran, G.; Agrawal, R. Optimization of a probiotic beverage based on Moringa leaves and beetroot. *LWT Food Sci. Technol.* **2015**, *63*, 1268–1273. [CrossRef]

12. Klewicka, E.; Nowak, A.; Zduńczyk, Z.; Juśkiewicz, J.; Cukrowska, B. Protective effect of lactofermented red beetroot juice against aberrant crypt foci formation, genotoxicity of fecal water and oxidative stress induced by 2-amino-1-methyl-6-phenylimidazo [4¨C-b] pyridine in rats model. *Environ. Toxicol. Pharmacol.* **2012**, *34*, 895–904. [CrossRef]

13. Rakin, M.; Vukasinovic, M.; Siler-Marinkovic, S.; Maksimovic, M. Contribution of lactic acid fermentation to improved nutritive quality vegetable juices enriched with brewer's yeast autolysate. *Food Chem.* **2007**, *100*, 599–602. [CrossRef]

14. Vaithilingam, M.; Chandrasekaran, S.; Mehra, A.; Prakash, S.; Agarwal, A.; Ethiraj, S.; Vaithiyanathan, S. Fermentation of beet juice using lactic acid bacteria and its cytotoxic activity against human liver cancer cell lines HepG2. *Curr. Bioact. Compd.* **2016**, *12*, 258–263. [CrossRef]

15. Gamage, S.; Mihirani, M.; Perera, O.; Weerahewa, H.D. Development of synbiotic beverage from beetroot juice using beneficial probiotic Lactobacillus Casei 431. *Ruhuna J. Sci.* **2016**, *7*, 64–69. [CrossRef]

16. Januário, J.; Da Silva, I.; De Oliveira, A.; De Oliveira, J.; Dionísio, J.; Klososki, S.; Pimentel, T. Probiotic yoghurt flavored with organic beet with carrot, cassava, sweet potato or corn juice: Physicochemical and texture evaluation, probiotic viability and acceptance. *Int. Food Res. J.* **2017**, *24*, 359–366.

17. Ferreira, M.S.; Santos, M.C.; Moro, T.M.; Basto, G.J.; Andrade, R.M.; Gonçalves, É.C. Formulation and characterization of functional foods based on fruit and vegetable residue flour. *J. Food Sci. Technol.* **2015**, *52*, 822–830. [CrossRef] [PubMed]

18. Plazzotta, S.; Manzocco, L.; Nicoli, M.C. Fruit and vegetable waste management and the challenge of fresh-cut salad. *Trends Food Sci. Technol.* **2017**, *63*, 51–59. [CrossRef]

19. Vodnar, D.C.; Calinoiu, L.F.; Dulf, F.V.; Stefanescu, B.E.; Crisan, G.; Socaciu, C. Identification of the bioactive compounds and antioxidant, antimutagenic and antimicrobial activities of thermally processed agro-industrial waste. *Food Chem.* **2017**, *231*, 131–140. [CrossRef]

20. Vulic, J.J.; Ebovic, T.N.C.; Anadanovic'-Bruneta, J.M.C.; Etkovic, G.S.C.; Anadanovic, V.M.C.; Djilasa, S.M.; Aponjac, V.T.T.S. In vivo and in vitro antioxidant effects of beetroot pomace extracts. *J. Funct. Foods* **2014**, *6*, 168–175. [CrossRef]

21. Shyamala, B.; Jamuna, P. Nutritional Content and Antioxidant Properties of Pulp Waste from Daucus carota and *Beta vulgaris*. *Malays. J. Nutr.* **2010**, *16*, 397–408.

22. Neha, P.; Jain, S.; Jain, N.; Jain, H.; Mittal, H. Chemical and functional properties of Beetroot (*Beta vulgaris* L.) for product development: A review. *Int. J. Chem. Stud.* **2018**, *6*, 3190–3194.

23. Kazimierczak, R.; Agata, S.; Ewelina, H.; Dominika, Ś.T.; Rembialkowska, E. Chemical Composition of Selected Beetroot Juices in Relation to Beetroot Production System and Processing Technology. *Not. Bot. Horti Agrobot. Cluj-Napoca* **2016**, *44*, 491–498. [CrossRef]

24. Domínguez, R.; Uenca, E.; Maté-Muñoz, J.L.; García-Fernández, P.; Serra-Paya, N.; Estevan, M.C.L.; Herreros, P.V.; Garnacho-Castaño, M.V. Effects of beetroot juice supplementation on cardiorespiratory endurance in athletes: A systematic review. *Nutrients* **2017**, *9*, 1–18.

25. Siervo, M.; Lara, J.; Ogbonmwan, I.; Mathers, J.C. Inorganic nitrate and beetroot juice supplementation reduces blood pressure in adults: A systematic review and meta-analysis. *J. Nutr.* **2013**, *143*, 818–826. [CrossRef] [PubMed]

26. Stalikas, C.D. Extraction, separation, and detection methods for phenolic acids and flavonoids. *J. Sep. Sci.* **2007**, *30*, 3268–3295. [CrossRef] [PubMed]

27. Čanadanović-Brunet, J.M.; Savatović, S.S.; Gordana, S.-Ć.; Vulić, J.J.; Ilas, S.M.D.; Markov, S.L.; Cvetković, D.D. Antioxidant and Antimicrobial Activities of Beet Root Pomace Extracts. *Czech J. Food Sci.* **2011**, *29*, 575–585.

28. Kujala, T.; Loponen, J.; Pihla, K. Betalains and Phenolics in Red Beetroot (*Beta vulgaris*) Peel Extracts: Extraction and Characterisation. *Z. Naturforsch. C J. Biosci.* **2001**, *56c*, 343–348. [CrossRef]

29. El-Beltagi, H.S.; Mohamed, H.I.; Megahed, B.M.H.; Gamal, M.; Safwat, G. Evaluation of some chemical constituents, antioxidanr, antibacterial, and anticancer activities of *Beta vulgaris* L. root. *Fresenius Environ. Bull.* **2018**, *27*, 6369–6378.

30. Prieto-Santiago, V.; Cavia, M.M.; Alonso-Torre, S.R.; Carrillo, C. Relationship between color and betalain content in different thermally treated beetroot products. *J. Food Sci. Technol.* **2020**, *57*, 3305–3313. [CrossRef]

31. Wruss, J.; Waldenberger, G.; Huemer, S.; Uygun, P.; Lanzerstorfer, P.; Müller, U.; Höglinger, O.; Weghuber, J. Compositional characteristics of commercial beetroot products and beetroot juice prepared from seven beetroot varieties grown in Upper Austria. *J. Food Compos. Anal.* **2015**, *42*, 46–55. [CrossRef]

32. Koubaier, H.B.H.; Snoussi, A.; Essaidi, I.; Chaabouni, M.M.; Bouzouita, P.T.N. Betalain and Phenolic Compositions, Antioxidant Activity of Tunisian Red Beet (*Beta vulgaris* L.conditiva) Roots and Stems Extracts. *Int. J. Food Prop.* **2014**, *17*, 1934–1945. [CrossRef]

33. Nantongo, J.S.; Odoi, J.B.; Abigaba, G.; Gwali, S. Variability of phenolic and alkaloid content in different plant parts of Carissa edulis Vahl and Zanthoxylum chalybeum Engl. *BMC Res. Notes* **2018**, *11*, 1–5. [CrossRef]

34. Do, Q.D.; Ngkawijaya, A.E.; Tran-Nguyen, P.L.; Huynh, L.H.; Soetaredjo, F.E.; Ismadji, S.; Ju, Y.-H. Effect of extraction solvent on total phenol content, total flavonoid content, and antioxidant activity of Limnophila aromatica. *J. Food Drug Anal.* **2014**, *22*, 296–302. [CrossRef]

35. Yoon, K.Y.; Woodams, E.E.; Hang, Y.D. Fermentation of beet juice by beneficial lactic acid bacteria. *LWT Food Sci. Technol.* **2005**, *38*, 73–75. [CrossRef]

36. Tang, A.; Shah, N.; Wilcox, G.; Walker, K.Z.; Stojanovska, L. Fermentation of calcium-fortified soymilk with Lactobacillus: Effects on calcium solubility, isoflavone conversion, and production of organic acids. *J. Food Sci.* **2007**, *72*, M431–M436. [CrossRef] [PubMed]

37. Obadina, A.; Akinola, O.; Shittu, T.; Bakare, H. Effect of natural fermentation on the chemical and nutritional composition of fermented soymilk nono. *Niger. Food J.* **2013**, *31*, 91–97. [CrossRef]

38. AOAC. *Official Methods of Analysis of the Association Official Analytical Chemist*, 15th ed.; Association Official Analytical Chemist, Inc.: Arlington, VA, USA, 1990; Volume 1, pp. 40–81.

39. Folch, L.M.; Stenly, G.S. A simple method for the isolation and purification of total lipids from animal tissues. *J. Biol. Chem.* **1957**, *226*, 497–509.

40. Peach, K.; Tracy, M. *Modern Methods of Plant Analysis*; Springer: Berlin/Heidelberg, Germany, 1956; 368p.

41. Lowther, J. Use of a single sulphuric acid-hydrogen peroxide digest for the analysis of Pinus radiata needles. *Commun. Soil Sci. Plant Anal.* **1980**, *11*, 175–188. [CrossRef]

42. Jackson, M.L. *Soil Chemical Analysis*; Prentice-Hall of India Pvt. Ltd.: New Delhi, India, 1973; 498p.

43. Vasconcellos, J.; Conte-Junior, C.; Silva, D.; Pierucci, A.P.; Paschoalin, V.; Alvares, T.S. Comparison of total antioxidant potential, and total phenolic, nitrate, sugar, and organic acid contents in beetroot juice, chips, powder, and cooked beetroot. *Food Sci. Biotechnol.* **2016**, *25*, 79–84. [CrossRef]

44. Raupp, D.D.S.; Rodrigues, E.; Rockenbach, I.I.; Carbonar, A.; Campos, P.F.D.; Borsato, A.V.; Fett, R. Effect of processing on antioxidant potential and total phenolics content in beet (*Beta vulgaris* L.). *Food Sci. Technol.* **2011**, *31*, 688–693. [CrossRef]

45. Baba, S.A.; Malik, S.A. Determination of total phenolic and flavonoid content, antimicrobial and antioxidant activity of a root extract of Arisaema jacquemontii Blume. *J. Taibah Univ. Sci.* **2015**, *9*, 449–454. [CrossRef]

46. Anand, P.; Singh, K.P.; Prasad, K.; Kaur, C.; Verma, A.K. Betalain Estimation and Callus Induction in Different Explants of Bougainvillea Spp. *Indian J. Agric. Sci.* **2016**, *87*, 191–196.

47. Castellanos-Santiago, E.; Yahia, E.M. Identification and quantification of betalains from the fruits of 10 Mexican prickly pear cultivars by high-performance liquid chromatography and electrospray ionization mass spectrometry. *J. Agric. Food Chem.* **2008**, *56*, 5758–5764. [CrossRef]

48. Buruleanu, L.C.; Bratu, M.G.; Manea, L.; Avram, D.; Nicolescu, C.L. Fermentation of vegetable juice by Lactobacillus acidophilus LA-5. In *Lactic Acid Bacteria-R & D for Food, Health and Livestock Purposes*; Kongo, M., Ed.; InTechOpen: London, UK, 2013; pp. 173–194.

49. BeMiller, J.N. Carbohydrate analysis. In *Food Analysis*; Springer: Berlin/Heidelberg, Germany, 2010; pp. 149–175.
50. Amerine, M.A.; Pangborn, R.M.; Roessler, B.E. Laboratory studies: Types and principles. In *Principles of Sensory Evaluation of Food*; Academic Press: New York, NY, USA, 1965; pp. 275–314.

Publisher's Note: MDPI stays neutral with regard to jurisdictional claims in published maps and institutional affiliations.

Article

Quality Evaluation, Storage Stability, and Sensory Characteristics of Wheat Noodles Incorporated with Isomaltodextrin

Da-Wei Huang [1,†], Yung-Jia Chan [2,†], Yuan-Chao Huang [3], Ya-Ju Chang [3], Jen-Chieh Tsai [4], Amanda Tresiliana Mulio [4], Zong-Ru Wu [4], Ya-Wen Hou [5], Wen-Chien Lu [6,*] and Po-Hsien Li [4,*]

1   Department of Biotechnology and Food Technology, Southern Taiwan University of Science and Technology, Tainan City 710, Taiwan; hdw0906@stust.edu.tw
2   College of Biotechnology and Bioresources, Da-Yeh University, Dacun, Changhua 51591, Taiwan; d0867601@cloud.dyu.edu.tw
3   Pharma Power Biotec Co., LTD., Taipei City 105412, Taiwan; snack997@ppbiotec.com.tw (Y.-C.H.); car-rie@ppbiotec.com.tw (Y.-J.C.)
4   Department of Medicinal Botanical and Health Applications, Da-Yeh University, Dacun, Changhua 51591, Taiwan; jenchieh@mail.dyu.edu.tw (J.-C.T.); tresiliana@gmail.com (A.T.M.); ian44989@gmail.com (Z.-R.W.)
5   Seafood technology division, Fisheries Research Institute, Council of Agriculture, Keelung 202008, Taiwan; ywhou@mail.tfrin.gov.tw
6   Department of Food and Beverage Management, Chung-Jen Junior College of Nursing, Health Sciences and Management, Hung-Mao-Pi, Chia-Yi City 60077, Taiwan
*   Correspondence: m104046@cjc.edu.tw (W.-C.L.); pohsien@mail.dyu.edu.tw (P.-H.L.); Tel.: +886-5-2772932 (ext. 860) (W.-C.L.); +886-4-8511888 (ext. 6233) (P.-H.L.)
†   These authors contributed equally to this work.

Citation: Huang, D.-W.; Chan, Y.-J.; Huang, Y.-C.; Chang, Y.-J.; Tsai, J.-C.; Mulio, A.T.; Wu, Z.-R.; Hou, Y.-W.; Lu, W.-C.; Li, P.-H. Quality Evaluation, Storage Stability, and Sensory Characteristics of Wheat Noodles Incorporated with Isomaltodextrin. *Plants* 2021, 10, 578. https://doi.org/10.3390/plants10030578

Academic Editor: Ivo Vaz de Oliveira

Received: 16 February 2021
Accepted: 9 March 2021
Published: 18 March 2021

Publisher's Note: MDPI stays neutral with regard to jurisdictional claims in published maps and institutional affiliations.

Abstract: Wheat noodles incorporated with isomaltodextrin were assessed in relation to physico-chemical properties (color), microstructure features, biochemical composition (fiber profile), cooking properties, textural attributes, and sensory evaluations during different storage temperatures (25, 4, −20 °C) and periods (0, 3, 6, 9, 12, 15, 18, 21, 24 months). Meanwhile, an accelerated study was also carried out at 40 °C storage conditions for 12 months to evaluate the fiber profile changes. Under different conditions, the overall quality of both raw and cooked noodle samples depended slightly on both the type and amount of added fiber isomaltodextrin, resistant starch (RS), insoluble high-molecular-weight dietary fiber (IHMWDF), and soluble high-molecular-weight dietary fiber (SHMWDF). However, this significantly changed for the fiber profile under 40 °C of storage for 12 months. Cooking quality, fiber profile, and color parameter did not differ by storage at −20 °C after 24 months than at 0 months, and noodles only slightly differed in texture and sensory characteristics. On sensory analysis, noodle samples were acceptable by panelists, with an acceptability score >5. In short, storage temperature is one of the most important factors in preserving food stability and retail properties. Isomaltodextrin noodles samples should be stored at low temperature to preserve the product functionality.

Keywords: wheat noodles; isomaltodextrin; storage stability; cooking quality; sensory evaluation; microstructure

1. Introduction

The glycemic index (GI), which is based on glycemic response, is a well-established indicator of the increase in blood glucose potential of a carbohydrate food [1]. Low GI diets play a vital role in the management and control of diabetes [2]. Previous research reported that the probable effects of a low-GI diet are lowering insulin secretion in type 2 diabetes and decreasing daily insulin requirements in type 1 diabetes [3]. Higher consumption of water-soluble dietary fiber, such as isomaltodextrin, above the level recommended by the

American Diabetes Association (ADA) enhances glycemic control, reduces hyperinsuline-mia, and lowers plasma lipid concentrations in people with type 2 diabetes [4]. However, noodles that are primarily formulated with wheat flour are not suitable for consumption by people with diabetes because of the high GI effects, which are evaluated for glycemic response (higher digestion rate or glucose absorption rate in the small intestine) [5,6].

In 2016, isomaltodextrin was assessed as a food that is generally recognized as safe (GRAS) by the US Food and Drug Administration (FDA) [7]. Isomaltodextrin presented health effects such as inhibition of fat absorption [8], anti-inflammatory properties, reduced risk of developing insulin resistance and associated metabolic diseases [9], diminished post-prandial blood glucose [10], positive effects on intestinal flora [11], and inhibition of hen egg ovalbumin allergic response by inducing immune tolerance in mice [12]. Isomaltodex-trin, a low-viscosity, water-soluble dietary fiber, which is enzymatically produced from corn starch, is also useful in reducing insulin secretion due to the reduction in sugar ab-sorption by inhibiting the disaccharidase-related transport system in both rats and healthy humans [13]. Furthermore, isomaltodextrin improved glucose tolerance after sucrose, maltose, and maltodextrin loading and also greatly lowered the accumulation of body fat, regardless of changes in body weight in male Sprague–Dawley rats [14]. Hence, because typical noodles will simply increase blood glucose level, fortification of low-GI ingredients (isomaltodextrin) in noodles may benefit people with diabetes and allow then to enjoy noodles as part of a healthy diet. However, few studies have investigated the stability and physicochemical of isomaltodextrin under different storage conditions.

Storage stability studies of food products is a vital characteristic for both manufac-turers and consumers, particularly in terms of food safety. Additionally important are physical and chemical features, appearances, and sensorial properties (organoleptic). Stor-age stability studies can provide important information to manufacturers and consumers to certify a high-quality product throughout the storage period [15]. Meanwhile, the products must be acceptable for consumers and comply with the nutritional value on the labeling. Choosing suitable storage-stability study models and data-analysis techniques are impor-tant to anticipate the shelf life, consider the changeability in environmental conditions, and guarantee real-time monitoring of products [16]. Appropriate scientific estimation and calculation of the storage stability of food commodities maintains the safety and quality of the products and can prevent food waste during the distribution and consumption stages of the food chain [17].

Therefore, this study aimed to investigate and evaluate the storage stability study of wheat noodles fortified with isomaltodextrin, in terms of physicochemical characteristics, microstructure features, dietary fiber profile, cooking quality, and sensory attributes during different storage temperature and periods. Meanwhile, it is hypothesized that isomaltodex-trin wheat noodles along the 24 months of storage at low temperature, which is 4 and $-20\ ^\circ$C, with only minor alterations in texture and sensory quality.

2. Materials and Methods

All chemical and reagents used in this study were purchased from Sigma-Aldrich Chemical (St. Louis, MO, USA) (ACS certified grade). Total dietary fiber assay kit (K-TDFR) was provided by Megazyme (Wicklow, Ireland). Isomaltodextrin (GRAS no. 610) was purchased from the Hayashibara (Okayama, Japan). All samples were established and analyzed in triplicate.

2.1. Noodles Sample

Wheat flour composed of hard red spring wheat and hard red winter wheat in a 2:3 proportion was provided by Chi-Fa Enterprise (Taichung, Taiwan). The analysis of mois-ture (Method 934.01), crude protein (Method 984.13), and ash content of corms followed the methods of the Association of Official Agricultural Chemists (1990). The moisture, crude protein, and ash content of wheat flour was 13.52, 11.74, and 0.35%, respectively. Isomaltodextrin was blended by using a domestic blender and passed through a 100-mesh

(150 µm) sieve. Wheat noodles were prepared as described [18] with modification. The formulated flours were prepared (5% of isomaltodextrin), and 2% salt was added; this mixture was mixed with water at low speeds for 3 min then rolled into a 2 mm thick sheet of dough. Folding and sheeting were repeated twice more. The dough sheet could rest for 30 min then was put through the sheeting rolls 3 times at progressively decreasing roll gaps of 2.60, 2.33, and 2.00 mm. From this dough sheet, the noodle strands were cut into strands of 18.1 × 0.12 × 0.11 cm, and the prepared noodles were dried in an oven at 55 °C for 12 h. Next, the dried noodle samples were packed in high-density polyethylene (HDPE) bags and stored in a dry box until further analysis use.

2.2. Storage Stability Study

To understand the effect of storage conditions on the product quality, and by referring the method of ICH Q1A (R2) Stability Testing of New Drug Substances and Products (European Medicines Agency), the noodle samples were stored under four different conditions, which is in general ambient condition (25 °C ± 2 °C/60% RH ± 5% RH), refrigerated storage temperature (5 °C ± 3 °C), freezing temperature (−20 °C ± 5 °C) for 24 months, and accelerated (40 °C ± 2 °C/75% RH ± 5% RH) conditions for 12 months (Figure S1). The noodle samples, packed in HDPE bags and stored under these conditions, were analyzed for physicochemical properties, such as cooking quality, color characteristics, texture analysis; changes in isomaltodextrin, resistant starch (RS), insoluble high-molecular-weight dietary fiber (IHMWDF), soluble high-molecular-weight dietary fiber (SHMWDF), sensory characteristics, as well as microstructure properties by scanning electron microscope (SEM). The noodle samples placed under different conditions were withdrawn at 3-month intervals for the above analyses. Still, the analyses were completed within 5 days of every single withdrawal, and the analyses were carried out in three replications.

2.3. Cooking Quality

The impact of storage conditions on the cooking characteristics of noodle samples was analyzed by the method of the American Association for Clinical Chemistry (AACC) (66–50) (AACC 2000). Cooking time (min) of the noodle samples was determined by starting the timer when adding one sample of 25 g to a beaker containing 300 mL boiling distilled water. The sample was stirred to ensure that noodles were separated. Cooking time corresponded to the disappearance of the opaque center core of the noodle when the noodle was squeezed between two clear glass plates. Cooking loss (%) was examined by drying the cooking water in an oven at 105 °C until constant weight was achieved. Cooking loss was calculated as follows:

$$\text{Cooking loss (\%)} = \frac{\text{weight of dried residues in noodles cooking water (g)}}{\text{weight of uncooked noodles (g)}} \times 100 \quad (1)$$

2.4. Dietary Fiber Profile Analysis

2.4.1. Determining Resistant Starch (RS)

The RS of noodle samples was analyzed by using the Megazyme kit (Megazyme K-TDFR, Wicklow, Ireland), which recognizes the method of the Association of Official Analytical Chemists (AOAC) (Method 2002.02/AACC Method 32-40.01).

2.4.2. Determining Isomaltodextrin

In this study, the content of isomaltodextrin usually was based on AOAC Method 991.43 "Total, Soluble, and Insoluble Dietary Fiber in Foods" (First Action 1991) and AACC Method 32-07.01 "Determination of Soluble, Insoluble, and Total Dietary Fiber in Foods and Food Products" (Final Approval 10-16-91).

2.4.3. Determining IHMWDF and SHMWDF

Analysis of IHMWDF was based on the AOAC Official Method 2011.25 "Insoluble, Soluble, and Total Dietary Fiber in Foods" (First Action 2011).

2.5. Color Measurement

Color of noodle samples was measured with a ZE2000 Color Measurement colorimeter (Nippon Denshoku) by using the CIE L*a*b* system. Color parameter was reported as lightness (L*); positive a* represents the redness, and negative a* is green; positive b* is yellowness, and negative b* is blue. A standard black and white ceramic tile was used to calibrate the instrument before every measurement. Color measurements were performed at room temperature in pentaplicates for every test sample.

2.6. Texture Profile Analysis

Texture analyzer model Brookfield CT3 Texture Analyzer (AMK CT3) specific for the food industry was used to measure noodle texture. Five noodle strands were arranged adjacent to each other on the fixture base tables and tested for hardness (g), adhesiveness (g.s), maximum tensile strength (g), and tensile fracture distance (mm) under the experimental conditions of 250 kg load cell and 10 mm/s cross head speed. The analyses were carried out with three repetitions and five replicates.

2.7. Sensory Evaluation

A total of 30 panelists (15 females and 15 males), aged from 25–35 years old, who were students from the College of Biotechnology and Bioresources, were randomly selected to participate in this sensory evaluation session. The sensory evaluation test was conducted in a sensory laboratory at room temperature and strictly followed the GB/T 13662-2008 and ISO 4121 criteria.

2.8. Microstructural Characteristics

The noodle samples stored under three different conditions 25, 4, −20 °C for 3, 12, and 24 months were coated with gold-palladium (Model JBS-ES 150, Ion sputter coater, Topon Corp., Japan), and the instrument was used with a model of the ABT-150S system (Topon Corp., Kyoto, Japan) equipped with an Olympus BX53 polarized light microscope (PLM) and a CCD camera to detect the collaboration and effects of the noodles matrix and variations during the storage study.

2.9. Statistical Analysis

Experimental assessments and analysis were carried out in triplicate. Data were reported as a mean score for each individual attribute. For statistical analysis, all data were assessed by using single-factor ANOVA. If the F-value was significant ($p < 0.05$) on ANOVA, then Duncan's new multiple range tests was used to correlate treatment means.

3. Results and Discussions

3.1. Effects of Storage on Cooking Quality

Cooking quality and characteristics of wheat noodles fortified with isomaltodextrin during storage periods of 0 to 24 months, at 25, 4, and −20 °C are summarized in Table 1. Determining the cooking quality aimed to understand the interrelationship between storage temperature and storage time in terms of noodle quality, to ensure consumer acceptancy and the mouthfeel of the products. The cooked weight gain of the noodles stored at 4 and −20 °C did not significantly differ from 0 to 24 months. However, the cooked weight gain significantly differed after 18 months for noodles stored at 25 °C (62.62 ± 0.15, 62.56 ± 0.16, 62.37 ± 0.16 g for 18, 21, and 24 months, respectively) as compared with 0 months (63.97 ± 0.60 g). The results agreed with a previous study [19], finding that the reduction in cooked weight was probably due to an increase in cooking loss. Cooking loss was to measure the number of solids remaining in the cooking water, which was caused by the leaking of amylose and solubilization of soluble proteins [20]. The cooking loss for noodles stored at 25 °C significantly differed from 3.71 ± 0.02 to 3.93 ± 0.04, 3.98 ± 0.05, 4.08 ± 0.07, and 4.13 ± 0.07 at 0, 15, 21, and 24 months. Appropriate cooking time and adequate cooking water may control the cooking loss, which deeply affects the texture and sensory attributes of the noodles.

Table 1. Cooking quality characteristics of wheat noodles fortified with isomaltodextrin during different storage periods and temperature.

Months	Temperature (°C)	Cooking Time (min)	Cooked Weight Gained (g)	Cooking Loss (%)
0	25	7.38 ± 0.05 [c]	63.97 ± 0.60 [ab]	3.71 ± 0.02 [bc]
	4	7.34 ± 0.07 [c]	64.12 ± 0.37 [a]	3.79 ± 0.03 [bc]
	−20	7.38 ± 0.06 [c]	63.99 ± 0.36 [ab]	3.87 ± 0.05 [ab]
3	25	7.43 ± 0.07 [bc]	63.58 ± 0.50 [bc]	3.78 ± 0.04 [bc]
	4	7.43 ± 0.06 [bc]	64.16 ± 0.56 [a]	3.75 ± 0.06 [bc]
	−20	7.41 ± 0.09 [bc]	63.93 ± 0.21 [ab]	3.81 ± 0.05 [bc]
6	25	7.50 ± 0.18 [a]	63.93 ± 1.08 [ab]	3.74 ± 0.09 [bc]
	4	7.39 ± 0.07 [c]	63.96 ± 0.26 [ab]	3.81 ± 0.08 [bc]
	−20	7.35 ± 0.04 [c]	63.62 ± 0.62 [bc]	3.77 ± 0.06 [bc]
9	25	7.43 ± 0.04 [bc]	64.19 ± 0.93 [a]	3.79 ± 0.08 [bc]
	4	7.47 ± 0.06 [ab]	63.39 ± 0.01 [bc]	3.85 ± 0.06 [bc]
	−20	7.43 ± 0.02 [bc]	64.11 ± 0.49 [a]	3.86 ± 0.02 [ab]
12	25	7.47 ± 0.09 [a]	63.49 ± 0.08 [bc]	3.89 ± 0.07 [ab]
	4	7.45 ± 0.06 [bc]	63.21 ± 0.06 [c]	3.74 ± 0.04 [bc]
	−20	7.46 ± 0.08 [ab]	64.53 ± 0.70 [a]	3.78 ± 0.04 [bc]
15	25	7.44 ± 0.02 [bc]	63.36 ± 0.26 [bc]	3.93 ± 0.04 [a]
	4	7.47 ± 0.15 [a]	63.58 ± 0.42 [bc]	3.88 ± 0.04 [ab]
	−20	7.38 ± 0.05 [c]	64.25 ± 0.84 [a]	3.86 ± 0.04 [ab]
18	25	7.55 ± 0.04 [a]	62.62 ± 0.15 [c]	3.98 ± 0.05 [a]
	4	7.45 ± 0.05 [bc]	63.72 ± 0.52 [ab]	3.87 ± 0.11 [ab]
	−20	7.45 ± 0.04 [bc]	64.29 ± 0.84 [a]	3.85 ± 0.02 [bc]
21	25	7.39 ± 0.03 [c]	62.56 ± 0.16 [c]	4.08 ± 0.07 [a]
	4	7.44 ± 0.03 [bc]	63.55 ± 0.46 [bc]	3.88 ± 0.07 [ab]
	−20	7.42 ± 0.05 [bc]	64.19 ± 0.87 [a]	3.87 ± 0.05 [ab]
24	25	7.36 ± 0.03 [c]	62.37 ± 0.16 [c]	4.13 ± 0.07 [a]
	4	7.43 ± 0.01 [bc]	63.85 ± 0.62 [ab]	3.97 ± 0.05 [a]
	−20	7.45 ± 0.08 [bc]	63.54 ± 0.55 [bc]	3.92 ± 0.03 [a]

Scores are presented as mean ± SD of triplicate analysis. The lowercase letters indicated significant difference in each column (*p < 0.05*).

3.2. Effects of Storage on Fiber Profile

Because of different structural characteristics of fibers, molecular weight distribution and spatial density of different dietary fibers added to noodles compounds can have positive or negative effects on noodles properties [21]. Therefore, assurances and preserved product quality control have become more crucial and have led to the need to determine the amount of dietary fiber composition in final products. Figure 1 illustrates the fiber profile of wheat noodles fortified with isomaltodextrin during the different storage periods at 25, 4, and −20 °C storage. Figure 2 shows the fiber profile of wheat noodles fortified with isomaltodextrin during the different storage periods at 40 °C storage. The isomaltodextrin content ranged from 9.47 ± 0.06 to 9.31 ± 0.03%, RS content from 0.24 ± 0.01 to 0.22 ± 0.02%, IHMWDF content from 7.96 ± 0.04 to 7.44 ± 0.04%, and SHMWDF content from 0.95 ± 0.02 to 0.61 ± 0.02%, from 0 to 24 months of storage at 25 °C (Table S1). However, storage at 4 °C and −20 °C slowed the degradation of SHMWDF, which was 1.01 ± 0.05 to 0.79 ± 0.03% and 1.01 ± 0.04 to 0.80 ± 0.03%, respectively, from 0 to 24 months of storage (Tables S2 and S3). The composition of isomaltodextrin, RS, and IHMWDF did not differ by storage and storage temperature. Regardless, SHMWDF content was degraded with storage at ambient temperature. Previous study has revealed that dietary fibers with high molecular weight have stronger viscosity than those with low molecular weight [22]. Moreover, high-molecular-weight dietary fiber, including cellulose, RS, and guar gum, originally from wheat flour, had a better effect on noodles than low-molecular-weight dietary fiber (Table S4) [23].

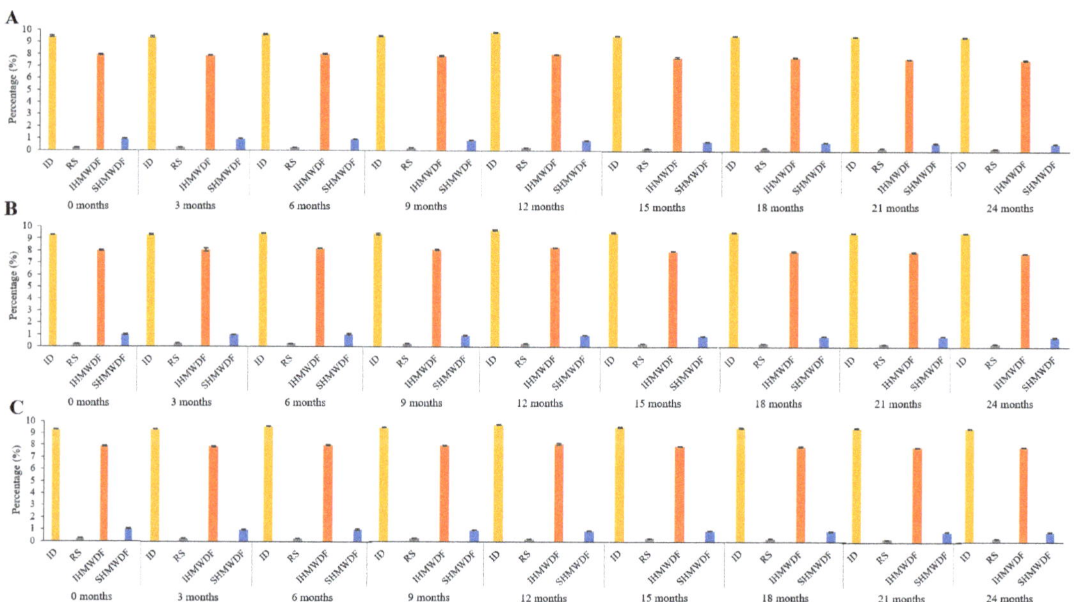

Figure 1. Fiber profile of wheat noodles fortified with isomaltodextrin during different storage period (0, 3, 6, 9, 12, 15, 18, 21, and 24 months) at 25 °C (**A**), 4 °C (**B**), and −20 °C (**C**). ID, isomaltodextrin; RS, resistant starch; IHMWDF, insoluble high-molecular-weight dietary fiber; SHMWDF, soluble high-molecular-weight dietary fiber.

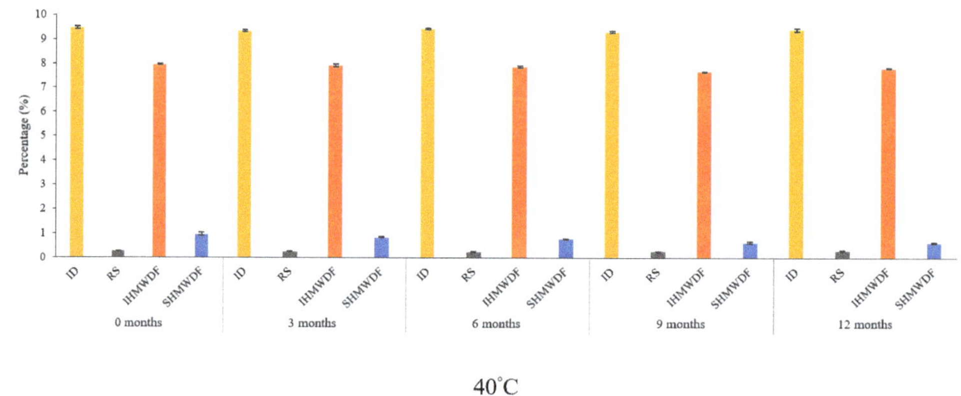

Figure 2. Fiber profile of wheat noodles fortified with isomaltodextrin during different storage periods (0, 3, 6, 9, and 12 months) at 40 °C storage. ID, isomaltodextrin; RS, resistant starch; IHMWDF, insoluble high-molecular-weight dietary fiber; SHMWDF, soluble high-molecular-weight dietary fiber.

Dietary fiber is proclaimed to help avoid obesity caused by high-fat diets [24], by reducing energy intake, increasing the viscosity of intestinal contents, delaying gastric exhaustion, and augmenting the short-chain fatty acid content in the large bowel, thus preventing central appetite and encouraging the release of anorexic intestinal hormones [25]. Meanwhile, dietary fiber as a food additive helps to maintain the level of fiber content in food and encourage the technological functionalities of food, such as gelatinization, product consistency, colloidal stability, texture, and viscosity [26]. As compared with insoluble dietary fiber, soluble dietary fiber is more common because of its critical physiological roles and benefits in physical and chemical properties [27], for example, the release of hyperlipidemia, controlling diabetes mellitus, effect on cardiovascular disease, and monitoring of colon cancer [28]. Because of the characteristics of water retention, gel formation, fat simulation, and thickening effects, the addition of soluble dietary fiber can significantly enhance the texture, shelf life, stability, and sensory properties of starch-based food; decrease the rate of aging; preserve the relative stability of the rheological and tissue properties [29].

3.3. Effects of Storage on Color Parameters

The impact of color characteristics on wheat noodles fortified with isomaltodextrin during different storage periods is in Table 2. The appearance of cooked wheat noodles fortified with isomaltodextrin for 0, 6, 12, 18, and 24 months of storage at 25, 4, and −20 °C is in Figure 3. Color parameters of L*, a*, and b* of noodle samples stored at 25 °C showed a significant deviation from 0 to 24 months. For noodle samples stored at 4 °C, the L* and a* value started to change beyond 9 months of storage. The a^* and b^* values of noodle samples stored at −20 °C did not differ; only the L* value differed after 6 months. Fortification of noodle samples with isomaltodextrin may alter the microstructure of the noodles, causing a disrupted starch, protein, and dietary fiber matrix. Changing the microstructure of the noodle samples may trigger a chemical reaction, such as browning effects and oxidation. Mohamed, Xu, and Singh (2010) [30] found that the Maillard reaction occurring in between reducing sugar and proteins may lead to the lightlessness of the cooked noodles. Color deterioration of food products during storage is mainly due to an enzymatic or chemical reaction [31]. Furthermore, during storage, the moisture is absorbed, and an oxidation reaction may also be responsible in part for the changes in color value [6]. Results of this color parameter study revealed that storage condition and duration of storage has a corresponding effect on the color of the noodle samples.

Table 2. Color characteristics of wheat noodles fortified with isomaltodextrin during different storage periods.

Months	Temperature (°C)	L*	a*	b*
0	25	53.90 ± 1.59 [a]	−2.95 ± 0.08 [b]	4.19 ± 0.07 [c]
	4	53.22 ± 1.96 [a]	−2.95 ± 0.09 [b]	4.25 ± 0.08 [bc]
	−20	53.18 ± 1.95 [a]	−2.88 ± 0.13 [ab]	4.36 ± 0.07 [bc]
3	25	53.60 ± 3.09 [a]	−2.94 ± 0.09 [b]	4.20 ± 0.08 [c]
	4	48.51 ± 3.78 [bc]	−2.86 ± 0.06 [ab]	4.27 ± 0.07 [bc]
	−20	50.59 ± 1.84 [a]	−2.78 ± 0.08 [a]	4.44 ± 0.09 [bc]
6	25	49.38 ± 0.89 [bc]	−2.96 ± 0.11 [b]	4.01 ± 0.58 [c]
	4	50.40 ± 1.16 [a]	−2.92 ± 0.11 [b]	4.27 ± 0.07 [bc]
	−20	49.87 ± 0.96 [bc]	−2.74 ± 0.04 [a]	4.52 ± 0.11 [ab]
9	25	49.13 ± 0.92 [bc]	−3.17 ± 0.09 [c]	4.59 ± 0.08 [ab]
	4	48.37 ± 0.44 [bc]	−2.83 ± 0.07 [a]	4.13 ± 0.11 [c]
	−20	48.65 ± 0.67 [bc]	−2.78 ± 0.07 [a]	4.28 ± 0.09 [bc]
12	25	47.60 ± 1.23 [bc]	−3.25 ± 0.09 [c]	4.78 ± 0.07 [a]
	4	49.01 ± 0.29 [bc]	−2.87 ± 0.07 [ab]	4.30 ± 0.05 [bc]
	−20	48.87 ± 0.57 [bc]	−2.77 ± 0.03 [a]	4.51 ± 0.11 [ab]
15	25	46.46 ± 1.06 [c]	−3.33 ± 0.07 [c]	4.68 ± 0.05 [a]
	4	48.65 ± 0.95 [bc]	−2.89 ± 0.06 [ab]	4.30 ± 0.03 [bc]
	−20	48.23 ± 0.31 [bc]	−2.81 ± 0.11 [a]	4.51 ± 0.06 [ab]

Table 2. *Cont.*

Months	Temperature (°C)	L*	a*	b*
18	25	46.39 ± 1.11 [c]	−3.38 ± 0.04 [c]	4.82 ± 0.04 [a]
	4	47.63 ± 0.95 [bc]	−3.08 ± 0.08 [b]	4.35 ± 0.06 [bc]
	−20	47.96 ± 0.66 [bc]	−2.83 ± 0.09 [a]	4.34 ± 0.13 [bc]
21	25	44.88 ± 1.25 [c]	−3.46 ± 0.04 [c]	4.93 ± 0.03 [a]
	4	47.23 ± 0.22 [bc]	−3.09 ± 0.08 [b]	4.34 ± 0.05 [bc]
	−20	47.29 ± 0.16 [bc]	−2.83 ± 0.09 [a]	4.26 ± 0.04 [bc]
24	25	42.83 ± 0.71 [c]	−3.52 ± 0.22 [c]	5.12 ± 0.04 [a]
	4	47.21 ± 0.21 [bc]	−3.19 ± 0.08 [c]	4.36 ± 0.15 [bc]
	−20	47.12 ± 0.05 [bc]	−2.74 ± 0.02 [a]	4.34 ± 0.04 [bc]

Scores are presented as mean ± SD of triplicate analysis. The lowercase letters indicated significant differences in each column ($p < 0.05$).

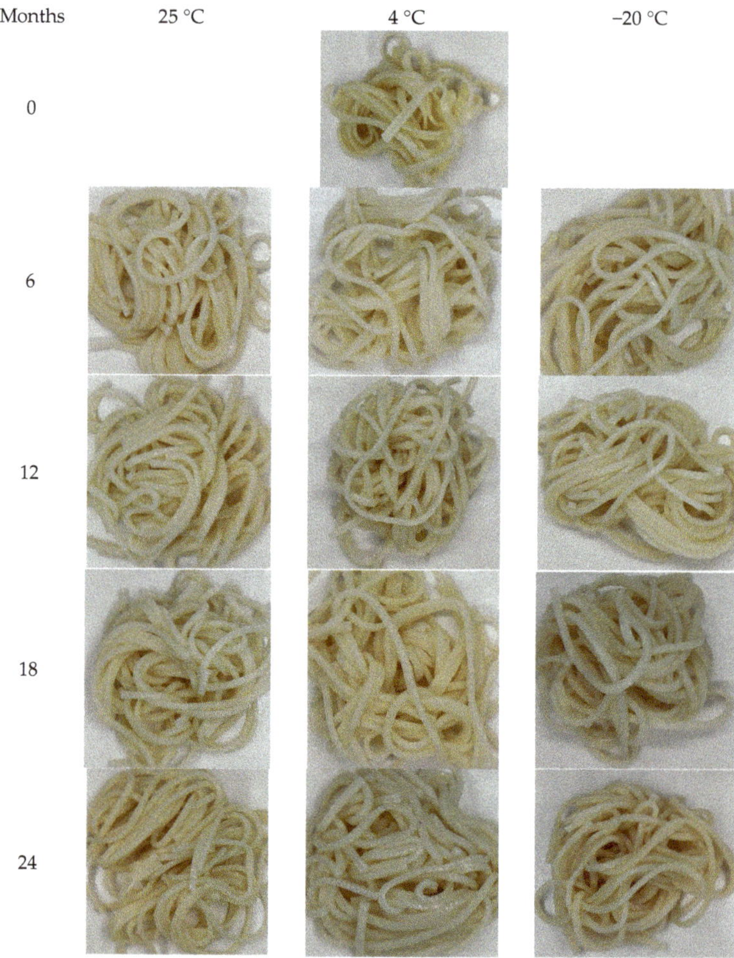

Figure 3. Appearance (after cooking) of cooked wheat noodles fortified with isomaltodextrin during different storage periods at different storage temperatures.

3.4. Effects of Storage on Texture Profile

Sensory attributes of fortified noodles rely on cooking qualities and texture of the cooked products, which also play an important role in affecting the acceptability of consumers. The texture profile of the cooked wheat noodles fortified with isomaltodextrin during different storage periods and temperatures is in Table 3. The hardness of the noodle samples stored at 25, 4, and $-20\,^{\circ}$C increased greatly from 4164.02 ± 49.37, 4172.81 ± 47.29, and 4189.19 ± 81.31 to 4271.44 ± 29.7, 4211.14 ± 53.34, and 4235.07 ± 47.84, respectively, at 24 months of storage. At 25 $^{\circ}$C storage, the adhesiveness, maximum tensile strength, and tensile fracture distance significantly differed after 6, 12, and 9 months of storage, respectively. However, noodles stored at 4 and $-20\,^{\circ}$C showed significant differences for only hardness and adhesiveness at 24 months of storage period. The increase in adhesiveness of isomaltodextrin noodles during storage may be due to the increase in cooking loss encouraging disintegration of more soluble proteins and starch during cooking. In contrast, the partially soluble proteins and starch are attached on the surface of noodles, causing aggravation of this sticky behavior [32]. The increase in hardness and adhesiveness of the isomaltodextrin noodles indicated the deterioration of the noodle samples during storage, especially at 25 $^{\circ}$C storage. The texture quality change in the isomaltodextrin noodle samples at different storage temperatures and periods may be due to the fiber supplements disturbing the fiber–starch matrix within the microstructure of noodles [33].

Table 3. Texture analysis of wheat noodles fortified with isomaltodextrin during different storage periods.

Months	Temperature ($^{\circ}$C)	Hardness (g)	Adhesiveness (g/s)	Maximum Tensile Strength (g)	Tensile Fracture Distance (mm)
0	25	4164.02 ± 49.37 [c]	77.36 ± 0.83 [c]	41.99 ± 0.65 [ab]	107.07 ± 2.66 [a]
	4	4172.81 ± 47.29 [bc]	76.78 ± 1.96 [c]	42.03 ± 0.27 [ab]	107.65 ± 1.18 [a]
	-20	4189.19 ± 81.31 [bc]	77.81 ± 0.31 [bc]	42.81 ± 0.31 [ab]	107.52 ± 1.22 [a]
3	25	4145.84 ± 50.49 [c]	77.79 ± 1.26 [c]	37.62 ± 0.51 [b]	101.17 ± 1.24 [ab]
	4	4217.83 ± 9.86 [a]	77.01 ± 1.55 [c]	40.94 ± 0.22 [ab]	101.29 ± 1.09 [ab]
	-20	4210.43 ± 54.29 [a]	79.20 ± 0.94 [ab]	42.36 ± 0.75 [ab]	105.52 ± 1.52 [a]
6	25	4154.76 ± 15.41 [c]	78.28 ± 1.08 [bc]	35.70 ± 0.42 [c]	94.82 ± 1.49 [ab]
	4	4180.45 ± 44.59 [bc]	77.59 ± 0.93 [c]	39.66 ± 0.25 [ab]	103.09 ± 1.49 [ab]
	-20	4198.39 ± 5.35 [bc]	76.98 ± 0.82 [c]	41.76 ± 0.59 [ab]	101.07 ± 0.76 [ab]
9	25	4187.61 ± 15.02 [bc]	77.91 ± 0.59 [bc]	34.06 ± 0.16 [c]	85.47 ± 1.25 [c]
	4	4204.14 ± 6.53 [bc]	76.09 ± 1.33 [c]	39.11 ± 0.15 [ab]	99.87 ± 0.24 [ab]
	-20	4196.64 ± 18.46 [bc]	77.97 ± 1.84 [bc]	41.71 ± 0.44 [ab]	99.61 ± 0.80 [ab]
12	25	4173.11 ± 31.86 [bc]	79.05 ± 1.23 [ab]	31.57 ± 0.27 [c]	80.28 ± 1.81 [c]
	4	4221.64 ± 91.03 [a]	76.61 ± 1.74 [c]	38.53 ± 0.39 [b]	98.44 ± 0.77 [ab]
	-20	4190.13 ± 48.19 [bc]	78.19 ± 0.97 [bc]	41.54 ± 0.99 [ab]	97.49 ± 1.02 [ab]
15	25	4186.11 ± 62.61 [bc]	79.31 ± 1.14 [ab]	29.01 ± 0.67 [c]	76.01 ± 1.54 [c]
	4	4157.65 ± 36.65 [c]	78.75 ± 0.45 [ab]	38.43 ± 0.26 [b]	97.46 ± 0.96 [ab]
	-20	4176.12 ± 63.16 [bc]	79.64 ± 0.56 [ab]	41.49 ± 1.61 [ab]	98.21 ± 0.93 [ab]
18	25	4210.48 ± 11.73 [a]	80.05 ± 1.52 [a]	26.94 ± 0.21 [c]	74.23 ± 0.94 [c]
	4	4128.51 ± 5.76 [c]	76.64 ± 1.28 [c]	36.61 ± 0.27 [b]	95.79 ± 0.44 [ab]
	-20	4243.74 ± 35.85 [a]	79.39 ± 0.82 [ab]	40.32 ± 0.56 [ab]	95.75 ± 2.04 [ab]
21	25	4190.08 ± 61.47 [bc]	80.86 ± 2.02 [a]	23.64 ± 0.48 [c]	70.19 ± 0.93 [c]
	4	4197.26 ± 41.38 [bc]	78.41 ± 0.72 [bc]	35.62 ± 0.31 [c]	94.59 ± 0.45 [ab]
	-20	4188.53 ± 10.96 [bc]	79.39 ± 0.82 [ab]	39.80 ± 0.52 [ab]	97.26 ± 0.88 [ab]
24	25	4271.44 ± 29.71 [a]	82.71 ± 3.46 [a]	21.84 ± 0.27 [c]	67.47 ± 0.94 [c]
	4	4211.14 ± 53.34 [a]	80.24 ± 1.06 [a]	35.65 ± 0.21 [c]	94.65 ± 0.48 [ab]
	-20	4235.07 ± 47.84 [a]	79.89 ± 0.51 [a]	39.05 ± 0.66 [ab]	95.58 ± 0.61 [ab]

Scores are presented as mean $\pm$ SD of triplicate analysis. The lowercase letters indicated significant differences in each column ($p < 0.05$).

3.5. Effects of Storage on Sensory Properties

Figure 4 illustrates the sensory evaluation of wheat noodles fortified with isomaltodextrin during different storage periods (0, 3, 6, 9, 12, 15, 18, 21, and 24 months) at 25, 4, and −20 °C storage. The cooked isomaltodextrin noodles were evaluated for sensory characteristics such as color, odor, taste, firmness, and overall acceptance. Long-term storage reduced the overall acceptance of the noodle samples, which may due to the odor changed during the long-term storage. However, taste, firmness, and color did not differ for the isomaltodextrin noodles stored at −20 °C. Sensory attributes, including color, odor, taste, firmness, and even overall acceptance of the isomaltodextrin noodles stored at ambient temperature slowly deteriorated during the storage study, from 0 to 24 months. Even though the sensory attributes and overall acceptance of the isomaltodextrin noodles stored at 25, 4, and −20 °C was slightly reduced by time, all the samples were acceptable to the panelists, with a score >5. The firmness of the isomaltodextrin noodles was increased slightly for the noodles stored at 25, 4, and −20 °C, from 0 to 24 months, which may due to the loss of moisture and breakdown of sample matrix interconnection during the long-term storage and directly affects the cooking characteristics of the noodle samples [6].

3.6. Effects of Storage on Microstructural Characteristics

Scanning electron microscope (SEM) was used to collect data on the structural integrity, size, shape, and arrangement of particles of both raw and cooked wheat noodles fortified with isomaltodextrin, which closely correlated with the cooking quality, texture profile, and sensory characteristics [34]. Figures 5–7 show the SEM micrographs of wheat noodles fortified with isomaltodextrin during different storage periods at 25, 4, and −20 °C, respectively. Micrographs of the raw isomaltodextrin noodle samples (before cooking) at 25 °C showed the well-formed fiber–starch matrix, with clearly seen isomaltodextrin particles attached to the starch granules. At 4 °C storage, the isomaltodextrin particles were partially adhered between each molecule and with the starch granules. The isomaltodextrin particles were frosted and held within starch granules to form a compact fiber–starch network because of the low temperature (storage temperature −20 °C). The starch granules within the noodle samples seemed to be a little swollen and with irregular size and shape, possibly demonstrating a gelatinization that occurred during the extrusion process [34]. Storage period did not affect the microstructure of the wheat noodles fortified with isomaltodextrin. SEM micrographs of raw noodle samples with isomaltodextrin showed a higher inclusion rate of soluble fiber in the binding relationship between starch and fiber (isomaltodextrin), which altered the general structural complexity of the noodle system.

Gelatinization of starch granules within the cooked noodle samples seems to be integrated into a developed fiber matrix to form a compacted noodles structure. The isomaltodextrin disturbs the structure with starch granules within the noodle matrix, causing loss of the regular structure and triggering the gelatinization process during cooking. The disruption to the isomaltodextrin noodles structures significantly affected the overall cooking quality and texture profiles of the isomaltodextrin noodles, which is described in Tables 1 and 3.

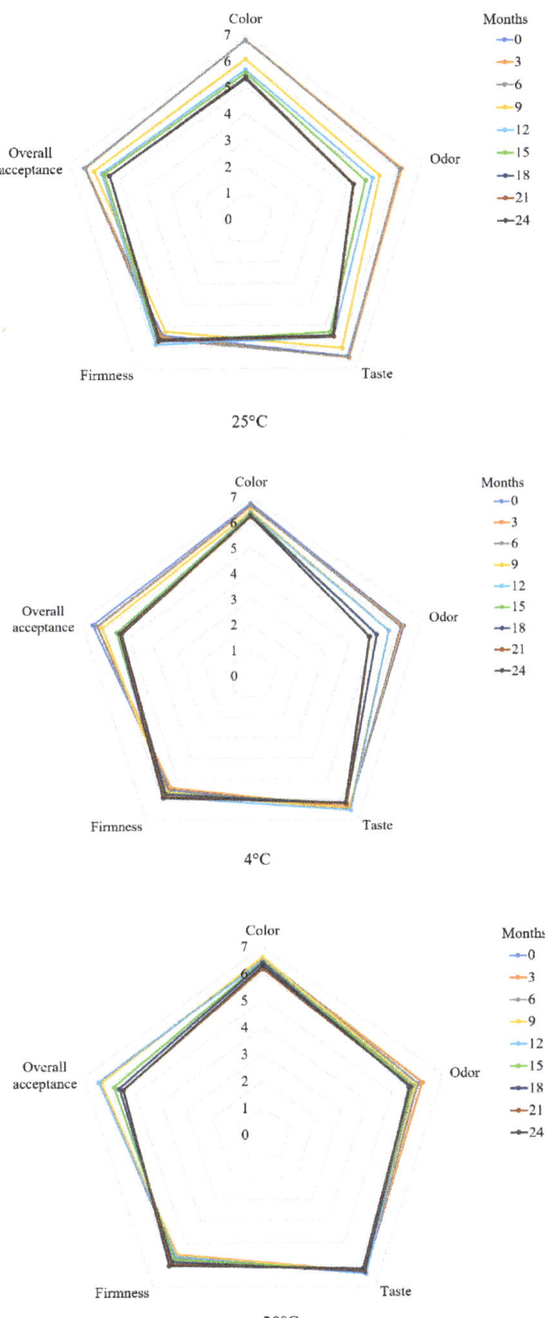

Figure 4. Sensory evaluation of wheat noodles fortified with isomaltodextrin during different storage periods at 25, 4, and −20 °C storage.

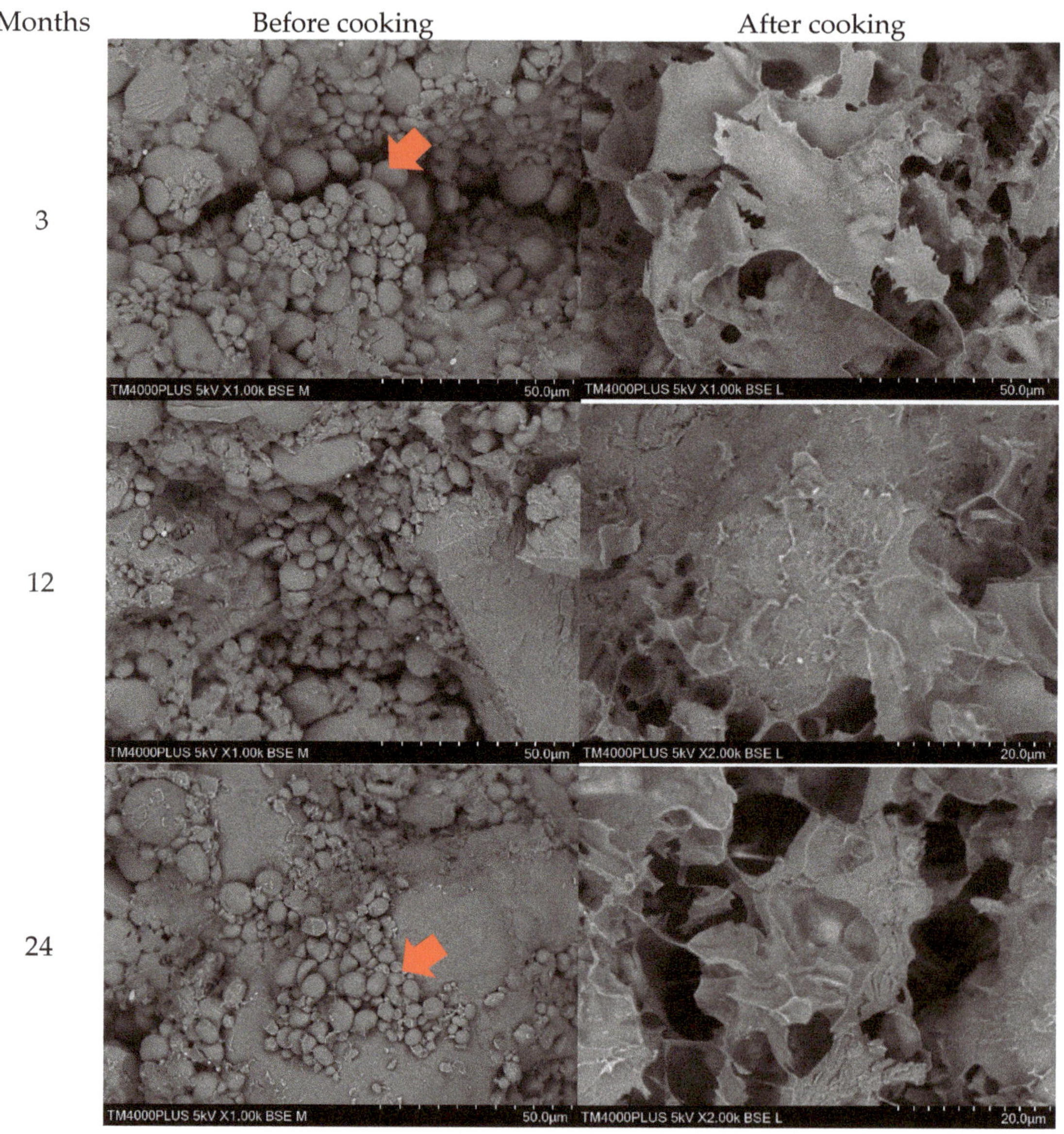

Figure 5. SEM of wheat noodles fortified with isomaltodextrin during different storage periods at 25 °C storage.

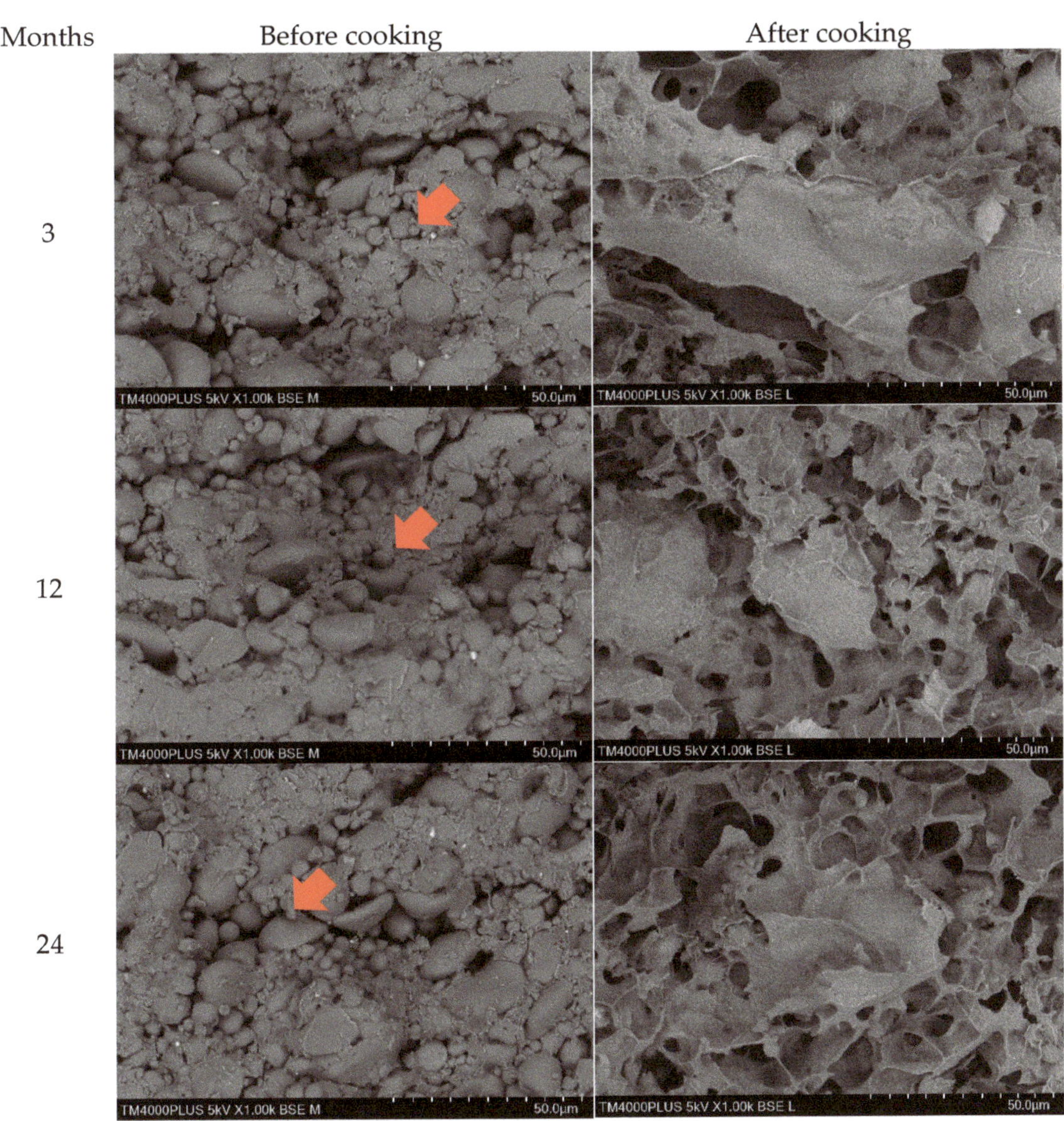

Figure 6. SEM of wheat noodles fortified with isomaltodextrin during different storage periods at 4 °C.

Months   Before cooking   After cooking

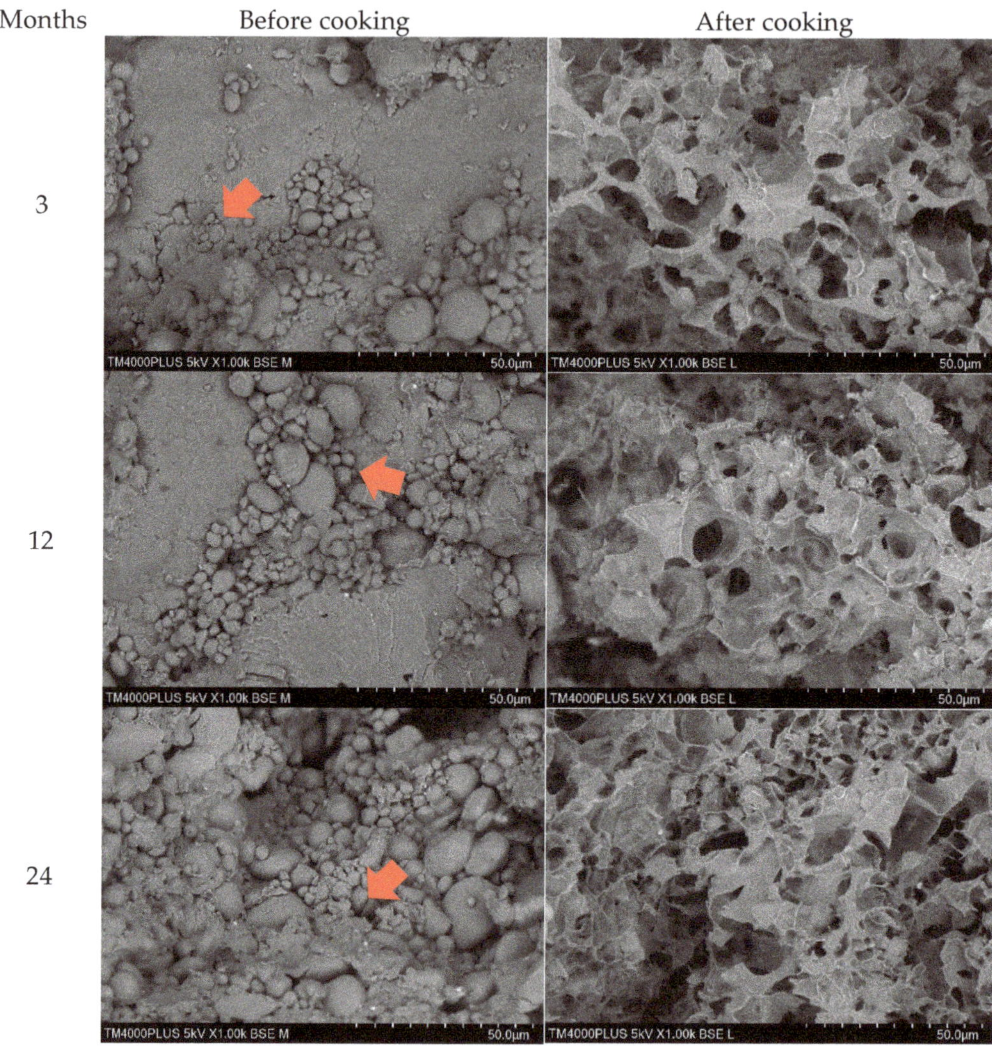

Figure 7. SEM of wheat noodles fortified with isomaltodextrin during different storage periods at −20 °C.

4. Conclusions

During a storage and stability study, food products may go through physical, chemical, and biochemical deterioration along with environmental condition, such as temperature and time range, which may cause changes in the product's physical, chemical, quality, and sensory attributes. These changes affect the safety of products and consumer acceptability. Hence, the quality and characteristics of the food products must be evaluated in a storage stability study to sustain the product's stability. In this study, during 24 months of storage at low temperatures (4 °C and −20 °C), cooking quality, fiber profile, and color did not differ from 0 months, with only minor differences in texture and sensory characteristics. Therefore, isomaltodextrin noodles could be stored at low temperature to preserve the product's functionality and quality.

Supplementary Materials: The following are available online at https://www.mdpi.com/2223-7747/10/3/578/s1, Figure S1. Storage stability study design of wheat noodles incorporated with isomaltodextrin, Table S1: Changes in indigestible dextrin (ID), resistant starch, IHMWDF, and SHMWDF of wheat noodles with maltodextrin during different storage period at 25 °C storage temperature, Table S2: Changes in indigestible dextrin (ID), resistant starch, IHMWDF, and SHMWDF of wheat noodles with maltodextrin during different storage period at 4 °C storage temperature. Table S3: Changes in indigestible dextrin (ID), resistant starch, IHMWDF, and SHMWDF of wheat noodles with maltodextrin during different storage period at −20 °C storage temperature. Table S4: Changes in indigestible dextrin (ID), resistant starch, IHMWDF, and SHMWDF of wheat noodles with maltodextrin during different storage period at 40 °C storage temperature.

Author Contributions: Conceptualization, Y.-W.H.; data curation, Y.-C.H.; formal analysis, D.-W.H.; funding acquisition, Y.-J.C. (Ya-Ju Chang); investigation, J.-C.T.; methodology, A.T.M.; project administration, P.-H.L.; validation, Z.-R.W.; writing—original draft, Y.-J.C. (Yung-Jia Chan); writing—review and editing, W.-C.L. All authors have read and agreed to the published version of the manuscript.

Funding: This study was supported by grants provided by the Pharma Power Biotech Company in Taiwan.

Conflicts of Interest: The authors declare no conflict of interest.

References

1. Kaur, B.; Ranawana, V.; Henry, J. The Glycemic Index of Rice and Rice Products: A Review, and Table of GI Values. *Crit. Rev. Food Sci. Nutr.* **2016**, *56*, 215–236. [CrossRef]
2. Thomas, D.E.; Elliott, E. The use of low-glycaemic index diets in diabetes control. *Br. J. Nutr.* **2010**, *104*, 797–802. [CrossRef]
3. Brand-Miller, J.; Hayne, S.; Petocz, P.; Colagiuri, S. Low-Glycemic Index Diets in the Management of Diabetes. *Diabetes Care* **2003**, *26*, 2261. [CrossRef]
4. Chandalia, M.; Garg, A.; Lutjohann, D.; von Bergmann, K.; Grundy, S.M.; Brinkley, L.J. Beneficial effects of high dietary fiber intake in patients with type 2 diabetes mellitus. *N. Engl. J. Med.* **2000**, *342*, 1392–1398. [CrossRef] [PubMed]
5. Shukla, K.; Srivastava, S. Evaluation of finger millet incorporated noodles for nutritive value and glycemic index. *J. Food Sci. Technol.* **2014**, *51*, 527–534. [CrossRef] [PubMed]
6. Srinivasan, B.K.; Prabhasankar, P. Shelf stability of low glycemic index noodles: Its physico-chemical evaluation. *J. Food Sci. Technol.* **2018**, *55*, 4811–4822.
7. Sadakiyo, T.; Inoue, S.I.; Ishida, Y.; Watanabe, H.; Mitsuzumi, H.; Ushio, S. Safety assessment of a soluble dietary fiber, isomaltodextrin, enzymatically produced from starch. *Fundam. Toxicol. Sci.* **2017**, *4*, 57–75. [CrossRef]
8. Takagaki, R.; Ishida, Y.; Sadakiyo, T.; Taniguchi, Y.; Sakurai, T.; Mitsuzumi, H.; Watanabe, H.; Fukuda, S.; Ushio, S. Effects of isomaltodextrin in postprandial lipid kinetics: Rat study and human randomized crossover study. *PLoS ONE* **2018**, *13*, e0196802. [CrossRef]
9. Hann, M.; Zeng, Y.; Zong, L.; Sakurai, T.; Taniguchi, Y.; Takagaki, R.; Watanabe, H.; Mitsuzumi, H.; Mine, Y. Anti-inflammatory activity of isomaltodextrin in a C57BL/6NCrl mouse model with lipopolysaccharide-induced low-grade chronic inflammation. *Nutrients* **2019**, *11*, 2791. [CrossRef] [PubMed]
10. Sadakiyo, T.; Ishida, Y.; Inoue, S.I.; Taniguchi, Y.; Sakurai, T.; Takagaki, R.; Kurose, M.; Mori, T.; Yasuda-Yamashita, A.; Mitsuzumi, H.; et al. Attenuation of postprandial blood glucose in humans consuming isomaltodextrin: Carbohydrate loading studies. *Food Nutr. Res.* **2017**, *61*, 1325306. [CrossRef]
11. Nishimura, N.; Tanabe, H.; Yamamoto, T. Isomaltodextrin, a highly branched α-glucan, increases rat colonic H_2 production as well as indigestible dextrin. *Biosci. Biotechnol. Biochem.* **2016**, *80*, 554–563. [CrossRef]
12. Mine, Y.; Jin, Y.; Zhang, H.; Rupa, P.; Majumder, K.; Sakurai, T.; Taniguchi, Y.; Takagaki, R.; Watanabe, H.; Mitsuzumi, H. Prophylactic effects of isomaltodextrin in a Balb/c mouse model of egg allergy. *NPJ Sci. Food* **2019**, *3*, 23. [CrossRef]
13. Wakabayashi, S. The effects of indigestible dextrin on sugar tolerance: I. Studies on digestion-absorption and sugar tolerance. *Nihon Naibunpi Gakkai Zasshi* **1992**, *68*, 623–635.
14. Wakabayashi, S.; Kishimoto, Y.; Matsuoka, A. Effects of indigestible dextrin on glucose tolerance in rats. *J. Endocrinol.* **1995**, *144*, 533–538. [CrossRef] [PubMed]
15. Phimolsiripol, Y.; Suppakul, P. Techniques in Shelf Life Evaluation of Food Products. *Ref. Modul. Food Sci.* **2016**, *1*, 1–8. [CrossRef]
16. Corradini, M. Shelf life of food products: From open labeling to real-time measurements. *Annu. Rev. Food Sci. Technol.* **2018**, *9*, 251–269. [CrossRef]
17. Beccaria, M.; Oteri, M.; Micalizzi, G.; Bonaccorsi, I.; Purcaro, G.; Dugo, P.; Mondello, L. Reuse of dairy product: Evaluation of the lipid profile evolution during and after their shelf-life. *Food Anal. Methods* **2016**, *9*, 3143–3154. [CrossRef]
18. Li, P.H.; Huang, C.C.; Yang, M.Y.; Wang, C.C.R. Textural and sensory properties of salted noodles containing purple yam flour. *Food Res. Int.* **2012**, *47*, 223–228. [CrossRef]

19. Reshmi, S.K.; Sudha, M.L.; Shashirekha, M.N. Noodles fortified with *Citrus maxima* (pomelo) fruit segments suiting the diabetic population. *Bioact. Carbohydr. Diet. Fibre* **2020**, *22*, 100213. [CrossRef]

20. Liang, Y.; Qu, Z.; Liu, M.; Wang, J.; Zhu, M.; Liu, Z.; Li, J.; Zhan, X.; Jia, F. Effect of curdlan on the quality of frozen-cooked noodles during frozen storage. *J. Cereal Sci.* **2020**, *95*, 103019. [CrossRef]

21. Nie, X.N.; She, D.; Liang, Z.S.; Geng, Z.C.; Wang, H.T.; Wang, K. Physicochemical Characterization of Hemicelluloses Obtained by Graded Ethanol Precipitation from Sweet Sorghum Stem. *J. Biobased Mater.* **2011**, *5*, 265–274. [CrossRef]

22. Xue, Z.; Chen, Y.; Jia, Y.; Wang, Y.; Lu, Y.; Chen, H.; Zhang, M. Structure, thermal and rheological properties of different soluble dietary fiber fractions from mushroom *Lentinula edodes* (Berk.) Pegler residues. *Food Hydrocoll.* **2019**, *95*, 10–18. [CrossRef]

23. Han, W.; Ma, S.; Li, L.; Wang, X.X.; Zheng, X.L. Application, and development prospects of dietary fibers in flour products. *J. Chem.* **2017**, 1–8. [CrossRef]

24. Hu, X.Z.; Sheng, X.L.; Li, X.P.; Liu, L.; Zheng, J.M.; Chen, X.Y. Effect of dietary oat β-glucan on high-fat diet induced obesity in HFA mice. *Bioact. Carbohydr. Diet. Fibre* **2015**, *5*, 79–85.

25. Bai, J.; Ren, Y.; Li, Y.; Fan, M.; Qian, H.; Wang, L.; Wu, G.; Zhang, H.; Qi, X.; Xu, M.; et al. Physiological functionalities and mechanisms of β-glucans. *Trends Food Sci. Technol.* **2019**, *88*, 57–66. [CrossRef]

26. Fabek, H.; Messerschmidt, S.; Brulport, V.; Goff, H.D. The effect of in vitro digestive processes on the viscosity of dietary fibres and their influence on glucose diffusion. *Food Hydrocoll.* **2014**, *35*, 718–726. [CrossRef]

27. Chen, H.; Zhao, C.; Li, J.; Hussain, S.; Yan, S.; Wang, Q. Effects of extrusion on structural and physicochemical properties of soluble dietary fiber from nodes of lotus root. *LWT* **2018**, *93*, 204–211. [CrossRef]

28. Sowbhagya, H.B.; Suma, P.F.; Mahadevamma, S.; Tharanathan, R.N. Spent residue from cumin—A potential source of dietary fiber. *Food Chem.* **2007**, *104*, 1220–1225. [CrossRef]

29. Wan, J.; Ding, Y.; Zhou, G.; Luo, S.; Liu, C.; Liu, F. Sorption isotherm and state diagram for indica rice starch with and without soluble dietary fiber. *J. Cereal Sci.* **2018**, *80*, 44–49. [CrossRef]

30. Mohamed, A.; Xu, J.; Singh, M. Yeast leavened banana-bread: Formulation, processing, colour and texture analysis. *Food Chem.* **2010**, *118*, 620–626. [CrossRef]

31. Karathanos, V.T.; Bakalis, S.; Kyritsi, A.; Rodis, P.S. Color Degradation of Beans During Storage. *Int. J. Food Prop.* **2006**, *9*, 61–71. [CrossRef]

32. Liu, Q.; Guo, X.N.; Zhu, K.X. Effects of frozen storage on the quality characteristics of frozen cooked noodles. *Food Chem.* **2019**, *283*, 522–529. [CrossRef] [PubMed]

33. Srinivasan, B.K.; Prabhasankar, P. A study on noodle dough rheology and product quality characteristics of fresh and dried noodles as influenced by low glycemic index ingredient. *J. Food Sci. Technol.* **2015**, *52*, 1404–1413.

34. Tudorica, C.M.; Kuri, V.; Brennan, C.S. Nutritional and physicochemical characteristics of dietary fiber enriched pasta. *J. Agric. Food Chem.* **2002**, *50*, 347–356. [CrossRef] [PubMed]

Article

Se-Enrichment Pattern, Composition, and Aroma Profile of Ripe Tomatoes after Sodium Selenate Foliar Spraying Performed at Different Plant Developmental Stages

Annalisa Meucci [1], Anton Shiriaev [1,*], Irene Rosellini [2], Fernando Malorgio [3] and Beatrice Pezzarossa [2]

1   Institute of Life Sciences, Sant'Anna School of Advanced Studies, 56127 Pisa, Italy; annalisa.meucci@santannapisa.it

2   Research Institute on Terrestrial Ecosystems, 56124 Pisa, Italy; irene.rosellini@cnr.it (I.R.); beatrice.pezzarossa@cnr.it (B.P.)

3   Department of Agriculture, Food and Environment, University of Pisa, 56124 Pisa, Italy; fernando.malorgio@unipi.it

*   Correspondence: anton.shiriaev@santannapisa.it

Abstract: Foliar spray with selenium salts can be used to fortify tomatoes, but the results vary in relation to the Se concentration and the plant developmental stage. The effects of foliar spraying with sodium selenate at concentrations of 0, 1, and 1.5 mg Se L^{-1} at flowering and fruit immature green stage on Se accumulation and quality traits of tomatoes at ripening were investigated. Selenium accumulated up to 0.95 µg 100 g FW^{-1}, with no significant difference between the two concentrations used in fruit of the first truss. The treatment performed at the flowering stage resulted in a higher selenium concentration compared to the immature green treatment in the fruit of the second truss. Cu, Zn, K, and Ca content was slightly modified by Se application, with no decrease in fruit quality. When applied at the immature green stage, Se reduced the incidence of blossom-end rot. A group of volatile organic compounds (2-phenylethyl alcohol, guaiacol, (E)-2-heptenal, 1-penten-3-one and (E)-2-pentenal), positively correlated with consumer liking and flavor intensity, increased following Se treatment. These findings indicate that foliar spraying, particularly if performed at flowering stage, is an efficient method to enrich tomatoes with Se, also resulting in positive changes in fruit aroma profile.

Keywords: biofortification; fruit quality; ripening; selenium; *Solanum lycopersicum*; volatile organic compounds (VOCs)

Citation: Meucci, A.; Shiriaev, A.; Rosellini, I.; Malorgio, F.; Pezzarossa, B. Se-Enrichment Pattern, Composition, and Aroma Profile of Ripe Tomatoes after Sodium Selenate Foliar Spraying Performed at Different Plant Developmental Stages. *Plants* **2021**, *10*, 1050. https:// doi.org/10.3390/plants10061050

Academic Editor: Ivo Vaz de Oliveira

Received: 13 May 2021
Accepted: 20 May 2021
Published: 23 May 2021

Publisher's Note: MDPI stays neutral with regard to jurisdictional claims in published maps and institutional affiliations.

1. Introduction

Selenium plays several crucial roles in human metabolism [1–3]. At the right concentrations, it positively affects DNA synthesis, fertility, reproduction, and muscle function. Selenium helps to slow down aging, prevent certain cancers, and reduce the incidence of viral infections, cardiovascular damage, arthritis, and alterations of the immune system [4]. Due to its ability to oxidize thiol groups in the virus protein disulfide isomerase, selenite may even prevent COVID-19 contagion [5].

The recommended dietary allowance (RDA) for selenium is 55 µg day^{-1}, and the tolerable upper intake for adults is 400 µg day^{-1} [6]. The European Food Safety Authority set the RDA for adults at 70 µg Se day^{-1} [7].

Worldwide, approximately one in eight people may have insufficient Se intake [8]. Those living in areas with low selenium in the soil, and consequently low levels in their daily diet, need to replenish a deficiency through medicines or nutraceuticals. Agronomic biofortification with Se fertilization is considered an effective way to produce Se-rich crops, thus improving the Se intake in the target population [9–11].

Selenium is not considered an essential element in plants, but it elicits the production of secondary metabolites and increases the enzymatic and non-enzymatic antioxidant capacity.

Thus, in addition to biofortification, Se enhances the levels of specific bio-active/health-promoting compounds that may also have positive effects on the plant physiology and metabolism [12,13]. In fact, plants supplemented with Se can better counteract oxidative stress as well as abiotic (salt, drought) and biotic stresses [14].

Although the exact mechanisms of Se action are still poorly understood [15], positive effects of Se on delaying senescence and, in crops bearing fleshy fruits, modulating ripening and metabolic profiling and composition of the produce have been reported [16–23].

Se-enrichment of fruits to be consumed as fresh produce is of great interest since some selenium is lost after cooking or processing [24]. Given that tomatoes (*Solanum lycopersicum*) are mostly consumed fresh and do not require heat processing, and since tomato is the second-most important vegetable crop worldwide, it is one of the best commodities for producing a selenium biofortified fruit for dietary supplementation.

Various strategies can be used for the Se-biofortification of different crops, including tomato [14]. Se supplementation (as sodium selenate or sodium selenite) to tomato plants via the root system (hydroponic, soils, or artificial substrate fertilization) enriches tomato fruit [13,18,20,25,26].

However, with the global focus on sustainability, ensuring that such strategies are safe for the environment is challenging. Excessive remaining chemicals in the growth substrate or solution, which are not taken up by plants, may cause water pollution [27] or increase soil salinity [28].

Foliar spraying is a feasible alternative, but few papers have investigated the effects of this method on Se-enrichment and on the metabolic, compositional, and physiological changes occurring in tomato fruit [13,19,22,29,30]. The results of these studies demonstrate that the enrichment effects of foliar treatments on tomato plants depend not only on the Se concentration and chemical form but also on the plant developmental stage when treatment is performed.

Since the range of Se concentration between essentiality and toxicity for humans and for plant tissues is quite narrow [31], the protocols need to be designed to enrich tomatoes with no risks for the consumers and no negative effects on the plants.

We investigated the responses to sodium selenate foliar spraying at flowering (FL) and fruit immature green stage (IG) in terms of Se accumulation in tomato fruit, and specific composition and quality-related traits at ripening, with a specific focus on volatile organic compound (VOC) profiles.

2. Results and Discussion

2.1. Selenium Concentration

Spraying tomato plants with sodium selenate resulted in a significant increase in selenium concentration in the fruit of both trusses (Table 1). The time of treatment did not affect the Se content in the fruit of the first truss, whereas in the second truss, the treatment performed at FL gave a higher selenium concentration compared to the IG treatment. The two doses of Se applied (1 and 1.5 mg L^{-1}) did not result in statistically different accumulations of Se in the fruit tissues.

The selenium concentration in the fruits of sprayed plants was much lower than the fruits of plants grown in the nutrient solution supplied with sodium selenate at the same concentration [18,20]. In fact, in those two studies, the uptake of selenium by roots, and its translocation to the aerial part, led to a higher accumulation of Se in fruit.

In plants sprayed with selenate, the composition of the cuticle (epicuticular wax or the deposition of wax platelets) and the leaf tomentosity could have hampered selenium absorption. The lower Se absorption could also be ascribed to the absence of adhesive substances in the spraying solution, which generally facilitate better penetration of the active ingredients and prevent their washout.

Table 1. Selenium concentration ($\mu g\ kg^{-1}$ DW) in tomato fruit sprayed with sodium selenate (Na_2SeO_4) at different selenium concentrations and distributed at FL and fruit IG stages.

Treatment Time	Se Spray mg L^{-1}	[Se] $\mu g\ kg\ DW^{-1}$	
		1st Truss	2nd Truss
	0	0 a	0 a
Flowering	1.0	108 b	123 c
	1.5	146 b	131 c
Immature green	0	0 a	0 a
	1.0	92 b	84 b
	1.5	105 b	95.2 bc
Variance analysis			
Treatment time (A)		ns	***
Se dosage (B)		***	***
Interaction A × B		ns	***

Significance is as follows: ns, not significant; *** significant at 0.1%. Different letters in each column correspond to significantly different values for $p < 0.05$ according to the LSD (least significant difference) test.

The selenium concentration in the fruit of plants treated with 1 mg Se L^{-1} was lower than that reported by Zhu et al. [29], i.e., 500 μg Se kg^{-1}, in plants treated with the same dose of selenium at the onset of flowering. Different growing and environmental conditions, as well as different genotypes, might explain the observed discrepancy of results.

The selenium content in 100 g of fresh tomatoes was calculated by multiplying the concentration of Se by the percentage of dry matter. Values ranged between 0.62 and 0.95 μg Se 100 g FW^{-1} in the fruit of the first truss and between 0.35 and 0.66 μg Se 100 g FW^{-1} in the fruit of the second truss (data not shown). Thus, the daily consumption of treated tomatoes appears to increase the selenium supplementation, with no risk of toxicity for humans.

2.2. Mineral Element Content

The enrichment with selenium affected the content of all the minerals, except for Fe in the fruit of the first truss, and for K in fruit of the second truss (Tables 2 and 3).

In the first truss, the supply with the higher dose of selenium increased the content of Ca when plants were treated at IG stage, whereas it decreased the content of Cu, Ca, and Zn in the fruit of plants sprayed at FL and of K in plants sprayed at the IG stage.

Selenium treatment decreased the content of all the minerals, except for Fe, in fruit of the second truss. To the best of our knowledge, only one study has been conducted on the effects of foliar spraying with selenium on the mineral content of tomato fruit [26]. The results of that study showed that foliar spraying at 10 and 20 mg Se L^{-1}, repeated every 20 days, did not modify the content of Ca, Mg, and K.

2.3. Effect of Selenium on Blossom-End Rot

A statistical analysis of data relating to blossom-end rot highlights that the biofortification with selenium at IG phase, at both concentrations, significantly reduced the incidence of this rot, but only in fruit of the second truss (Table 4).

Blossom-end rot is believed to be caused by a calcium imbalance within the plant, but in our trial, the selenium treatment did not increase the calcium content in fruit. Factors other than Se-concentration and/or Ca content should thus be considered in terms of the incidence of this physiological disorder.

Table 2. Macro- and microelements concentration (mg kg^{-1}) in tomato fruit of the first truss treated with sodium selenate (Na$_2$SeO$_4$) at different concentrations of Se at FL and fruit IG stages.

Treatment Time	Se Spray mg L^{-1}	Cu	Zn	Mn	Fe	K	Ca	Mg
					mg kg^{-1}			
Flowering	0	14.4 a	60.4 a	79.7 b	484 a	1672 a	429 ab	978 a
	1.0	13.5 a	65.5 a	88.6 b	493 a	1076 a	511 a	1007 a
	1.5	7.4 a	35.5 c	51.3 c	363 a	1540 ab	375 b	781 b
Immature green	0	14.1 a	60.2 a	80.0 b	479 a	1668 a	431 ab	980 a
	1.0	9.9 a	50.0 b	77.1 b	437 a	1374 b	458 a	909 a
	1.5	13.4 a	61.6 a	106 a	401 a	1005 c	540 a	1053 a
Variance analysis								
Treatment time (A)		ns	***	**	ns	ns	ns	**
Se dosage (B)		**	***	ns	ns	***	**	ns
Interaction A × B		**	***	***	ns	**	***	***

Significance is as follows: ns, not significant; ** significant at 1%; *** significant at 0.1%. Different letters in each column correspond to significantly different values for $p < 0.05$ according to the LSD (least significant difference) test.

Table 3. Macro and microelements concentration (mg kg^{-1}) in tomato fruit of the second truss treated with sodium selenate (Na$_2$SeO$_4$) at different concentrations of Se at FL and fruit IG stages.

Treatment Time	Se Spray mg L^{-1}	Cu	Zn	Mn	Fe	K	Ca	Mg
					mg kg^{-1}			
Flowering	0	14.3 a	67.1 a	95.0 a	550 a	1320 a	628 a	1074 a
	1.0	15.4 a	67.0 a	94.3 a	580 a	402 a	594 a	1054 a
	1.5	9.8 b	44.3 c	68.9 a	320 b	1355 a	516 b	851 b
Immature green phase	0	14.5 a	67.4 a	94.9 a	556 a	1327 a	630 a	1072 a
	1.0	8.9 b	39.8 c	55.2 a	392 b	1469 a	462 b	785 b
	1.5	10.5 b	53.8 c	84.1 a	366 b	1508 a	603 a	908 b
Variance analysis								
Treatment time (A)		***	***	ns	ns	ns	ns	ns
Se dosage (B)		*	***	**	**	ns	***	***
Interaction A × B		**	***	**	*	ns	***	**

Significance is as follows: ns, not significant; * significant at 5%; ** significant at 1%; *** significant at 0.1%. Different letters in each column correspond to significantly different values for $p < 0.05$ according to the LSD (least significant difference) test.

Table 4. Incidence of blossom-end rot on fruits of plants treated with sodium selenate (Na$_2$SeO$_4$), at different selenium concentrations and distributed at FL and fruit IG stages.

Treatment Time	Se Spray mg L^{-1}	Blossom−End Rot Incidence Affected/Total Fruit Ratio	
		1st Truss	2nd Truss
Flowering	0	25.3 a	40.4 a
	1.0	19.4 ab	33.0 ab
	1.5	14.4 ab	31.2 bc
Immature green	0	18.2 ab	36.8 a
	1.0	9.2 b	31.4 b
	1.5	9.6 b	20.2 c
Variance analysis			
Treatment time (A)		*	ns
Se dosage (B)		ns	*
Interaction A × B		ns	ns

Significance is as follows: ns, not significant; * significant at 5%. Different letters in each column correspond to significantly different values for $p < 0.05$ according to the LSD (least significant difference) test.

2.4. Qualitative Characteristics of Fruit

2.4.1. Fruit Composition

Biofortification with selenium, both at FL and IG, affected the compositional traits of fruit collected at the red-ripe stage. In the fruit of the first truss, Se enrichment decreased titratable acidity and increased dry weight compared to the control (Table 5), while SSC (soluble solid content) and maturity index were not affected. In the fruit of the second truss (Table 6), dry weight and titratable acidity showed the same effects as Se-enrichment, as observed in the first truss, whereas SSC increased. Thus, the maturity index was significantly affected by Se-enrichment.

Table 5. Qualitative characteristics of tomato fruits of the first truss treated with sodium selenate (Na_2SeO_4) at different concentrations of selenium and distributed at FL and fruit IG stages.

Treatment Time	Se Spray mg L^{-1}	DW %	SSC °Brix	Titrable Acidity g Citric Acid 100 mL^{-1}	Maturity Index	Taste Index
	0	5.0 c	6.5 a	0.8 a	8.0 a	1.1 a
Flowering	1.0	7.1 a	6.1 a	0.7 b	8.5 a	1 b
	1.5	6.5 a	6.4 a	0.7 b	9.3 a	0.9 b
	0	5.1 c	6.3 a	0.9 a	8.1 a	1 a
Immature green	1.0	6.7 a	6.4 a	0.8 ab	8.4 a	1 ab
	1.5	5.9 b	6.2 a	0.7 b	8.9 a	0.9 b
Variance Analysis Treatment time (A)		***	ns	ns	ns	ns
Se dosage (B)		***	ns	***	ns	**
Interaction A × B		*	ns	ns	ns	ns

Significance is as follows: ns, not significant; *, significant at 5%; **, significant at 1%; ***, significant at 0.1%. Different letters in each column correspond to significantly different values for $p < 0.05$ according to the LSD (least significant difference) test.

Table 6. Qualitative characteristics of tomato fruits of the second truss treated with sodium selenate (Na_2SeO_4) at different concentrations of selenium and distributed at FL and fruit IG stages.

Treatment Time	Se Spray mg L^{-1}	DW %	SSC °Brix	Titrable Acidity g Citric Acid 100 mL^{-1}	Maturity Index	Taste Index
	0	4.5 bc	6.2 c	0.8 a	7.5 b	1 a
Flowering	1.0	5.4 a	7.0 a	0.7 a	9.4 a	1 a
	1.5	4.8 b	6.3 bc	0.6 b	10.2 a	0.8 c
	0	4.6 bc	6.0 c	0.8 a	7.6 b	1 a
Immature green	1.0	4.2 c	6.3 bc	0.7 a	8.8 a	0.9 b
	1.5	5.3 a	6.5 b	0.7 a	9.4 a	0.9 b
Variance Analysis Treatment time (A)		***	ns	ns	ns	ns
Se dosage (B)		***	**	**	*	**
Interaction A × B		***	***	***	ns	***

Significance is as follows: ns, not significant; *, significant at 5%; **, significant at 1%; ***, significant at 0.1%. Different letters in each column correspond to significantly different values for $p < 0.05$ according to the LSD (least significant difference) test.

The taste index, based on brix and titratable values, slightly decreased in fruit treated with 1.5 mg Se L^{-1} of both protocols and in fruit treated with 1 mg Se L^{-1} at the IG stage. Considering that a taste index of <0.7 is associated with low quality [32], plants treated with selenium selenate produced tasty fruit. These results on fruit composition agree with previous studies conducted by our research group [18,20] in which tomato plants grown in hydroponics were enriched with the same concentration of Se, as sodium selenate, which was added to the nutrient solution.

2.4.2. Aroma Profiles

Fresh tomato fruit produces several hundred VOCs, some of which are major contributors to the overall aroma and consequently affect consumer experience [33]. Specific pathways are involved in the metabolism of the VOCs, and their concentration in ripe fruit depends on factors affecting fruit development and physiology. To assess whether Se-enrichment affects this important quality trait, we carried out specific VOC analyses on red-ripe fruit of the second truss from the control and from plants treated with 1.0 and 1.5 Se L^{-1} at FL. A total of 39 VOCs were identified in the three sets of samples (Supplementary material, Table S1).

Partial least squares regression analysis indicated that sodium selenate treatment affects the volatile profile of the tomatoes at ripening (Figure 1). Overall, the validated PLSDA (Partial least squares discriminant analysis) model explains about 90% (first and second factors together) of the variability recorded between sample groups. In fact, a clear segregation of the different treatments was observed, with Se-enriched samples clustering separately from the control tomatoes and from each other. This clustering indicates important differences in terms of the numbers of VOCs specifically induced by the different levels of Se applied at the flowering stage. In addition, the fact that the analysis was repeatable was highlighted by the correct positioning of the different analyzed fruit replicates, with samples from the same treatment being close together.

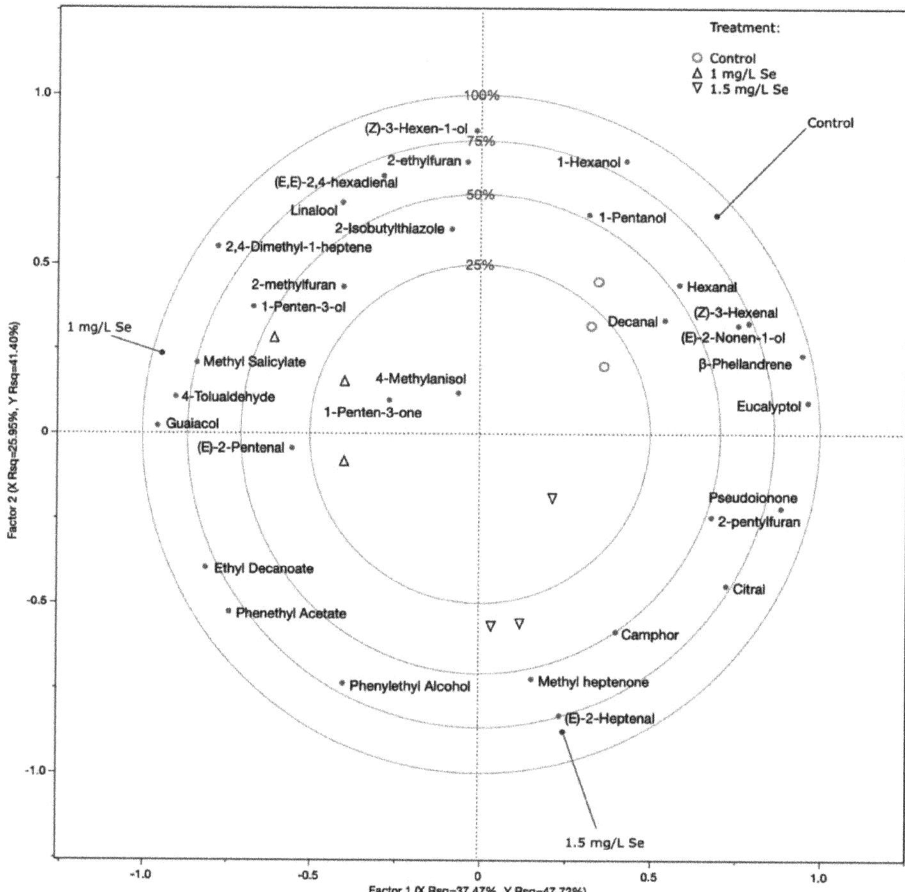

Figure 1. Partial least squares discriminant analysis (PLSDA). Treatment was employed as response variable, while the identified VOCs were used as predictor variables. Variable importance in projection (VIP) scores were used to filter compounds varying the most between different theses, and only compounds with a VIP score higher than 0.8 were included in the analysis. Circles, triangles, and inverted triangles were used to depict control fruit, 1 mg L^{-1}, and 1.5 mg L^{-1} sodium selenate-treated fruit, respectively.

Lipoxygenase pathway-related compounds, such as hexanal, (Z)-3-hexenal and 1-hexanol, were associated with control fruit. Specific C5 and C7 compounds and derivatives instead appeared to be more abundantly accumulated in Se-biofortified fruit. In particular, (E)-2-heptenal and methyl heptenone, together with phenylethyl alcohol, characterized the 1.5 mg Se L^{-1}-treated fruit. On the other hand, 1-penten-3-ol, 1-penten-3-one, and (E)-2-pentenal sit, in the PLSDA score plot, closer to the 1.0 mg Se L^{-1}-treated samples, which also showed a strict association with guaiacol and methyl salicylate.

2-phenylethyl alcohol, guaiacol, and (E)-2-heptenal, together with 1-penten-3-one and (E)-2-pentenal, were identified by Tieman et al. [33] as compounds positively correlated with consumer liking and overall flavor intensity, with decreased accumulation in modern cultivars compared to heirloom *S. lycopersicum* varieties. Thus, Se-enriched tomatoes may possibly also benefit from the treatment in terms of organoleptic properties and more attractive traits for consumers. Of particular interest appears to be the increase in methyl salicylate characterizing the fruit treated with 1 mg Se L^{-1}. This compound is one of several phenylpropanoids that significantly contribute to the unique flavor of tomato fruit and is also considered to be better than salicylic acid as the signal molecule transmitted in the systemic acquired resistance (SAR) response [34]. Together with linalool, methyl salicylate is involved in the immune response against *Pseudomonas syringae* bacterial attack [35].

The association of both compounds with selenium treatments rather than with control clusters may indicate a better immunity response of fruit treated with selenium against some bacterial diseases. The association of methylated compounds (methyl salicylate and methyl heptanone) with Se-treated fruit may show how selenium affects tomato methyltransferases. In fact, in a proteomics study, Zeng et al. [36] demonstrated that methyl transferases are among the differentially expressed proteins in naturally selenium-enriched rice.

3. Materials and Methods

3.1. Experimental Set-Up

The experiment was conducted in a greenhouse located at the Department of Agricultural, Food, and Agro-Environmental Sciences (DISAAAa) of the University of Pisa (PI), characterized by a controlled temperature system (13–30 °C) and a ventilation air temperature of 27 °C. The seedlings of *Solanum lycopersicum* var. Kreos were provided in rock wool cubes (GRODAN, ROCKWOOL Group, 75 × 75 × 65 mm) supplied with a nutrient solution for about a week before being transferred to rock wool slabs (1 × 0.15 × 0.75 m).

Plants were grown in a hydroponic system with a density of 3 plants m^{-2}. They were trimmed above the second truss. Pollination was performed by mechanical vibration of the inflorescences. The nutrient solution, characterized by E.C. of 2.6 mS cm^{-1} and pH value of 5.6, was distributed through a drip irrigation system, three times a day in the early phenological phases of plant growth, and five times a day in the final phases. The composition of the nutrient solution was the following: 14 mM L^{-1} N-NO$_3$, 1 mM L^{-1} N-NH$_4$, 1 mM L^{-1} P, 8 mM L^{-1} K, 5 mM L^{-1} Ca, 1.6 mM L^{-1} Mg, 5.23 mM L^{-1} Na, 1.71 mM L^{-1} S-SO$_4$, 3.43 mM L^{-1} Cl, 15 μM L^{-1} Fe, 20 μM L^{-1} B, 1 mM L^{-1} Cu, 5 μM L^{-1} Zn, 10 μM L^{-1} Mn, and 1 μM L^{-1} Mo.

The experimental scheme was organized in randomized blocks. Selenium was sprayed as sodium selenate solution (Sigma Aldrich, colorless crystalline powder) at concentrations of 0.1 and 1.5 mg Se L^{-1}. Each plant was treated with 250 mL of Se-enriched solution, whereas each control plant was sprayed with 250 mL of distilled water. When the solution was supplied, the surrounding plants were shielded with plastic sheets to avoid contamination. For each selenium dosage, half of the plants were treated at flowering (FL) (June), and the other half at the immature green (IG) fruit stage (July). Fruits were harvested for all the analyses at the red ripening stage.

After harvesting, the selenium content in the fruit of both trusses was analyzed. Composition, taste index, micro- and macro element concentrations, and VOC profiles of fruit were determined.

3.2. Selenium and Mineral Content Analyses

Total selenium and micro- and macro- element concentrations were determined in the fruit of Se-enriched and control plants. Collected fruit samples were oven-dried at 50 °C and then ground in a mortar. A total of 0.5 g of powder for each replicate was digested with nitric and perchloric acids. The mineral elements (Cu, Zn, Mn, Fe, K, Ca, Mg) were determined with a fast sequential atomic absorption spectrometer (AA240FS, Agilent). To calculate the selenium content, the digests were reduced by hydrochloric acid, following Zasoski and Burau [37], and the atomic absorption spectrometer was coupled to a hydride generator (Varian VGA 77) [38].

3.3. Fruit Composition and Quality Parameters

Fresh and dry weights of harvested fruit were determined. Titratable acidity was measured on 10 mL of tomato juice with an automatic titrator, using sodium hydroxide (0.1 M NaOH) as the titling and phenophthalein as the indicator. The result was expressed in g of citric acid per 100 mL of juice. Soluble solid content, expressed in °Brix, was investigated using a digital refractometer (model 53011, Turoni, Forlì, Italy). Maturation and taste indexes were calculated using Navez et al.'s [32] formula, on the basis of °Brix, and titratable acidity as proposed by Hernandez Suarez et al. [39]:

$$\text{Maturity Index} = ° \text{ Brix acidity} \tag{1}$$

$$\text{Taste Index} = (° \text{ Brix}/20 \times \text{acidity}) + \text{acidity} \tag{2}$$

We also tried to detect the possible effect of Se dosage and time of treatment on the incidence of blossom-end rot. The percentage of fruit affected by blossom-end rot was evaluated by calculating the percentage of fruit that showed the symptoms on the total of fruit present at each flowering stage. Percentage data were then transformed into angular values before statistical analysis.

3.4. HS-SPME-GC-MS Analysis

For the VOC analysis, a slightly modified method optimized by Brizzolara et al. [40,41] was used. A total of 15 g of pericarp tissue (without seeds) were homogenized with 15 mL 1M NaCl water solution in a 50 mL plastic tube using T25 Ultra-Turrax (IKA, Königswinter, Germany) homogenizer. The obtained puree was frozen in liquid nitrogen and stored at −80 °C. Three replicates per treatment were analyzed for both the red ripe stage with a different level of Se accumulation in fruit.

VOCs were identified using a gas-chromatograph (Perkin Elmer Clarus 680) coupled with a mass spectrometer (Perkin Elmer Clarus 600 S). A total of 5 g of tomato puree was transferred into a 20 mL crimp vial (Sigma Aldrich, Milano, Italy) and was thawed at 15 °C for 15 min. Samples were incubated for 60 min at 40 °C, and VOCs were extracted for 45 min at the same temperature using an SPME Fiber (50/30 μm, DVB/CAR/PDMS, Sigma Aldrich, Italy).

We used the following GC temperature program: the initial temperature was 40 °C, which was maintained for 1 min; the temperature was then increased to 250 °C at a rate of 4 °C min^{-1} and held for 1 min; finally, the temperature was increased to 280 °C at a rate of 15 °C min^{-1} and held for 2 min.

Desorption was performed in spitless mode using a PSSI injector at 260 °C. A fused silica 30 m × 0.25 mm × 0.25 μm film thickness SLB-5MS column was used for the analysis, and helium was used as a carrier gas at a constant flow of 1 mL min^{-1}. Chromatograms were analyzed using AMDIS (National Institute of Standards and Technology, Gaithersburg, MD, United States), and compound identification was carried out by comparing detected spectra with NIST v. 2 (National Institute of Standards and Technology, United States). Only compounds with 80%, or higher, levels of matching were considered. The retention index (RI) information was used to increase the identification efficiency. VOCs were also compared with molecules in the literature on tomato fruit [35,42]. In order to reduce variability due to the SPME fiber decay, the raw peak area was normalized on the sum of the areas for each specific chromatogram.

3.5. Statistical Analysis

Data were processed by means of two-way ANOVA, using Statgraphics and considering the selenium concentration and time of its administration as variables. The results were then analyzed with the least significant difference (LSD) test ($p < 0.05$). Statistically significant values were indicated with different letters. VOCs were analyzed by partial least squares discriminant analysis (PLSDA) using JMP®, v. 16 (SAS Institute Inc., Cary, NC, USA).

4. Conclusions

Although less effective than supplementation via nutrient solution, selenium treatments performed by spraying tomato plants with 1 and 1.5 mg L^{-1} sodium selenate are effective in enriching tomatoes with Se concentrations, with no risks of toxicity for the human diet. In addition to the Se dosage, the developmental stage of the plant also seems to affect the biofortification effectiveness of the treatment with limited effects on fruit composition. Our approach to studying the physiological effect of Se-biofortification of fleshy fruit supports the evidence that the uptake of selenium causes metabolic changes in ripening fruit. This, then, results in increases in specific VOCs putatively associated with higher consumer preference scores. These changes in the aroma profile induced by Se represent a new and interesting aspect to study in future research, considering that the poor aroma of the commercial tomato varieties is one of the major causes of consumer complaints.

Supplementary Materials: The following are available online at https://www.mdpi.com/article/10.3390/plants10061050/s1, Table S1: List of VOCs identified in control and Se-enriched fruit samples.

Author Contributions: Conceptualization, A.M., F.M. and B.P.; Methodology, A.M., F.M., I.R. and B.P.; Software, A.M., A.S. and F.M.; Validation, A.M., B.P. and F.M.; Formal Analysis, A.M., I.R. and A.S; Writing—Original Draft Preparation, A.M., B.P., F.M. and A.S.; Writing—Review and Editing, B.P. and A.S. All authors have read and agreed to the published version of the manuscript.

Funding: This research received no external funding.

Institutional Review Board Statement: Not applicable.

Informed Consent Statement: Not applicable.

Data Availability Statement: The data presented in this study are available on request from thecorresponding author.

Acknowledgments: The authors wish to thank Stefano Brizzolara, Institute of Life Sciences, Scuola Superiore Sant'Anna, Pisa (Italy), for his technical assistance in performing the VOC analyses. A special thanks to Pietro Tonutti, Institute of Life Sciences, Scuola Superiore Sant'Anna, Pisa (Italy), for reading the text and adding support through valuable comments and suggestions.

Conflicts of Interest: The authors declare no conflict of interest.

References

1. Surai, P.F. *Selenium in Nutrition and Health*; Nottingham University Press: Nottingham, UK, 2006; pp. 151–260. [CrossRef]
2. Fairweather-Tait, B.Y.; Broadley, M.; Berry, R.; Ford, D.; Hesketh, J.; Hurst, R. Selenium in Human Health and Disease. *Antioxid Redox Signal.* **2011**, *14*, 1337–1383. [CrossRef]
3. Tóth., R.; Csapó, J. Tóth. R.; Csapó, J.The role of selenium in nutrition—A review. *Acta Univ. Sapientiae Aliment.* **2018**, *11*, 128–144. [CrossRef]
4. Rayman, M.P. Selenium and Human Health. *Lancet* **2012**, *379*, 1256–1268. [CrossRef]
5. Kieliszek, M.; Lipinski, B. Selenium Supplementation in the Prevention of Coronavirus Infections (COVID-19). *Med. Hypotheses* **2020**, *143*. [CrossRef] [PubMed]
6. Institute of Medicine (US). *Panel on Dietary Antioxidants and Related Compounds. Dietary Reference Intakes for Vitamin C, Vitamin E, Selenium, and Carotenoids*; National Academies Press (US): Washington, DC, USA, 2000. [CrossRef]

7. EFSA NDA Panel (EFSA Panel on Dietetic Products, Nutrition and Allergies). Scientific Opinion on Dietary Reference Values for selenium. *EFSA J.* **2014**, *12*. [CrossRef]

8. Combs, G.F. Selenium in Global Food Systems. *Br. J. Nutr.* **2001**, *85*, 517–547. [CrossRef]

9. Puccinelli, M.; Malorgio, F.; Pezzarossa, B. Selenium Enrichment of Horticultural Crops. *Molecules* **2017**, *22*, 933. [CrossRef]

10. D'Amato, R.; Regni, L.; Falcinelli, B.; Mattioli, S.; Benincasa, P.; Dal Bosco, A.; Pacheco, P.; Proietti, P.; Troni, E.; Santi, C.; et al. Current Knowledge on Selenium Biofortification to Improve the Nutraceutical Profile of Food: A Comprehensive Review. *J. Agric. Food Chem.* **2020**, *68*, 4075–4097. [CrossRef]

11. Izydorczyk, G.; Ligas, B.; Mikula, K.; Witek-Krowiak, A.; Moustakas, K.; Chojnacka, K. Biofortification of edible plants with selenium and iodine—A systematic literature review. *Sci. Total Environ.* **2021**, *754*. [CrossRef]

12. Newman, R.; Waterland, N.; Moon, Y.; Tou, J. Selenium Biofortification of Agricultural Crops and Effects on Plant Nutrients and Bioactive Compounds Important for Human Health and Disease Prevention—A Review. *Plant Foods Hum. Nutr.* **2019**, *74*. [CrossRef]

13. Schiavon, M.; Dall'Acqua, S.; Mietto, A.; Pilon-Smits, E.A.H.; Sambo, P.; Masi, A.; Malagoli, M. Selenium Fertilization Alters the Chemical Composition and Antioxidant Constituents of Tomato (*Solanum Lycopersicon* L.). *J. Agric. Food Chem.* **2013**, *61*, 10542–10554. [CrossRef]

14. Schiavon, M.; Nardi, S.; Vecchia, F.; Ertani, A. Selenium biofortification in the 21 century: Status and challenges for healthy human nutrition. *Plant Soil* **2020**, *453*, 245–270. [CrossRef]

15. Malerba, M.; Cerana, R. Effect of Selenium on the Responses Induced by Heat Stress in Plant Cell Cultures. *Plants* **2018**, *7*, 64. [CrossRef] [PubMed]

16. Pezzarossa, B.; Malorgio, F.; Tonutti, P. Effects of selenium uptake by tomato plants on senescence, fruit ripening and ethylene evolution. In *Biology and Biotechnology of The Plant Hormone Ethylene*, 2nd ed.; Kanellis, A.K., Chang, C., Klee, H., Bleecker, A.B., Pech, J.C., Grierson, D., Eds.; Kluwer Academic Publishers: Dordrecht, The Netherlands, 1999; pp. 275–276. [CrossRef]

17. Pezzarossa, B.; Remorini, D.; Gentile, M.L.; Massai, R. Effects of foliar and fruit addition of sodium selenate on selenium accumulation and fruit quality. *J. Sci. Food Agric.* **2012**, *92*, 781–786. [CrossRef] [PubMed]

18. Pezzarossa, B.; Rosellini, I.; Borghesi, E.; Tonutti, P.; Malorgio, F. Effects of Se-Enrichment on Yield, Fruit Composition and Ripening of Tomato (Solanum Lycopersicum) Plants Grown in Hydroponics. *Sci. Hortic.* **2014**, *165*, 106–110. [CrossRef]

19. Zhu, Z.; Chen, Y.; Shi, G.; Zhang, X. Selenium delays tomato fruit ripening by inhibiting ethylene biosynthesis and enhancing the antioxidant defense system. *Food Chem.* **2017**, *219*, 179–184. [CrossRef]

20. Puccinelli, M.; Malorgio, F.; Terry, L.A.; Tosetti, R.; Rosellini, I.; Pezzarossa, B. Effect of Selenium Enrichment on Metabolism of Tomato (Solanum Lycopersicum) Fruit during Postharvest Ripening. *J. Sci. Food Agric.* **2019**, *99*, 2463–2472. [CrossRef] [PubMed]

21. Babalar, M.; Mohebbi, S.; Zamani, Z.; Askari, M.A. Effect of foliar application with sodium selenate on selenium biofortification and fruit quality maintenance of 'Starking Delicious' apple during storage. *J. Sci. Food Agric.* **2019**, *99*, 5149–5156. [CrossRef] [PubMed]

22. Neysanian, M.; Iranbakhsh, A.; Ahmadvand, R.; Ardebili, Z.O.; Ebadi, M. Comparative efficacy of selenate and selenium nanoparticles for improving growth, productivity, fruit quality, and postharvest longevity through modifying nutrition, metabolism, and gene expression in tomato; potential benefits and risk assessment. *PLoS ONE* **2021**, *16*. [CrossRef] [PubMed]

23. Businelli, D.; D'Amato, R.; Onofri, A.; Tedeschini, E.; Tei, F. Se-enrichment of cucumber (Cucumis sativus L.), lettuce (Lactuca sativa L.) and tomato (Solanum lycopersicum L. Karst) through fortification in pre-transplanting. *Sci. Hortic.* **2015**, *197*, 697–704. [CrossRef]

24. Lu, X.; He, Z.; Zhiqing, L.; Yuanyuan, Z.; Linxi, Y.; Ying, L.; Xuebin, Y. Effects of Chinese Cooking Methods on the Content and Speciation of Selenium in Selenium Bio-Fortified Cereals and Soybeans. *Nutrients* **2018**, *10*, 317. [CrossRef]

25. Keskinen, R.; Yli-Halla, M.; Hartikainen, H. Retention and Uptake by Plants of Added Selenium in Peat Soils. *Commun. Soil Sci. Plant Anal.* **2013**, *2013 44*, 3465–3482. [CrossRef]

26. Narvaez-Ortiz, W.; Becvort-Azcurra, A.; Fuentes-Lara, L.; Benavides-Mendoza, A.; Valenzuela-García, J.; Gonzalez, F.J. Mineral Composition and Antioxidant Status of Tomato with Application of Selenium. *Agronomy* **2018**, *8*, 185. [CrossRef]

27. Gupta, M.; Gupta, S. An Overview of Selenium Uptake, Metabolism, and Toxicity in Plants. An Overview of Selenium Uptake, Metabolism, and Toxicity in Plants. *Front. Plant Sci.* **2017**, *7*, 2074. [CrossRef] [PubMed]

28. Kaur, N.; Shuchi, S.; Simranjeet, K.; Harsh, N. Selenium in Agriculture: A Nutrient or Contaminant for Crops? *Arch. Agron. Soil Sci.* **2014**, *60*, 1593–1624. [CrossRef]

29. Zhu, Z.; Yanli, C.; Xueji, Z.; Miao, L. Effect of foliar treatment of sodium selenate on postharvest decay and quality of tomato fruits. *Sci. Hortic.* **2016**, *198*, 304–310. [CrossRef]

30. Zhu, Z.; Zhang, Y.; Liu, J.; Chen, Y.; Zhang, X. Exploring the effects of selenium treatment on the nutritional quality of tomato fruit. *Food Chem.* **2018**, *25*, 9–15. [CrossRef]

31. Natasha, M.S.; Nabeel, K.N.; Sana, K.; Behzad, M.; Irshad, B.; Muhammad, I.R. A critical review of selenium biogeochemical behavior in soil-plant system with an inference to human health. *Environ. Pollut.* **2018**, *234*, 915–934. [CrossRef] [PubMed]

32. Navez, B.; Letard, M.; Graselly, D.; Jost, M. Les criteres de qualite de la tomate. *Infos-Ctifl* **1999**, *155*, 41–47.

33. Tieman, D.; Zhu, G.; Resende, M.F.R., Jr.; Lin, T.; Nguyen, C.; Bies, D.; Rambla, J.L.; Ortiz Beltran, K.S.; Taylor, M.; Zhang, B.; et al. A chemical genetic roadmap to improved tomato flavor. *Science* **2017**, *355*, 391–394. [CrossRef]

34. Tieman, D.; Zeigler, M.; Schmelz, E.; Taylor, M.G.; Rushing, S.; Jones, J.B.; Klee, H.J. Functional analysis of a tomato salicylic acid methyl transferase and its role in synthesis of the flavor volatile methyl salicylate. *Plant J.* **2010**, *62*, 113–123. [CrossRef]

35. López-Gresa, M.P.; Purificación, L.; Laura, C.; Ismael, R.; José, L.R.; Antonio, G.; Vicente, C.; José, M.B. A Non-Targeted Metabolomics Approach Unravels the VOCs Associated with the Tomato Immune Response against Pseudomonas Syringae. *Front. Plant Sci.* **2017**, *8*, 1–15. [CrossRef]

36. Zeng, R.; Farooq, M.U.; Wang, L.; Su, Y.; Zheng, T.; Ye, X.; Jia, X.; Zhu, J. Study on Differential Protein Expression in Natural Selenium-Enriched and Non-Selenium-Enriched Rice Based on iTRAQ Quantitative Proteomics. *Biomolecules* **2019**, *9*, 130. [CrossRef] [PubMed]

37. Zasoski, R.J.; Burau, R.G. A rapid nitric-percloric acid digestion method for multi-element tissue analysis. *Commun. Soil Sci. Plant Anal.* **1977**, *8*, 425–436. [CrossRef]

38. Huang, P.M.; Fujii, R. Selenium and Arsenic. In *Methods of Soil Analysis: Part 3 Chemical Methods*; Sparks, D.L., Page, A.L., Helmke, P.A., Loeppert, R.H., Soltanpour, P.N., Tabatabai, M.A., Johnston, C.T., Sumner, M.E., Eds.; Soil Science Society of America: Madison, WI, USA, 1996; pp. 793–831. [CrossRef]

39. Hernández Suárez, M.; Rodríguez Rodríguez, E.M.; Díaz Romero, C. Chemical composition of tomato (*Lycopersicon esculentum*) from Tenerife, the Canary Islands. *Food Chem.* **2008**, *106*, 1046–1056. [CrossRef]

40. Brizzolara, S.; Santucci, C.; Tenori, L.; Hertog, M.; Nicolai, B.; Stürz, S.; Zanella, A.; Tonutti, P. A metabolomics approach to elucidate apple fruit responses to static and dynamic controlled atmosphere storage. *Postharvest Biol. Technol.* **2017**, *127*, 76–87. [CrossRef]

41. Brizzolara, S.; Hertog, M.; Tosetti, R.; Nicolai, B.; Tonutti, P. Metabolic responses to low temperature of three peach fruit cultivars differently sensitive to cold storage. *Front. Plant Sci.* **2018**, *9*, 706. [CrossRef]

42. Cortina, P.R.; Asis, R.; Peralta, I.E.; Asprelli, P.D.; Santiago, A.N. Determination of Volatile Organic Compounds in Andean Tomato Landraces by Headspace Solid Phase Microextraction-Gas Chromatography-Mass Spectrometry. *J. Braz. Chem. Soc.* **2017**, *28*, 30–41. [CrossRef]

Article

The Effect of the Cultivar and Harvest Term on the Yield and Nutritional Value of Rhubarb Juice

Ivana Mezeyová [1,*], Ján Mezey [2] and Alena Andrejiová [1]

[1] Department of Vegetable Production, Faculty of Horticulture and Landscape Engineering, Slovak University of Agriculture, Dražovská 4, 94901 Nitra, Slovakia; alena.andrejiova@uniag.sk
[2] Department of Fruit Growing, Viticulture and Enology, Faculty of Horticulture and Landscape Engineering, Slovak University of Agriculture, Dražovská 4, 94901 Nitra, Slovakia; jan.mezey@uniag.sk
*   Correspondence: ivana.mezeyova@uniag.sk

Abstract: Since scientific interest in rhubarb from a culinary point of view is a relatively new issue, the aim of this study was to test five edible cultivars of *Rheum rhabarbarum* L. ('Poncho', 'Canadian Red', 'Valentine', 'Red Champagne', and 'Victoria') from a specific culinary perspective, i.e., processing into juice. Total yields (t/ha) were established in six harvests during a two-year field experiment. For juice production and subsequent laboratory analysis, rhubarb petioles from two different harvest terms were used (i.e., harvest term A (HTA) and harvest term B (HTB)). Analyses of total sugar, glucose, fructose, total soluble solids (TSS), total acidity, malic acid, and pH level were determined by FT-IR spectrophotometer. Total yields of petioles varied between 28.77 t/ha ('Canadian Red') and 45.58 t/ha ('Red Champagne') at a density of 11,000 pl/ha. 'Red Champagne' significantly ($p < 0.05$) reached the highest juice yield potential (85%) and the highest values of glucose (9.97 g/L), total soluble solids (4.37 g/L), and total sugars (54.96 g/L).

Keywords: rhubarb; sugars; organic acids; juices; yield of petioles

Citation: Mezeyová, I.; Mezey, J.; Andrejiová, A. The Effect of the Cultivar and Harvest Term on the Yield and Nutritional Value of Rhubarb Juice. *Plants* **2021**, *10*, 1244. https://doi.org/10.3390/plants10061244

Academic Editor: Ivo Vaz de Oliveira

Received: 21 May 2021
Accepted: 15 June 2021
Published: 18 June 2021

Publisher's Note: MDPI stays neutral with regard to jurisdictional claims in published maps and institutional affiliations.

1. Introduction

Rhubarb (*Rheum rhabarbarum* L.) is a perennial plant from the family Polygonacea with valuable nutritional and medicinal properties [1] and high potential for cultivation as it provides early yields when the vegetable supply to market is deficient. Rhubarb is grown for its large, thick leaf stalks or petioles that are consumed as food [2]. The leaf stalks are of various widths, with a range of 30–50 mm in diameter. The stems range in length from 300 to 500 mm and can be green to anthocyanin in color. The position of the stems can be upright or semi-upright to horizontal [3]. The enlarged petioles develop from a central crown of the rhubarb plant. Petiole color is associated with rhubarb quality and the order of preference is red, pink, and green [2]. Although rhubarb is a well-known vegetable, scientific interest in this plant is a relatively new issue; most of the evidence of its biological activities and therapeutic potential derives from the last 15 years [4]. It shows high levels of both polyphenol content and antioxidant capacity in edible parts [5], which are petioles characterized by very high antioxidant properties and rich in many compounds that have a pro-health effect on the human body [6]. There is a wide variety of rhubarb cultivars that contain bioactive compounds such as flavonoids, anthraquinone, glycosides, tannins, volatile oils, and saponins [1]. The chemical composition of rhubarb includes anthraquinones, anthrones, stilbenes, tannins, polysaccharides, etc., which show extensive pharmacological activities including gastrointestinal regulation, anticancer and antimicrobial properties, hepatoprotection, cardiovascular and cerebrovascular anti-inflammatory protection, etc. [7,8]. The main characteristics of rhubarb quality are taste and aroma, which depend on the chemical composition [9]. Studies on the phytochemical composition of different species of rhubarb have provided information on the presence of a variety of inorganic and organic acids (including tartaric, oxalic, citric, malic, and ascorbic acids) [4].

Increased interest in the application of natural biologically active substances in human nutrition has led scientists to develop functional foods. The production of functional drinks is an ever-changing aspect of the beverage industry, as there are still new and prospective trends in this area [10]. Some examples of these kinds of foods include juices that are not obtained from concentrate (NFC) and freshly squeezed non-pasteurized juices (FS). These juices are obtained from the fruit tissue by pressing and centrifugation of the pulp [11]. Nutritionists recommend the consumption of fruit or vegetable juices instead of replacing natural products with synthetic vitamin and mineral supplements [12]. Fruit juices as functional drinks offer promoting good health by reducing the risk of serious illness. These beverages contain ingredients that provide specific benefits and are enriched with vitamins, minerals, amino acids, fiber, or antioxidants [13]. Products made from rhubarb have a favorable taste due to a high content of organic acids and rhubarb stalks taste best in early spring when they are ripe. Because rhubarb leaves are toxic due to their content of oxalic acid [14], the petioles are used for processing juices.

It should be noted that the scientific literature lacks studies on this plant and its application in the food industry. Therefore, for constant development and consumers' search for new solutions, it is necessary to focus attention on this plant [6]. It has been utilized for thousands of years for medicinal purposes, but only recently identified for its culinary use [15]; it was not until the 18th century that the culinary use of petioles was first reported [16]. In culinary rhubarb, the high oxalate content is a major drawback. A renewed interest in rhubarb production is now directed towards the use of stalks from low-oxalate cultivars as a cheap filler for industrial production of marmalade, jam, and syrup [17]. It has potential importance in the food and pharmaceutical industries to expand the range of products in the future by adding a new product with outstanding antioxidant properties [6]. Therefore, the aim of this study was to test selected rhubarb cultivars from a specific culinary perspective, i.e., processing into juice. From qualitative parameters, the sugars and acids were established in rhubarb juices which are important from a sensory analysis point of view. In addition, the juice yield of the plant was tested as an important parameter in the case of practical uses.

2. Materials and Methods

This 2-year vegetation field experiment was carried out at the Botanical Garden (BG), Slovak University of Agriculture (SUA), in 2018 and 2019 (Nitra, Slovak Republic, 48°18′ N, 18°05′ E, 144 masl).

2.1. Rhubarb Cultivars' Characterization

Morphological characterization and quantitative parameters of petioles of selected cultivars of *Rheum rhabarbarum* L. are summarized in Tables 1 and 2.

Table 1. Rhubarb cultivars' characterization (morphological description according to UPOV (1999)) [18].

	Morphological Description of Petiole	Picture of the Petioles *
'Poncho'	Semi-erect attitude, type of cross-section 1, green ground color of skin, entire distribution of skin superimposed color at base, absent distribution of skin superimposed color at middle, absent distribution of skin superimposed color just below leaf blade, present hairiness just below leaf blade, absent or very weak ribbing of dorsal side, green color of flesh	

Table 1. *Cont.*

	Morphological Description of Petiole	Picture of the Petioles *
'Victoria'	Erect attitude, type of cross-section 2, green ground color of skin, entire distribution of skin superimposed color at base, speckled distribution of skin superimposed color at middle, absent distribution of skin superimposed color just below leaf blade, present hairiness just below leaf blade, weak ribbing of dorsal side, green color of flesh	
'Valentine'	Erect attitude, type of cross-section 3, red ground color of skin, entire distribution of skin superimposed color at base, speckled distribution of skin superimposed color at middle, speckled distribution of skin superimposed color just below leaf blade, absent hairiness just below leaf blade, medium ribbing of dorsal side, pink color of flesh	
'Red Champagne'	Erect attitude, type of cross-section 2, red ground color of skin, entire distribution of skin superimposed color at base, speckled distribution of skin superimposed color at middle, speckled distribution of skin superimposed color just below leaf blade, present hairiness just below leaf blade, weak ribbing of dorsal side, green color of flesh	
'Canadian Red'	Semi-erect attitude, type of cross-section 7, red ground color of skin, entire distribution of skin superimposed color at base, entire distribution of skin superimposed color at middle, speckled distribution of skin superimposed color just below leaf blade, present hairiness just below leaf blade, strong ribbing of dorsal side, green color of flesh	

* Photo by Andrejiová.

Table 2. Quantitative parameters of *Rheum rhabarbarum* L. petioles ($n = 100$).

Cultivar	Length (cm)	Width (cm)	Thickness (cm)	Weight (g)
'Poncho'	29.42 ± 6.21 a	1.79 ± 0.33 a	1.16 ± 0.29 cd	49.35 ± 21.88 a
'Victoria'	33.99 ± 6.39 b	1.78 ± 0.28 a	0.97 ± 0.19 a	54.38 ± 15.80 a
'Valentine'	33.38 ± 6.64 b	2.31 ± 0.34 c	1.29 ± 0.30 d	77.61 ± 33.60 c
'Red Champagne'	37.03 ± 7.14 c	1.96 ± 0.48 b	1.02 ± 0.22 ab	63.04 ± 25.58 b
'Canadian Red'	29.55 ± 4.09 a	2.27 ± 0.30 c	1.07 ± 0.28 bc	62.47 ± 20.43 b

Values with different letters are significantly different at $p < 0.05$ by LSD test in ANOVA (Statgraphic XVII).

2.2. Soil and Climate Characteristics

The soil type of the experimental area is brown soil to chernozem on loess and loess loams and the part along the river Nitra belongs to the area of fluvial soils, where the original soil type was fluvial and fluvial glue. The plants were grown in growing substrate suitable for *Rheum rhabarbarum* L. In terms of climate classification of the region, Nitra is situated in a warm and dry area of Slovakia. The evaluations of experimental years according to climate normals are given in Tables 3 and 4.

Table 3. Evaluation of months according to air temperature climate normals 1961–1990.

Month	Normal 1961–1990 (°C)	T (°C) 2018	Characteristic— 2018	T (°C) 2019	Characteristic— 2019
III	5.0	3.4	cold	8.1	very hot
IV	10.4	15.4	extremely hot	9.4	normal
V	15.1	18.8	very hot	9.3	very cold
VI	18.0	20.7	hot	18.7	normal
VII	19.8	21.7	hot	21.9	hot

Table 4. Evaluation of months according to annual precipitation climate normals 1961–1990.

Month	Normal 1961–1990 (mm)	PRC (mm) 2018	Characteristic— 2018	PRC (mm) 2019	Characteristic— 2019
III	30	36	normal	16	very dry
IV	39	16	very dry	21	dry
V	58	29	very dry	135	extremely wet
VI	66	44	dry	29	very dry
VII	52	13	very dry	21	very dry

2.3. Organization of the Experiment

The cultivation of plant material was carried out in accordance with modern agrotechnical practices of rhubarb cultivation. The plants were purchased from the perennial nursery VICTORIA (Čab, Slovakia) and were planted on 5 May 2017, in spacing 1.0 × 0.9 m (density of 11,000 pl/ha). A total of 100 petioles from each cultivar were monitored. The petioles were harvested at the stage of consumption maturity at BBCH 39/49 (rosette development completed/typical leaf mass reached). When collecting the petioles, the minimum 20 cm length and 1 cm width (diameter) was respected. The harvest was conducted manually and gradually over two months (6 realized dates in the case of total yields and morphological descriptions) from the middle of April to the middle of June, at intervals of 1 time per week. For juice production and subsequent laboratory analysis, the rhubarb petioles from two different harvest terms were used, in a time span of 1 month, i.e., harvest term A (HTA) carried out on 18 May 2018 and harvest term B (HTB) on 17 June 2018. In 2019, HTA was carried out on May 13 and HTB was carried out on June 13.

2.4. Sample Preparation

Subsequent washing and analyses were conducted in the Laboratory of Beverages of Research Centre AgroBioTech SAU in Nitra. A total weight of 500 g of petioles was washed and harvested in 100 mm long pieces. They were processed on a commercial fruit juicer with 1200 rpm without heating (Magimix Duo Le Plus XL, Vincennes, France). The total volume of obtained juice was approximately 400 mL from each rhubarb cultivar replication. Immediately after juicing, the average juice yield from each cultivar was calculated. Following that, juice was strained through sieves of 0.2 mm diameter and processed through a centrifuge (Hettich BV-20, Tuttlingen, Germany) for 120 s at 6000 RPM in 8 × 10 mL plastic test tubes.

2.5. Juice Yield

The efficiency of juicing was calculated using the formula according to Wilczyński, 2019 [11]:

$$Wj\ (\%) = Mj/Mi \times 100$$

where Wj is the efficiency of juicing (%), Mj is the mass of juice after juicing (kg), and Mi is the mass of input material (kg).

2.6. Testing of Selected Parameters in Juice

Analyses of total sugar, glucose, fructose, total soluble solids, total acidity, malic acid, and pH level were analyzed by FT-IR spectrophotometer (Alpha Wine Analyzer, module for juices, Bruker Optics, Billerica, MA, USA) [19] which uses invisible infrared light of which certain wave lengths are absorbed by the sample, depending on its characteristics. It analyzes the sample by utilizing the attenuated total reflection (ATR) measurement technique. The core of the ATR technique is a diamond crystal that reflects the incoming light at a right angle to the detector. A sample is placed on top of the diamond and a fraction of the light penetrates the sample and is attenuated according to the absorption characteristics of the sample. This attenuation is measured by the detector and transformed into a spectrum [20,21]. The parameters were determined in three replicates for each sample of extracted juice.

2.7. Statistical Analyses

The analysis of variance (ANOVA), the multifactor analysis of variance (ANOVA), and the multiple Range test were done using the Statgraphic Centurion XVII (Stat Point Inc., Warrenton, Virginia, USA). Differences were tested depending on cultivar, year and term of harvest, and interactions by LSD test with significance: non-significant (NS) or significant at $p \leq 0.05$ (*), $p \leq 0.01$ (**), or $p \leq 0.001$ (***), respectively.

3. Results and Discussion

3.1. Yield Per Hectare

The total yield of petioles was estimated from data of six harvests per every tested cultivar grown at a density of 11,000 pl/ha. According to Table 5, the highest average values reached 'Red Champagne' (45.58 t/ha) and the lowest 'Canadian Red' (28.77 t/ha). As shown in Figure 1, the influence of the cultivar effect was significant ($p < 0.001$), as well as the year influence ($p < 0.0025$), when the average yield was higher in 2019.

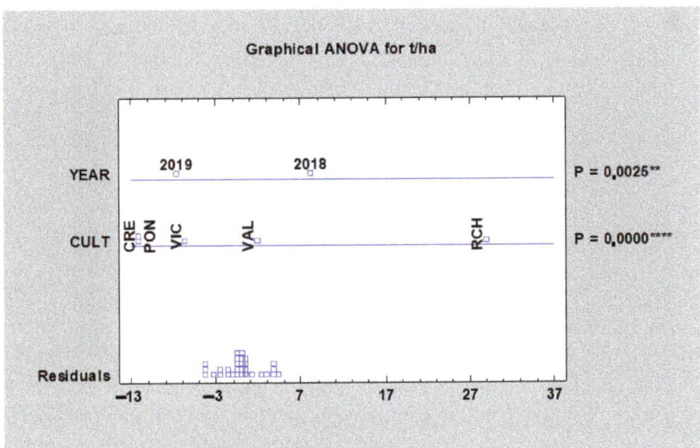

Figure 1. Effect of the experimental year and cultivar on the mean juice yield of all rhubarb cultivars. Abbreviations: PON, 'Poncho'; CRE, 'Canadian Red'; VAL, 'Valentine'; RCH, 'Red Champagne'; VIC, 'Victoria'. LSD test with significance: significant at $p \leq 0.01$ (**), or $p = 0.0000$ (****).

Table 5. The yields of rhubarb petioles.

Cultivar	Yield (t/ha)		
	2018	**2019**	**2018–2019**
PON	30.48 ± 4.22 a	28.40 ± 1.10 ab	29.44 ± 1.47 a
CRE	29.87 ± 1.12 a	27.66 ± 3.20 a	28.77 ± 1.56 a
VAL	36.61 ± 0.36 b	32.35 ± 3.32 b	34.48 ± 3.01 b
RCH	47.46 ± 3.60 c	43.70 ± 1.24 c	45.58 ± 2.65 c
VIC	32.70 ± 4.14 ab	29.26 ± 2.36 ab	30.98 ± 2.43 ab

Values with different letters are significantly different at $p < 0.05$ by LSD test in ANOVA (Statgraphic XVII). Abbreviations: PON, 'Poncho'; CRE, 'Canadian Red'; VAL, 'Valentine'; RCH, 'Red Champagne'; VIC, 'Victoria'.

The yield of the rhubarb cultivars is influenced both by the cultivar and the density according to Cojocaru et al. [14], where the total yield varied between 27.20 t/ha in the case of the 'Glanskin's perpetual' cultivar, at a reduced density, up to 39.02 t/ ha in the case of the 'Moldova Local population' cultivar, at a high density. A study by Stoleru et al. [9] on the establishment of the rhubarb crop highlighted the fact that in the first year after a crop was established, there were obtained yields of 0.7 kg/pl. at a density of 8000 pl/ha. In the second and third year of a plantation, intensive plant growth was observed according to Salata and Kozak [22]. Our plants were harvested in the second and third years after the planting. The results correspond with the study by Stoleru et al. [9] who reported that, at a density of 13,300 pl/ha, the Victoria cultivar obtained an overall yield of 46.64 t/ha as compared with at a density of 8000 pl/ha, where the obtained yield was 21.72 t/ha. Lepse [23] observed no differences in total yield and yield quality in the second year of a plantation in an experimental study on rhubarb propagation by different methods. Very good commercial yields for rhubarb are approximately 1.5–3 kg of petioles per plant according to Welbaum [2], which corresponds with our results as shown in Figure 2. The average yields per plant varied between 2489.3 ('Victoria') and 3933.4 g/plant ('Red Champagne').

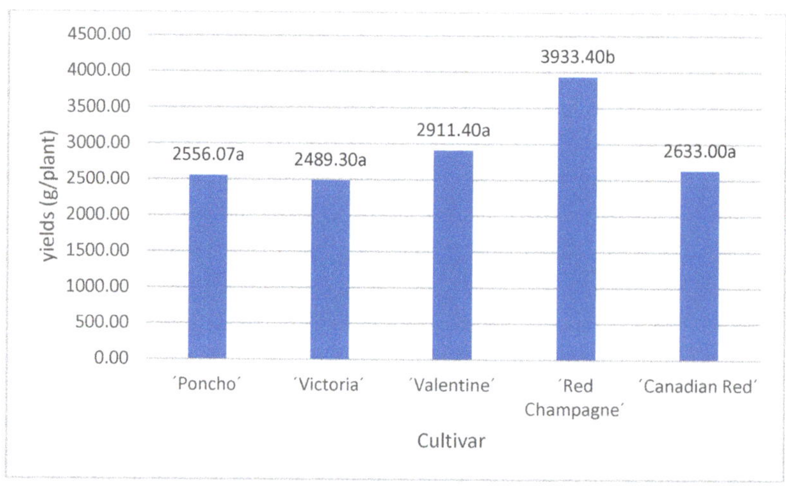

Figure 2. Average petiole yield for tested rhubarb cultivars. Values with different letters are significantly different at $p < 0.05$ by LSD test in ANOVA (Statgraphic XVII).

3.2. Juice Yield

As shown in Table 6, the juice yield significantly ($p < 0.05$) differed among the tested rhubarb cultivars. The average juice yield of the rhubarb cultivars ranged from 75% ('Poncho', HTA) to 85% ('Red Champagne', HTB). Generally, 'Red Champagne' was characterized by the highest juice yield potential for both harvests, as it significantly reached the highest values as compared with other tested cultivars. The statistical analysis of all tested results (Figure 3) showed significant differences ($p < 0.001$) in the case of morphological variability and influence of tested years on rhubarb juice yield. The term of harvests has not been shown to be significantly important to juice yield. The yield and composition of petioles are clearly influenced by the technology applied to a crop, especially the cultivar, the planting distance, and the nutritional regimen [14]. There were no data about juice yield of rhubarb found in the available literature. A comparison to similar celery petioles showed that high yield of celery juice was easy with a household juicer according to Donaldson, 2020 [24]. The juice yield and enzyme activity were tested in carrot, apples, spinach, and celery. The petiole juice yield of celery varied between 67.3% and 87.5% depending on juicer type. The yield of pressing ranged from 61.9% to 71.6% in the case of three apple cultivars according to Wilczyński et al. [11]. In addition, the dry matter of roots, in the case of carrot, were estimated to be from 41.1% to 65.2% depending on the juicer type [24].

Table 6. Influence of cultivar and term of harvest on juice yield of rhubarb petioles.

Cultivar	2018 (HTA)	2019 (HTA)	2018–2019 (HTA)
	JY (%)	JY (%)	JY (%)
PON	77 ± 2 a	74 ± 3 abc	75 ± 1 a
CRE	79 ± 2 ab	80 ± 3 cd	79 ± 2 abc
VAL	88 ± 2 c	72 ± 2 ab	80 ± 3 abc
RCH	90 ± 3 c	79 ± 3 bcd	84 ± 2 c
VIC	87 ± 3 c	78 ± 2 bcd	82 ± 2 bc
	2018 (HTB)	2019 (HTB)	2018–2019 (HTB)
PON	84 ± 2 bc	76 ± 2 abc	80 ± 3 abc
CRE	83 ± 3 abc	70 ± 2 a	77 ± 2 ab
VAL	84 ± 3 bc	80 ± 2 cd	82 ± 2 bc
RCH	87 ± 1 c	84 ± 3 d	85 ± 1 c
VIC	84 ± 2 abc	75 ± 1 abc	79 ± 2 abc

Values with different letters are significantly different at $p < 0.05$ by LSD test in ANOVA (Statgraphic XVII). Abbreviations: PON, 'Poncho'; CRE, 'Canadian Red'; VAL, 'Valentine'; RCH, 'Red Champagne'; VIC, 'Victoria'.

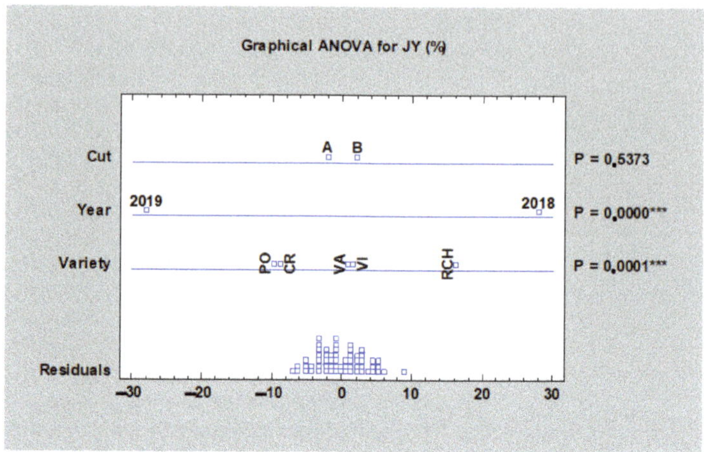

Figure 3. Effect of the experimental year, cultivar, and term of the harvest on the mean juice yield of all rhubarb cultivars. Abbreviations: PON, 'Poncho'; CRE, 'Canadian Red'; VAL, 'Valentine'; RCH, 'Red Champagne'; VIC, 'Victoria'. LSD test with significance: $p \leq 0.001$ (***).

3.3. Sugar Content

The average fructose content ranged from 33.93 ('Valentine', HTA) to 37.93 g/L ('Valentine', HTB), but cultivars significantly differed ($p < 0.05$) in fructose content only in the later harvest term (HTB), as shown in Table 7. When comparing all data from the cultivar variability point of view, the differences had not been proved (Figure 4). Fructose is highly represented in rhubarb. Its percentage representation is from 67% to 78% (as compared with total sugar content) according to the average from all results (Figure 5); therefore, it has high sensorial value in juice production. On the one hand, a statistically significant effect ($p < 0.05$) of the experimental year on fructose content was not found among rhubarb cultivars.

Table 7. Effect of cultivar on fructose and glucose content in rhubarb juice.

Cultivar	2018 (term HTA)		2019 (HTA)		2018–2019 (HTA)	
	Fructose (g/L)	Glucose (g/L)	Fructose (g/L)	Glucose (g/L)	Fructose (g/L)	Glucose (g/L)
PON	35.40 ± 0.34 cd	7.88 ± 0.19 d	35.03 ± 0.27 c	7.77 ± 0.18 e	35.22 ± 0.26 abc	7.83 ± 0.08 cde
CRE	34.58 ± 0.20 b	7.47 ± 0.16 c	34.40 ± 0.36 bc	5.69 ± 0.21 c	34.49 ± 0.13 ab	6.58 ± 1.26 ab
VAL	33.35 ± 0.45 a	6.34 ± 0.28 a	34.50 ± 0.21 bc	6.65 ± 0.12 d	33.93 ± 0.82 ab	6.50 ± 0.22 ab
RCH	35.50 ± 0.80 cd	9.48 ± 0.08 f	34.05 ± 0.71 b	5.33 ± 0.14 b	34.78 ± 1.03 ab	7.40 ± 2.93 bcd
VIC	36.03 ± 0.43 d	7.78 ± 0.07 d	32.80 ± 0.53 a	4.74 ± 0.07 a	34.41 ± 2.28 a	6.26 ± 2.15 a
	2018 (HTB)		2019 (HTB)		2018–2019 (HTB)	
PON	33.58 ± 0.49 a	6.86 ± 0.10 b	38.18 ± 0.58 e	7.81 ± 0.15 e	35.88 ± 3.25 bc	7.34 ± 0.67 abcd
CRE	38.20 ± 0.41 ef	6.90 ± 0.21 b	36.63 ± 0.50 d	6.82 ± 0.15 d	37.42 ± 1.11 de	6.86 ± 0.05 abc
VAL	38.46 ± 0.30 f	8.71 ± 0.12 e	37.40 ± 0.27 de	8.89 ± 0.18 g	37.93 ± 0.75 e	8.80 ± 0.12 e
RCH	37.56 ± 0.09 e	11.74 ± 0.13 g	34.87 ± 1.00 bc	8.20 ± 0.21 f	36.21 ± 1.90 cd	9.97 ± 2.51 f
VIC	34.89 ± 0.44 bc	8.48 ± 0.13 e	33.99 ± 0.19 b	8.19 ± 0.10 f	34.44 ± 0.64 a	8.33 ± 0.21 de

Values with different letters in columns are significantly different at $p < 0.05$ by LSD test in ANOVA (Statgraphic XVII). Abbreviations: PON, 'Poncho'; CRE, 'Canadian Red'; VAL, 'Valentine'; RCH, 'Red Champagne'; VIC, 'Victoria'.

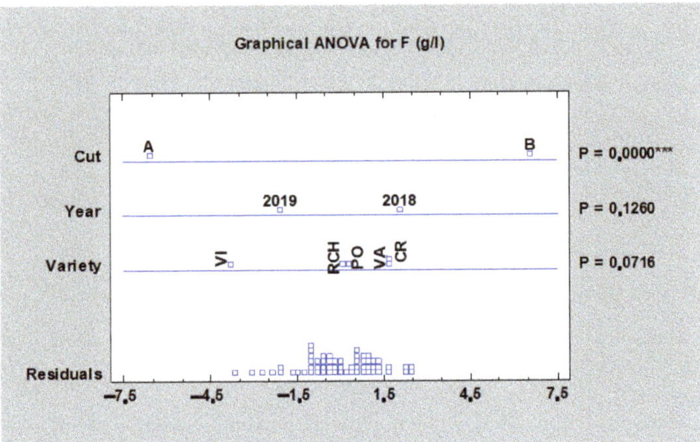

Figure 4. Effect of the experimental year, cultivar, and term of the harvest on the mean fructose content of all rhubarb cultivars. Abbreviations: PON, 'Poncho'; CRE, 'Canadian Red'; VAL, 'Valentine'; RCH, 'Red Champagne'; VIC, 'Victoria'. LSD test with significance $p \leq 0.001$ (***).

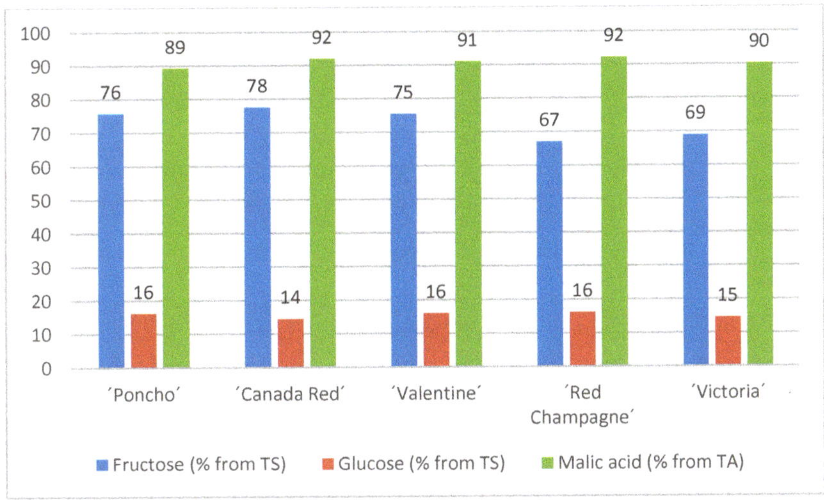

Figure 5. Share of fructose, glucose, and malic acid in total sugar and total acid content in the evaluated rhubarb cultivars. TS, total sugars and TA, total acid. Values are average from both harvests and years 2018–2019.

Values with different letters in columns are significantly different at $p < 0.05$ by LSD test in ANOVA (Statgraphic XVII). Abbreviations: PON, 'Poncho'; CRE, 'Canadian Red'; VAL, 'Valentine'; RCH, 'Red Champagne'; VIC, 'Victoria'.

On the other hand, the term of harvest had a significant effect ($p < 0.001$) on the fructose content (Figure 4), which was higher in the second harvest.

According to Table 7, the glucose content in the tested cultivars varied from 6.26 ('Victoria', HTA) to 9.97 g/L ('Red Champagne', HTB), which was 14–16% of the total sugar content of the evaluated rhubarb cultivars (Figure 5). The obtained results confirmed the statistically significant impact of cultivars on the glucose content of rhubarb (Table 7). The higher values were found in the HTB and, like the case of fructose, those differences were proven at the $p < 0.001$ level (Figure 6). The impact of the year on glucose content was significantly important when following all data (for both harvests and all cultivars).

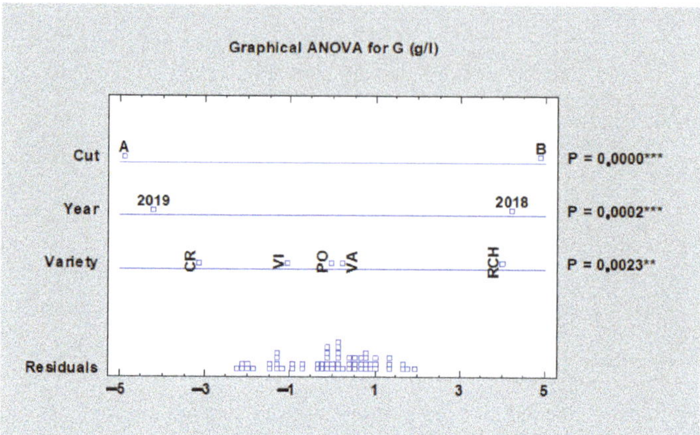

Figure 6. Effect of the experimental year, cultivar, and term of the harvest on the mean glucose content of all rhubarb cultivars. Abbreviations: PON, 'Poncho'; CRE, 'Canadian Red'; VAL, 'Valentine'; RCH, 'Red Champagne'; VIC, 'Victoria'. LSD test with significance: $p \leq 0.01$ (**), or $p \leq 0.001$ (***).

The total soluble solids, defined by Hegedűsová et al. [25] as additive quantity that expresses the content of dissolved substances, mainly sugars, in vegetable or fruit extracts, were also tested and are summarized in Table 8. The values for total soluble solids ranged from 3.54 ('Valentine', HTA) to 4.37 °BRIX ('Red Champagne', HTB). Significant differences ($p < 0.05$) were found between the tested cultivars and the values were higher in the second harvest (Table 8). When comparing all data in Figure 7, the significant effect ($p < 0.001$) of the cultivar, term of the harvest, and experimental year on the mean total soluble solids values of all rhubarb cultivars were found. The higher values in the late term of harvest could relate to the total soluble solid content increasing during ripening. This parameter is a very practical index of internal fruit quality and an accurate criterion for the decision to harvest in the field [26]. The mean value 3.8 °Brix was observed, with a range from 2.2 to 6.1° according to Pantoja and Kuhl [15], where they evaluated fifteen morphological characteristics to differentiate rhubarb cultivars.

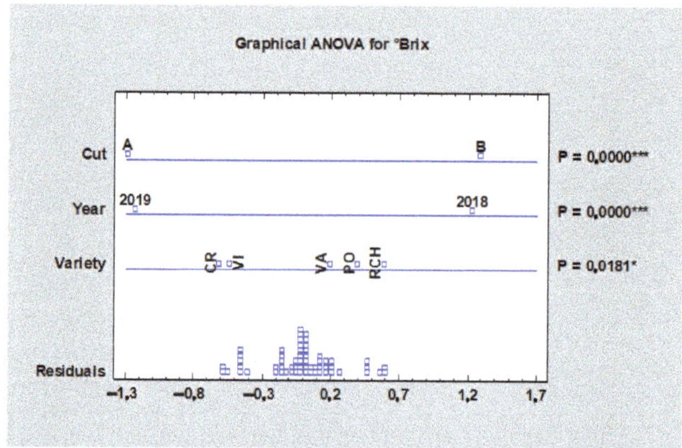

Figure 7. Effect of the experimental year, cultivar, and term of the harvest on the mean total soluble solid content of all rhubarb cultivars. Abbreviations: PON, 'Poncho'; CRE, 'Canadian Red'; VAL, 'Valentine'; RCH, 'Red Champagne'; VIC, 'Victoria'. LSD test with significance: significant at $p \leq 0.05$ (*) or $p \leq 0.001$ (***).

Table 8. Effect of cultivar on total soluble solid content and total sugar content in rhubarb juice.

Cultivar	2018 (HTA)		2019 (HTA)		2018–2019 (HTA)	
	TSS	Total sugar	TSS	Total sugar	TSS	Total sugar
	°BRIX	(g/L)	°BRIX	(g/L)	°BRIX	(g/L)
PON	3.98 ± 0.03 d	44.89 ± 0.40 c	4.22 ± 0.02 h	52.82 ± 0.80 f	4.10 ± 0.17 cd	48.86 ± 5.61 bc
CRE	3.74 ± 0.02 b	45.35 ± 0.14 c	3.44 ± 0.04 b	43.22 ± 0.11 a	3.59 ± 0.21 ab	44.29 ± 1.50 a
VAL	3.48 ± 0.03 a	43.32 ± 0.12 a	3.60 ± 0.02 c	45.06 ± 0.15 b	3.54 ± 0.08 a	44.19 ± 1.23 a
RCH	4.15 ± 0.02 f	52.59 ± 0.10 ef	3.25 ± 0.01 a	50.14 ± 0.37 e	3.70 ± 0.63 ab	51.36 ± 1.73 c
VIC	3.87 ± 0.04 c	50.69 ± 0.22 d	3.23 ± 0.01 a	47.90 ± 0.48 d	3.55 ± 0.45 a	49.30 ± 1.91 c
	2018 (HTB)		2019 (HTB)		2018–2019 (HTB)	
PON	3.76 ± 0.03 b	43.99 ± 0.16 b	3.95 ± 0.02 f	46.82 ± 0.11 c	3.86 ± 0.13 abc	45.40 ± 2.00 ab
CRE	4.04 ± 0.05 e	52.12 ± 0.57 e	3.60 ± 0.03 c	45.25 ± 0.19 b	3.82 ± 0.32 abc	48.69 ± 4.86 bc
VAL	4.50 ± 0.08 g	54.43 ± 0.59 h	4.12 ± 0.02 g	48.20 ± 0.23 d	4.31 ± 0.27 d	51.31 ± 4.41 c
RCH	4.86 ± 0.06 h	60.14 ± 0.43 i	3.88 ± 0.03 e	49.78 ± 0.26 e	4.37 ± 0.70 d	54.96 ± 7.33 d
VIC	4.03 ± 0.12 e	52.85 ± 0.33 g	3.76 ± 0.03 d	48.51 ± 0.43 d	3.90 ± 0.20 bc	50.68 ± 3.07 c

Values with different letters in columns are significantly different at $p < 0.05$ by LSD test in ANOVA (Statgraphic XVII). Abbreviations: PON, 'Poncho'; CRE, 'Canadian Red'; VAL, 'Valentine'; RCH, 'Red Champagne'; VIC, 'Victoria'.

The total sugar content varied between 44.19 ('Valentine', HTA) and 54.96 g/L ('Red Champagne', HTB) and according to Table 8 the differences between the tested cultivars were significant ($p < 0.05$). The values were higher at the later harvest (HTB), in which differences between the two harvests were also significantly proven at $p < 0.001$ (Figure 8). The year impact also significantly ($p < 0.01$) influenced the total sugar content parameter when comparing all cultivars and both harvests. The significant differences ($p < 0.001$) in terms of harvest influence on tested sugars as well as in the case of agrotechnical and soil/climate conditions through the tested years are in accordance with the study by Kalisz et al. [6] where the chemical composition of organic acids, minerals, carbohydrates, proteins, and vitamins highly depended on the cultivation and harvesting period.

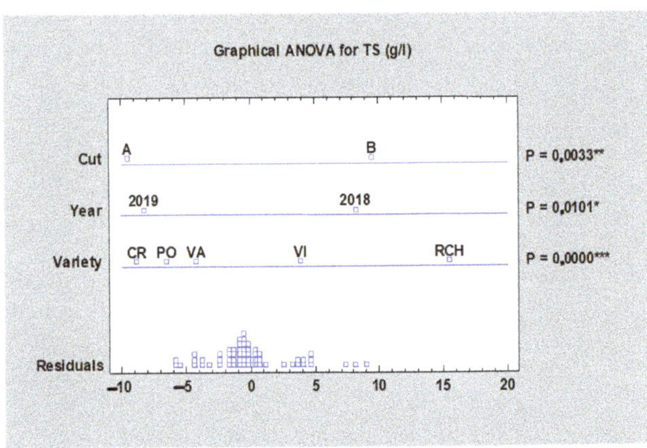

Figure 8. Effect of the experimental year, cultivar, and term of the harvest on the total sugar content of all rhubarb cultivars. Abbreviations: PON, 'Poncho'; CRE, 'Canadian Red'; VAL, 'Valentine'; RCH, 'Red Champagne'; VIC, 'Victoria'. LSD test with significance: non-significant (NS) or significant at $p \leq 0.05$ (*), $p \leq 0.01$ (**), or $p \leq 0.001$ (***).

3.4. Organic Acids and pH

Malic acid is the primary acid in rhubarb. Its value ranged from 14.06 ('Red Champagne', HTA) to 21.03 g/L ('Valentine', HTB). 'Valentine' reached the highest values for both harvests and the morphological variability was statistically improved at $p < 0.05$ (Table 9) in the frame of each harvest as well as in the statistical analyses of all data at $p < 0.001$ (Figure 9). Any significant differences were found in terms of harvest and the year influence on malic acid content (Figure 9). A similar situation was observed in the case of total acid content. In Table 9, cultivar variability on total acid content was significantly proven, where total acid content values were the highest in the case of the cultivar 'Valentine' for both harvests (21.88 g/L for HTA, and 22.95 g/L for HTB). The lowest value was reached by 'Canadian Red' at the second harvest (15.29 g/L) followed by 'Red Champagne' with a value of 15.45 g/L at the first harvest. As shown in Figure 10, cultivar variability had a significant impact ($p < 0.001$) on total acids content, as well as the year impact ($p < 0.01$). The term of harvest was not significantly important ($p = 0.8747$).

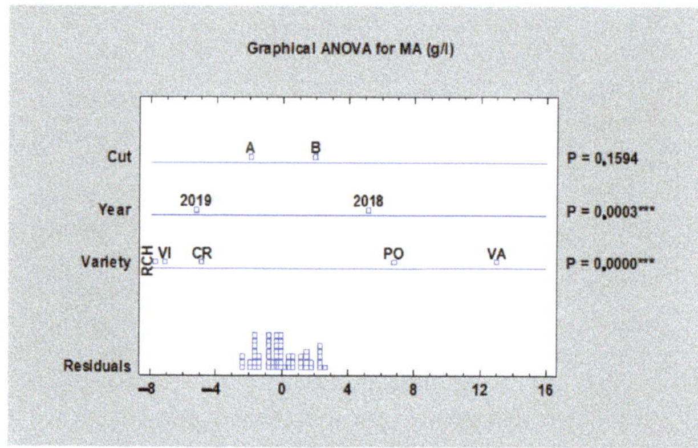

Figure 9. Effect of the experimental year, cultivar, and term of the harvest on the malic acid content in rhubarb cultivars. Abbreviations: PON, 'Poncho'; CRE, 'Canadian Red'; VAL, 'Valentine'; RCH, 'Red Champagne'; VIC, 'Victoria'. LSD test with significance: $p \leq 0.001$ (***).

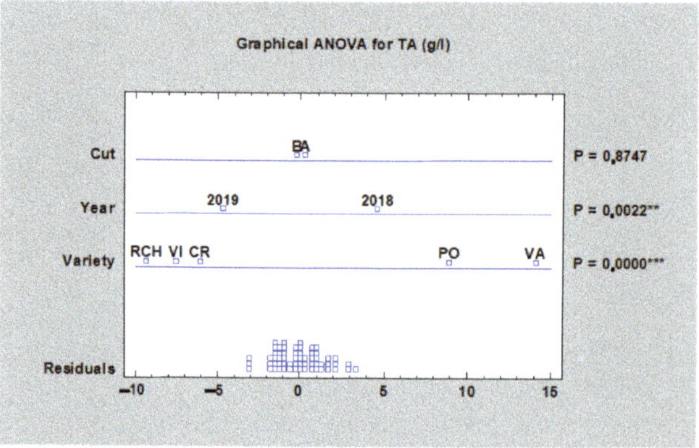

Figure 10. Effect of the experimental year, cultivar, and term of the harvest on the total acid content in rhubarb cultivars. Abbreviations: PON, 'Poncho'; CRE, 'Canadian Red'; VAL, 'Valentine'; RCH, 'Red Champagne'; VIC, 'Victoria'. LSD test with significance: $p \leq 0.01$ (**) or $p \leq 0.001$ (***).

Table 9. Effect of cultivar on malic acid and total acid content in rhubarb juice.

Cultivar	2018 (HTA)		2019 (HTA)		2018–2019 (HTA)	
	Malic Acid	Total Acids	Malic Acid	Total Acids	MALIC ACID	Total Acids
	(g/L)	(g/L)	(g/L)	(g/L)	(g/L)	(g/L)
PON	21.62 ± 0.16 h	24.73 ± 2.23 i	15.34 ± 0.59 d	17.43 ± 0.55 e	18.48 ± 4.45 c	21.08 ± 5.16 c
CRE	16.47 ± 1.06 e	18.49 ± 0.10 e	16.99 ± 1.04 g	18.42 ± 1.04 f	16.73 ± 0.37 b	18.46 ± 0.25 b
VAL	19.40 ± 2.03 f	21.92 ± 1.05 f	20.19 ± 2.06 j	21.84 ± 0.88 h	19.80 ± 0.56 cd	21.88 ± 0.46 cd
RCH	14.43 ± 1.07 a	15.17 ± 0.03 a	13.69 ± 0.16 c	15.72 ± 1.06 b	14.06 ± 0.52 a	15.45 ± 0.39 a
VIC	15.49 ± 0.94 c	16.14 ± 0.65 c	12.13 ± 0.17 a	15.89 ± 2.14 c	13.81 ± 2.38 a	16.01 ± 0.17 a
	2018 (HTB)		2019 (HTB)		2018–2019 (HTB)	
PON	19.79 ± 1.03 g	22.39 ± 1.06 g	18.03 ± 1.05 h	19.44 ± 2.07 g	18.91 ± 1.25 c	20.92 ± 2.08 c
CRE	15.08 ± 0.25 b	16.16 ± 1.03 c	13.44 ± 1.09 b	14.42 ± 0.22 a	14.26 ± 1.16 a	15.29 ± 1.23 a
VAL	22.63 ± 2.08 i	24.16 ± 2.13 h	19.43 ± 1.06 i	21.74 ± 0.83 h	21.03 ± 2.27 d	22.95 ± 1.71 d
RCH	14.97 ± 1.06 b	15.90 ± 1.09 b	15.81 ± 0.25 e	17.10 ± 1.05 d	15.39 ± 0.60 ab	16.50 ± 0.85 a
VIC	15.71 ± 1.12 d	16.75 ± 0.14 d	16.21 ± 0.39 f	17.06 ± 2.08 d	15.96 ± 0.35 b	16.90 ± 0.22 ab

Values with different letters in columns are significantly different at $p < 0.05$ by LSD test in ANOVA (Statgraphic XVII). Abbreviations: PON, 'Poncho'; CRE, 'Canadian Red'; VAL, 'Valentine'; RCH, 'Red Champagne'; VIC, 'Victoria'.

Organic acids play a biochemical role in maintaining the nutritional value and the quality of a vegetable species, and therefore they are among the frequently quantified compounds [27]. According to Welbaum [2], rhubarb leaves contain oxalic acid and should not be eaten. The petiole juice has a sharp acidic flavor. The oxalic acid in the petioles is much lower, the proportion of oxalic acid is about 10% of the total 2–2.5% acidity, which is dominated by nontoxic malic acid [7]. This means that the stalks are not hazardous to eat but have a very tart taste [2]. Malic acid is the predominant one, and citric and oxalic acids are also present in a smaller quantity in rhubarb plants according to Will and Dietrich [28]. The share of malic acid in total acid content varied between 89% and 92% in the evaluated rhubarb cultivars, as shown in Figure 5. Among the organic acids, the highest content was found in the case of malic acid, with average values of 679 ± 2.88 mg.100 g^{-1} fresh weight, according to Stoleru et al. [9]. Our results showed higher values because, in early petiole harvest, malic acid content is higher and decreases with vegetation period advancement in which the malic acid is naturally decomposed. According to the same study, the total acid content was 2239 mg.100 g^{-1} fresh weight. The chemical composition in organic acids highly depends on the cultivation and harvesting period [6].

The flavor of fruits and vegetable can also be determined by the sugar/acid (S/A) ratio that, in our case, expresses the ratio between total sugars and total acids; when the ratio is higher, the juice is subjectively less acidic. As shown in Table 10, the lowest S/A ratio of 2.02 was detected in the 'Valentine' cultivar, which means that there are 2.02 parts of sugars per one part of acid and the 'Valentine' is sensorially sweeter as compared with the 'Red Champagne' with the highest value of S/A ratio (3.35). The year and term of harvest did not have a significant effect on the S/A ratio according to the statistical analyses (Figure 11). Although the SSC/TA ratio is currently used as a maturity index for some types of fruit, it has been recognized that this measurement does not always correlate well with the perception of sweetness or tartness in other fruits [29].

Table 10. Effect of cultivar on pH level and the S/A ratio in rhubarb juice.

Cultivar	2018 (HTA)		2019 (HTA)		2018–2019 (HTA)	
	pH	S/A Ratio	pH	S/A Ratio	pH	S/A Ratio
PON	2.60 ± 0.02 a	1.82 ± 0.03 a	2.95 ± 0.04 d	3.03 ± 0.05 g	2.77 ± 0.08 b	2.42 ± 0.86 b
CRE	2.86 ± 0.22 bc	2.45 ± 0.01 d	2.93 ± 0.12 d	2.35 ± 0.00 c	2.90 ± 0.24 b	2.40 ± 0.07 b
VAL	2.73 ± 0.13 ab	1.98 ± 0.00 b	2.76 ± 0.14 a	2.06 ± 0.00 a	2.74 ± 0.25 a	2.02 ± 0.06 a
RCH	2.99 ± 0.21 c	3.47 ± 0.01 g	3.18 ± 0.18 g	3.19 ± 0.03 i	3.09 ± 0.06 cd	3.33 ± 0.20 cd
VIC	3.00 ± 0.32 c	3.14 ± 0.00 e	3.31 ± 0.11 h	3.01 ± 0.05 g	3.16 ± 0.04 de	3.08 ± 0.09 cd
	2018 (B)		2019 (B)		2018–2019 (B)	
PON	2.66 ± 0.04 a	1.96 ± 0.01 b	3.08 ± 0.18 e	2.41 ± 0.00 d	2.87 ± 0.29 b	2.19 ± 0.31 ab
CRE	3.39 ± 0.28 d	3.23 ± 0.04 f	3.11 ± 0.16 f	3.14 ± 0.02 h	3.25 ± 0.20 e	3.18 ± 0.06 cd
VAL	2.88 ± 0.03 c	2.25 ± 0.03 c	2.80 ± 0.06 b	2.22 ± 0.01 b	2.84 ± 0.09 b	2.24 ± 0.03 ab
RCH	2.99 ± 0.02 c	3.78 ± 0.05 h	2.88 ± 0.09 c	2.91 ± 0.02 f	2.93 ± 0.06 bc	3.35 ± 0.62 d
VIC	2.93 ± 0.01 c	3.16 ± 0.04 e	2.90 ± 0.06 c	2.84 ± 0.03 e	2.92 ± 0.11 b	3.00 ± 0.22 c

Values with different letters in columns are significantly different at P < 0.05 by LSD test in ANOVA (Statgraphic XVII). Abbreviations: PON, 'Poncho'; CRE, 'Canadian Red'; VAL, 'Valentine'; RCH, 'Red Champagne'; VIC, 'Victoria'.

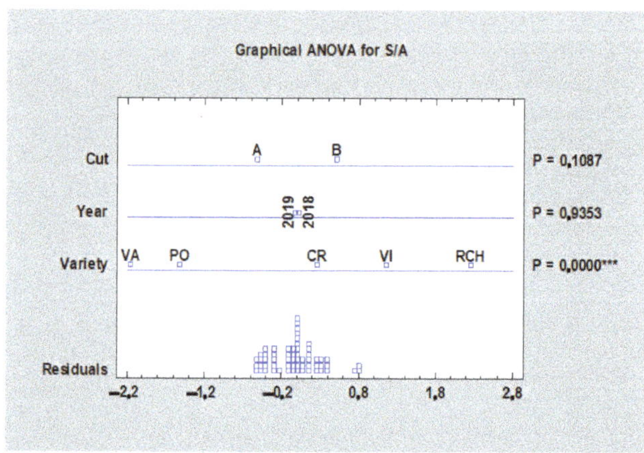

Figure 11. Effect of the experimental year, cultivar, and term of the harvest on the sugar/acid ratio in the rhubarb cultivars. Abbreviations: PON, 'Poncho'; CRE, 'Canadian Red'; VAL, 'Valentine'; RCH, 'Red Champagne'; VIC, 'Victoria'. LSD test with significance: $p \leq 0.001$ (***).

The pH of rhubarb juice varied between 2.77 ('Poncho', THA) and 3.25 ('Canadian Red', HTB), as shown in Table 10, which corresponded with Welbaum [2], where the pH of rhubarb juice was measured to be 3.2; therefore, large quantities of sugar are often used to prepare rhubarb dishes. The cultivar had a significant influence ($p < 0.001$) on pH, as shown in Figure 12, while the year and term of harvest did not have a significant influence.

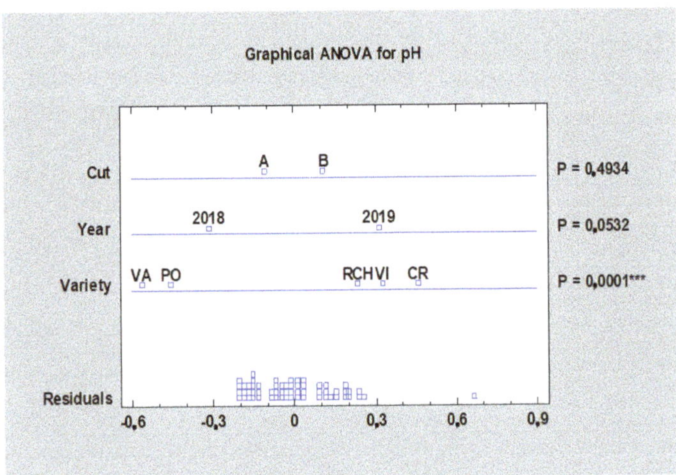

Figure 12. Effect of the experimental year, cultivar, and term of the harvest on the pH in the rhubarb cultivars. Abbreviations: PON, 'Poncho'; CRE, 'Canadian Red'; VAL, 'Valentine'; RCH, 'Red Champagne'; VIC, 'Victoria'. LSD test with significance: $p \leq 0.001$ (***).

4. Conclusions

Since fresh rhubarb is considered to be not tasty, i.e., it is very sour and slightly sweet, the ideal solution for using its concentrated nutrients and antioxidants is to use it as a juice, or as an additive to mixed juices. Among the given cultivars, the highest yield was achieved by the cultivar 'Red Champagne' (45.58 t/ha), with the highest juice yield potential (85%), which is also interesting from the tested parameters point of view. It reached the highest values of glucose (9.97 g/L), total soluble solids (4.37 g/L), and total sugars (54.96 g/L). In contrast, 'Valentine' had glucose (6.50 g/L) and total sugar content (44.19 g/L) at the lowest level with the highest total acids (22.95 g/L). The juice yield of the plant was tested as an important parameter in the case of practical uses and its value varied in the following order: 'Poncho' < 'Canadian Red' < 'Valentine' < 'Victoria' < 'Red Champagne'. The quality of the juice was significantly affected by the harvest term, where in the later term (HTB) the values were higher in the case of sugars (total sugars, TSS, fructose, and glucose). The acids were not changed by the term of harvest. In the case of some parameters, the significant influence of the year was also amplified, as there were significant differences in the agroclimatic characteristics tested for the two years. Since studying the sensory analysis of rhubarb juice is a new scientific issue, the results can serve as a basis for widespread use of this health-promoting crop, especially in the context of gastronomic use. We recommend adding rhubarb juice to vitalize, for example, apples harvested too late, where the share of malic acid is too low and achieves a balanced ratio between sugars and acids. The results show that rhubarb juice is a very suitable fortifier for such fruits, especially apples, which are unsaleable after long-term storage and juices made from such apples are often faint because malic acid has been degraded during the storage process. By enriching such juices with a rhubarb component, we achieve an optimal level of acidity in relation to the content of total sugars, which creates a very refreshing drink.

Author Contributions: Conceptualization—I.M.; methodology—I.M., J.M., and A.A.; resources—J.M. and A.A.; writing—original draft preparation—I.M.; writing—review and editing—I.M., J.M., and A.A. All authors have read and agreed to the published version of the manuscript.

Funding: This research was funded by the project, Development of Theoretical Knowledge and Practical Skills of Students for Teaching of Subject Vegetable Production, KEGA 018SPU-4/2020.

Conflicts of Interest: The authors declare no conflict of interest.

References

1. Clapa, D.; Borsai, O.; Hârţa, M.; Bonta, V.; Szabo, K.; Coman, V.; Bobiş, O. Micropropagation, Genetic Fidelity and Phenolic Compound Production of Rheum rhabarbarum L. *Plants* **2020**, *9*, 656. [CrossRef] [PubMed]
2. Welbaum, G.E. *Vegetable Production and Practices*, 1st ed.; CABI: Virginia, WV, USA, 2015; 476p.
3. Andrejiová, A.; Šlosár, M. *Instructions for Exercises in Vegetable Growing*, 1st ed.; Slovak University of Agriculture: Nitra, Slovakia, 2015; 124p.
4. Kolodziejczyk-Czepas, J.; Liudvytska, O. Rheum rhaponticum and Rheum rhabarbarum: A review of phytochemistry, biological activities and therapeutic potential. *Phytochem. Rev.* **2021**, *20*, 589–607. [CrossRef]
5. Cojocaru, A.; Vlase, L.; Munteanu, N.; Stan, T.; Teliban, G.C.; Burducea, M.; Stoleru, V. Dynamic of Phenolic Compounds, Antioxidant Activity, and Yield of Rhubarb under Chemical, Organic and Biological Fertilization. *Plants* **2020**, *9*, 355. [CrossRef] [PubMed]
6. Kalisz, S.; Oszmiański, J.; Kolniak-Ostek, J.; Grobelna, A.; Kieliszek, M.; Cendrowski, A. Effect of a variety of polyphenols compounds and antioxidant properties of rhubarb (Rheum rhabarbarum). *LWT* **2020**, *118*, 108775. [CrossRef]
7. Cao, Y.-J.; Pu, Z.-J.; Tang, Y.-P.; Shen, J.; Chen, Y.-Y.; Kang, A.; Zhou, G.-S.; Duan, J.-A. Advances in bio-active constituents, pharmacology and clinical applications of rhubarb. *Chin. Med.* **2017**, *12*, 1–12. [CrossRef] [PubMed]
8. Xiang, H.; Zuo, J.; Guo, F.; Dong, D. What we already know about rhubarb: A comprehensive review. *Chin. Med.* **2020**, *15*, 1–22. [CrossRef] [PubMed]
9. Stoleru, V.; Munteanu, N.; Stan, T.; Ipătioaie, C.; Cojocaru, A.; Butnariu, M. Effects of production system on the content of organic acids in Bio rhubarb (Rheum rhabarbarum L.). *Romanian Biotechnol. Lett.* **2019**, *24*, 184–192. [CrossRef]
10. Turkmen, N.; Akal, C.; Özer, B. Probiotic dairy-based beverages: A review. *J. Funct. Foods* **2019**, *53*, 62–75. [CrossRef]
11. Wilczyński, K.; Kobus, Z.; Dziki, D. Effect of Press Construction on Yield and Quality of Apple Juice. *Sustainability* **2019**, *11*, 3630. [CrossRef]
12. Liu, R.H. Health-Promoting Components of Fruits and Vegetables in the Diet. *Adv. Nutr.* **2013**, *4*, 384S–392S. [CrossRef] [PubMed]
13. Grumezescu, A.M.; Holban, A.M. *Production and Management of Beverages*; Woodhead Publishing, Elsevier: Amsterdam, The Netherlands, 2019; 504p. [CrossRef]
14. Cojocaru, A.; Munteanu, N.; Petre, B.A.; Stan, T.; Teliban, G.C.; Vintu, C.; Stoleru, V. Biochemical and Production of Rhubarb Under Growing Technological Factors. *Rev. Chim.* **2019**, *70*, 2000–2003. [CrossRef]
15. Pantoja, A.; Kuhl, J.C. Morphologic variation in the USDA/ARS rhubarb germplasm collection. *Plant Genet. Resour.* **2009**, *8*, 35–41. [CrossRef]
16. Kuhl, J.C.; DeBoer, V.L. Genetic Diversity of Rhubarb Cultivars. *J. Am. Soc. Hortic. Sci.* **2008**, *133*, 587–592. [CrossRef]
17. Helena, P. Estimating Genetic Variability in Horticultural Crop Species at Different Stages of Domestication. Ph.D. Thesis, Swedish University of Agricultural Sciences, Alnarp, Sweden, 2001.
18. UPOV. Geneva: Guidelines for the Conduct of Tests for Distinctness, Uniformity and Stability. 1999. Available online: https://www.upov.int/test_guidelines/en/fulltext_tgdocs.jsp?lang_code=EN&q=rheum (accessed on 7 June 2021).
19. ALPHA II, FTIR Must & Wine Analyzer, Brochure. Available online: https://takimya.com/storage/pdf/Wine_analyzer_brochure_EN.pdf (accessed on 7 June 2021).
20. Ján, M.; Ivana, M. Changes in the levels of selected organic acids and sugars in apple juice after cold storage. *Czech J. Food Sci.* **2018**, *36*, 175–180. [CrossRef]
21. Mezey, J.; Petrík, M.; Bajčan, D.; Harangozó, L.; Mezeyová, I. Monitoring of Heavy Metals, Bioactive Substances and Nutritional Composition of Cranberry (Vaccinium Vitis-Idaea) Fruits in Tatra National Park Forest Ecosystem High-Altitude Transects. In Proceedings of the 19th International Multidisciplinary Scientific Geoconference, SGEM 2019, Albena, Bulgaria, 30 June–6 July 2019; pp. 761–768. [CrossRef]
22. Salata, A.; Kozak, D. Comparison of Yield and Morphological Characters of Rhubarb (Rheum Rhaponticum L.) "Karpow Lipskiego" Plants Propagated in Vitro and By Conventional Methods. *Acta Sci. Polonorum. Hortorum Cultus* **2013**, *12*, 115–128.
23. Lepse, L. Comparison of in vitro and traditional propagation methods of rhubarb (rheum rhabarbarum) according to morphological features and yield. *Acta Hortic.* **2009**, 265–270. [CrossRef]
24. Donaldson, M. Household Juice Extractor Comparison and Optimization. *J. Food Process. Technol.* **2020**, *11*, 14. Available online: https://www.longdom.org/open-access/household-juice-extractor-comparison-and-optimization-53006.html (accessed on 1 April 2021).
25. Hegedusova, A.; Slosar, M.; Mezeyova, I.; Hegedus, O.; Andrejiova, A.; Szarka, K. *Methods for Estimation of Selected Biologically Active Substances*, 1st ed.; Slovak university of Agriculture: Nitra, Slovakia, 2018; 95p.
26. Lado, J.; Rodrigo, M.J.; Zacarías, L.; Review, S.P. Maturity Indicators and Citrus Fruit Quality. Stewart Postharvest Review. 2014. Available online: https://www.researchgate.net/profile/Lorenzo-Zacarias/publication/265858476_Maturity_indicators_and_citrus_fruit_quality/links/5747f02308ae14040e28d94c/Maturity-indicators-and-citrus-fruit-quality.pdf (accessed on 7 June 2021).
27. Kačániová, M.; Kňazovická, V.; Melich, M.; Fikselová, M.; Massanyi, P.; Stawarz, R.; Haščík, P.; Pechociak, T.; Kuczkowska, A.; Putała, A. Environmental concentration of selected elements and relation to physicochemical parameters in honey. *J. Environ. Sci. Heal. Part A* **2009**, *44*, 414–422. [CrossRef] [PubMed]

28. Will, F.; Dietrich, H. Processing and chemical composition of rhubarb (Rheum rhabarbarum) juice. *LWT* **2013**, *50*, 673–678. [CrossRef]
29. Obenland, D.; Collin, S.; Mackey, B.; Sievert, J.; Fjeld, K.; Arpaia, M.L. Determinants of flavor acceptability during the maturation of navel oranges. *Postharvest Biol. Technol.* **2009**, *52*, 156–163. [CrossRef]

Article

Use of Grape Peels By-Product for Wheat Pasta Manufacturing

Mădălina Iuga * and Silvia Mironeasa *

Faculty of Food Engineering, Ștefan cel Mare University of Suceava, 13 Universitatii Street,
720229 Suceava, Romania
* Correspondence: madalina.iuga@usm.ro (M.I.); silviam@fia.usv.ro (S.M.)

Abstract: Grape peels (GP) use in pasta formulation represents an economic and eco-friendly way to create value-added products with multiple nutritional benefits. This study aimed to evaluate the effect of the GP by-product on common wheat flour (*Triticum aestivum*), dough and pasta properties in order to achieve the optimal level that can be incorporated. Response surface methodology (RSM) was performed taking into account the influence of GP level on flour viscosity, dough cohesiveness and complex modulus, pasta color, fracturability, chewiness, cooking loss, total polyphenols, dietary fibers and resistant starch amounts. The result show that 4.62% GP can be added to wheat flour to obtain higher total polyphenols, resistant starch and dietary fiber contents with minimum negative effects on pasta quality. Flour viscosity, dough cohesiveness, complex modulus and pasta fracturability of the optimal sample were higher compared to the control, while chewiness was lower. Proteins' secondary structures were influenced by GP addition, while starch was not affected. Smooth starch grains embedded in a compact protein structure containing GP fiber was observed. These results show that GP can be successfully incorporated in wheat pasta, offering nutritional benefits by their antioxidants and fiber contents, without many negative effects on the final product's properties.

Keywords: grape peels; pasta; wheat flour; fibers; antioxidants

Citation: Iuga, M.; Mironeasa, S. Use of Grape Peels By-Product for Wheat Pasta Manufacturing. *Plants* **2021**, *10*, 926. https://doi.org/10.3390/plants10050926

Academic Editor: Ivo Vaz de Oliveira

Received: 4 April 2021
Accepted: 4 May 2021
Published: 6 May 2021

Publisher's Note: MDPI stays neutral with regard to jurisdictional claims in published maps and institutional affiliations.

1. Introduction

Nowadays, consumer behavior is changing and functional foods with significant health benefits are gaining increasing attention. The food industry generates high amounts of by-products that may be considered an opportunity for a sustainable valorization in order to minimize the waste impact and to ensure environmental protection. Winery by-products' conversion into value-added products is of high interest for producers, consumers and researchers. One method of valorization is the inclusion of byproducts in food products in order to increase their nutritional value. Pasta is consumed worldwide and can be considered a good matrix for bioactive compounds' incorporation [1].

About 50% of the grape pomace is composed of grape peels (GP), depending on the grape variety and pedo-climatic conditions [2,3]. Some health problems such as cardio-vascular diseases, stroke and some cancer types can be prevented by an adequate intake of fruits and vegetables due to their high amounts of bioactive compounds [4]. GP are a source of polyphenolic compounds and dietary fiber [5–7], components that can exert antioxidant an antimicrobial action [8,9]. The most important GP components with antioxi-dant characters are anthocyanins, hydroxycinnamic acids, catechins and flavonols, which can determine the inhibition of oxidative processes of low-density lipoproteins [10,11].

The potential applications of polyphenols and dietary fiber to preserve foods and prolong their shelf-life were demonstrated in some previous studies [12,13]. GP contain up to 60% (dm) dietary fiber, the insoluble fraction prevailing, followed by sugars, which can total up to 70%, depending on the vinery process applied [10,14]. Due to the essential role played by dietary fiber for human health, such as improvement of gastrointestinal activity, reducing glycemic responses and cholesterol levels in the blood [13], it is necessary to take alternative sources of dietary fiber to achieve the recommended consumption, which is about 25–30 g per day [1].

There are some studies revealing the possibility to create value-added products by incorporating grape by-products in bakery products or pasta [15–18]. Grape by-products can be used as semolina replacer in pasta with many positive effects on the technological properties, such as firmness and adhesiveness, but also on the physico-chemical and functional characteristics of the final product, such as increased total polyphenolics content and antioxidant activity, and lowering of the glycaemic index trough resistant starch content increase [1]. Simonato et al. [19] studied the effects of *Moringa oleifera leaf* powder on wheat fresh pasta and observed an increase of cooking loss and a reduced firmness along with nutritional value enhancement given by higher phenols and mineral contents. The addition of coconut by-products to wheat pasta led to lower firmness and color changes, while the fiber, protein and lipid contents increased [20]. Sobota et al. [21] underlined the possibility to increase durum wheat pasta fiber content by incorporating different vegetables powders (beet, carrot, kale), along with significant changes of color. According to the results presented by Xu et al. [22], the incorporation of apple pomace in noodles caused a cohesiveness and tensile strength reduction and a cooking loss increase, with the hardness and adhesiveness of the noodles not being changed, while gumminess, chewiness, and springiness recorded differences. Fortification of durum wheat pasta with onion skin by-products resulted in increased dietary fiber, ash, total phenolic compounds, flavonoids content and antioxidant activity, while cooking loss, water solubility index and redness were higher and the optimal cooking time lower [23]. Zarzycki et al. [24] showed that the addition of Moldavian dragonhead seeds residue in pasta resulted in higher nutritional value by increasing proteins, dietary fiber and mineral contents, without negative effects on the cooking and sensory characteristics of the pasta. Another study made by Simonato et al. [25] underlined the opportunity to increase total polyphenol contents and antioxidant capacity of pasta by supplementation with olive pomace. The authors found a decrease of rapidly digestible starch and an increase of slowly digestible starch, resistant starch, swelling index, water absorption, cooking loss and pasta firmness [25].

The addition of fiber-rich ingredients can have significant effects on dough rheological properties and on the final product texture, microstructure and color. The effects of grape by-products on the composite flour and final product properties are proportional to the addition level [3]. Mironeasa et al. [16] revealed some negative effects of GP on dough rheology caused by gluten dilution, which can be minimized by particle size reduction. Food texture, volume and color are strongly affected by high levels of grape by-products, Gaita et al. [17] suggesting that amounts up to 6% GP can be incorporated into pasta containing eggs without significant negative effects on the sensory acceptance. On the other hand, fortification of bakery and pasta products with grape by-products led to a nutritional value increase due to the intake of fiber and polyphenolics with antioxidant activity [4,5,15]. Grape peels' phenolics, such as phenolic acids, tannins and flavonoids, could have reducing effects on starch digestibility due to their abilities to inhibit enzyme activity or by the formation of starch-polyphenol complexes with resistance to enzyme attacks [26]. Furthermore, polyphenols can slow down starch gelatinization via the inter-action through hydrogen bonds with amylose molecules [27]. Saad et al. [28] found that wheat pasta dough rheological properties in terms of extensibility and water absorption increased, while elasticity decreased, when cucumber pomace was added. The mineral and polyphenols content of noodles were improved, while a reduction of protein and carbohydrate contents was observed [28].

In countries where durum wheat is not widely cultivated, common wheat usually represents the basic ingredient for pasta production. There are some studies revealing the possibility to use fiber and polyphenols-rich ingredients in pasta formulation, but to our knowledge, there are no studies revealing the effects of GP on common white wheat flour for pasta production. The approach of this study is complex, evidencing the technological, nutritional, molecular and structural changes of flour, dough and pasta. The knowledge of the interactions of grape peels with other biopolymers from wheat is very important for the development of novel pasta products. Thus, the aim of this study was to underline the impact of GP components on white wheat flour (WWF), dough and pasta properties, and to optimize the addition level in order to obtain the best product quality. Furthermore, a characterization of the optimum and control products was made.

2. Results

2.1. Diagnostic Checking of the Models

The ANOVA results for the model fitting presented in Table 1 show that all of the mathematic models chosen were significant and predicted accurately the responses, the F-value being significant ($p < 0.01$) in all cases and R^2 values being more than 0.76.

Table 1. ANOVA results for the fitted models for different characteristics of flour, dough and pasta.

Response	Model	F-Value	p-Value	R^2	Adj.-R^2
η_{max} (Pa·s)	quadratic	36.23	<0.01	0.83	0.81
G* (Pa)	quartic	31.27	<0.01	0.91	0.88
Co	quadratic	24.14	<0.01	0.76	0.73
C*	quartic	32.72	<0.01	0.91	0.88
F (g)	quartic	179.30	<0.01	0.98	0.98
Ch (J)	quadratic	26.38	<0.01	0.78	0.75
CL (%)	quartic	91.58	<0.01	0.97	0.96
RS (%)	cubic	181.00	<0.01	0.98	0.97
TPC (µg GAE/g)	quartic	85.84	<0.01	0.96	0.95
TDF (%)	fifth	1503.06	<0.01	0.99	0.99

η_{max}—peak viscosity, G*—complex modulus, Co—cohesiveness, C*—chroma, F—fracturability, Ch—chewiness, CL—cooking loss, RS—resistant starch, TPC—total polyphenols content, TDF—total dietary fiber.

Flour peak viscosity (η_{max}), dough cohesiveness (Co) and boiled pasta chewiness (Ch) data were fitted to the quadratic model, which described 83%, 76% and 78% of the data variation, respectively. Dough complex modulus (G*), dry pasta chroma (C*), fracturability (F), cooking loss (CL) and polyphenolic content (TPC) results were fitted to the quartic model, with 91%, 91%, 98%, 97% and 96%, respectively, of the data variation being explained. The cubic model explained 98% of data variation for resistant starch content (RS), while the results for the dietary fiber content (TDF) were fitted to the fifth model, which explained 99% of the variation.

2.2. Effects of GP on Flour and Pasta Characteristics

A Supplementary Materials section containing the graphics for the variation of the responses with GP level is provided.

Peak viscosity increased with GP level increase (Figure S1), the biggest significant ($p < 0.01$) positive influence being obtained for the linear term Equation (1).

$$\eta_{max}(Pa·s) = 0.42 + 0.14A^{**} - 0.01A^2 \tag{1}$$

where η_{max}—peak viscosity, A—GP level, ** significant at $p < 0.01$.

An increase of G* as the GP level was higher was observed (Figure S2a), the linear and quadratic terms having significant influence ($p < 0.05$) on the response Equation (2).

$$G^*(Pa) = 96083.29 + 23997.76A^{**} + 38352.65A^{2*} + 808.26A^3 - 27020.94A^4 \tag{2}$$

where G*—complex modulus, A—GP level, ** significant at $p < 0.01$, * significant at $p < 0.05$.

More cohesive dough was obtained as the GP addition level was higher (Figure S2b), the linear term having the greatest significant ($p < 0.01$) influence Equation (3). The negative effects of GP, which are a fiber-rich ingredient, were minimized because small particle size (<180 µm) was used.

$$Co = 0.40 + 0.02A^{**} + 0.01A^2 \tag{3}$$

where Co—cohesiveness, A—GP level, ** significant at $p < 0.01$.

Pasta color was affected by GP incorporation, a significant ($p < 0.01$) decrease of C^* being observed with the addition level increase (Figure S3a). The biggest negative influence was obtained for the linear term of the factor Equation (4).

$$C^* = 23.08 - 2.72A^{**} + 1.42A^2 - 0.35A^3 + 0.03A^4 \tag{4}$$

where C^*—chroma, A—GP level, ** significant at $p < 0.01$.

Dry pasta fracturability expressed as the maximum force needed to break the sample can be an indicator of pasta resistance to transport and manipulation. The addition of GP caused an increase of F with the level increase (Figure S3b). The linear and cubic terms presented significant positive influence ($p < 0.05$), while the quadratic term had a negative effect on the response Equation (5).

$$F(g) = 3158.72 + 1837.63A^{**} - 1052.11A^{2**} + 256.32A^{3*} - 19.84A^{4*} \tag{5}$$

where F—fracturability, A—GP level, ** significant at $p < 0.01$, * significant at $p < 0.05$.

GP addition caused a CL rise with the level increase (Figure S4a), the linear, quadratic and quartic terms presenting significant ($p < 0.05$) influences based on Equation (6). An acceptable cooking loss value should be less than 12% [5].

$$CL(\%) = 0.66 + 6.33A^{**} - 3.02A^{2*} + 0.61A^3 - 0.04A^{4*} \tag{6}$$

where CL—cooking loss, A—GP level, ** significant at $p < 0.01$, * significant at $p < 0.05$.

Pasta chewiness is expressed as the energy required to chop the sample until it is ready to swallow [29]. GP level increase caused a proportional decrease of pasta chewiness (Figure S4b), the highest influence being observed for the linear term Equation (7).

$$Ch = 3696.18 - 287.20A^{**} + 79.49A^2 \tag{7}$$

where Ch—chewiness, A—GP level, ** significant at $p < 0.01$.

Resistant starch content was significantly influenced ($p < 0.01$) by the linear, quadratic and cubic terms in Equation (8). A rise in RS with GP addition level increase was observed (Figure S5a).

$$RS(\%) = 4.49 + 0.34A^{**} - 0.43A^{2**} + 0.50A^{3**} \tag{8}$$

where RS—resistant starch content, A—GP level, ** significant at $p < 0.01$.

A significant positive influence ($p < 0.01$) was obtained for the linear term of GP level, while the quadratic and quartic terms exhibited a negative and significant ($p < 0.05$) effect on TPC response (Equation (9). TPC showed higher values with GP level increase (Figure S5b), as a result of polyphenols present in the added ingredient.

$$TPC(\%) = 29.25 + 117.50A^{**} - 55.362A^{2*} + 11.17A^3 - 0.77A^{4*} \tag{9}$$

where TPC—total polyphenols content, A—GP level, ** significant at $p < 0.01$, * significant at $p < 0.05$.

GP are a rich source of soluble and insoluble dietary fibers, their incorporation in wheat pasta determining an increase of TDF with the level increase (Figure S5c). Significant positive influences ($p < 0.01$) were obtained for the linear, cubic and fifth terms, while the quadratic and quartic terms presented negative effects on the response (Equation (10)).

$$TDF(\%) = -4.58 + 9.17A^{**} - 6.31A^{2**} + 2.01A^{3**} - 0.30A^{4**} + 0.02A^{5**} \tag{10}$$

where TDF—total dietary fiber content, A—GP level, ** significant at $p < 0.01$.

2.3. Optimization of GP Level and Models Validation

To obtain the maximum nutritional benefits with minimum quality characteristics' impairment, the optimization of GP level as a function of the considered responses showed that wheat flour can be supplemented with 4.62% GP (Table 2), with a desirability of 0.57.

Table 2. Confirmation of the optimized parameters and control sample characteristics.

Parameter	OGP			Control
	Predicted Value	**Experimental Value**	**Relative Deviation * (%)**	
A-GP (%)	4.62 ± 0.00	4.62 ± 0.00	-	-
η_{max} (Pa·s)	0.83 ± 0.06 [x]	0.85 ± 0.06 [xa]	2.35	0.42 ± 0.02 [b]
G* (Pa)	$113{,}595.55 \pm 6604.45$ [x]	$113{,}733.33 \pm 950.44$ [xa]	0.12	$51{,}170.00 \pm 1822.44$ [b]
Co	0.41 ± 0.01 [x]	0.41 ± 0.01 [xa]	0.00	0.36 ± 0.01 [b]
C*	19.07 ± 0.40 [x]	19.26 ± 0.06 [ya]	0.99	21.89 ± 0.06 [b]
F (g)	5432.15 ± 106.15 [x]	5659.67 ± 159.22 [xa]	4.02	4207.33 ± 123.18 [b]
Ch (J)	3583.12 ± 116.08 [x]	3353.11 ± 162.77 [xa]	−6.86	4910.27 ± 72.29 [b]
CL (%)	6.81 ± 0.25 [x]	7.03 ± 0.24 [xa]	3.13	5.53 ± 0.19 [b]
RS (%)	4.60 ± 0.10 [x]	4.79 ± 0.01 [ya]	3.97	2.58 ± 0.10 [b]
TPC (µg GAE/g)	141.48 ± 4.21 [x]	144.99 ± 2.78 [xa]	2.42	106.75 ± 4.18 [b]
TDF (%)	1.43 ± 0.03 [x]	1.38 ± 0.03 [xa]	−3.62	0.02 ± 0.00 [b]

OGP—optimal formulation of wheat flour with grape peels, A—GP (grape peels) level, η_{max}—peak viscosity, G*—complex modulus, Co—cohesiveness, C*—chroma, F—fracturability, Ch—chewiness, CL—cooking loss, RS—resistant starch, TPC—total polyphenols content, TDF—total dietary fiber, means in the same row followed by different letters (x–y for differences among predicted and observed values, a–b for differences between OGP and control) are significantly different ($p < 0.05$), * relative deviation = [(experimental value − predicted value)/experimental value] × 100.

For the model's validation, a pasta sample was made using the optimal level of GP that resulted after optimization. The responses were checked in triplicate and the experimental values were less than 5% different from the predicted ones (Table 2), except for chewiness, which was lower by 6.86% than the predicted value. Compared to the control, significantly ($p < 0.01$) higher G*, η_{max}, dough Co, F, CL, RS, TPC and TDF contents were obtained, while C* and boiled pasta Ch were smaller (Table 2). Consequently, the nutritional and functional values of the optimized pasta were enhanced compared to the control and the quality parameters were kept. Even if higher CL was obtained (6.81%), the value was less than 12%, the limit recommended for acceptable pasta. The OGP sample presented a more cohesive, elastic and viscous dough, which was probably related to the higher resistance to break (F) of pasta, which was desirable.

2.4. Determination of Control and Optimal Product Properties

2.4.1. FTIR Analysis of Flours

FTIR analysis allowed the identification of changes in bonding and possible interactions between composite flour components, underlying the impact of GP addition to WWF. The representative spectra of OGP and control samples is presented in Figure 1 and shows the peaks of the functional groups and the vibration ways of the compounds. Several peaks were identified in the analyzed spectra (650–4000 cm^{-1}) and were attributed to the molecular linkages of some chemical components such as starch, proteins and polyphenols.

The deconvoluted spectra in the range of 800–1300 cm^{-1} (Figure 1b1) showed the characteristics of starch grains. The amount of hydrated starch structures was identified at 995 cm^{-1}, the amorphous starch at 1022 cm^{-1} and the short-ordered starch structures at 1047 cm^{-1} [30]. No significant changes ($p > 0.05$) were observed in starch structure between OGP and control samples (Table 3). The intermolecular associations were identified at 1613–1620 cm^{-1}, the intramolecular associations at 1627–1635 cm^{-1}, β-sheets structures at 1620–1644 cm^{-1}, α-helix at 1650–1660 cm^{-1} and β-turn at 1660–1680 cm^{-1} [31,32]. The OGP sample presented higher inter- and intra-molecular associations compared to the control, which lacked absorbance for intramolecular associations. Significant lower ($p < 0.01$) α-helix conformations were identified for OGP compared to the control, while β-turn and antiparallel β-sheets structures were higher (Table 3). The presence of fibers could be observed from the peaks at 1149 and 1077 cm^{-1} [30], while the phenolic compounds could be identified at 1609–1608 and 1519–1516 cm^{-1} [33] (Figure 1b2) and 1747 cm^{-1}.

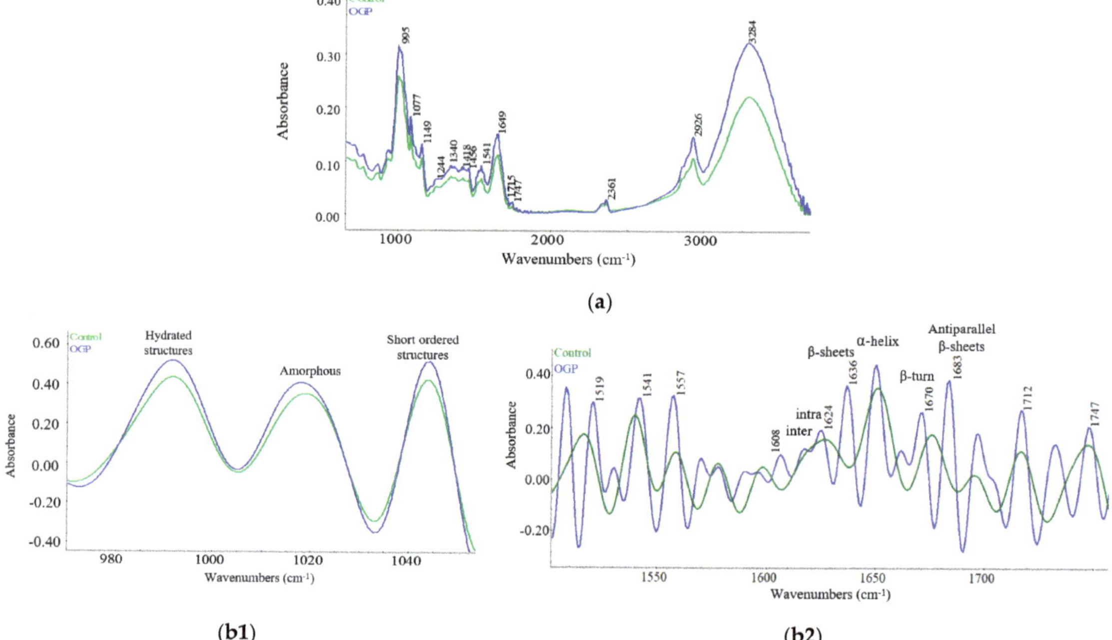

Figure 1. (**a**) Average spectra of optimal wheat-grape peels (OGP) and control samples in mid-infrared region; (**b1**) starch components' deconvoluted spectra; (**b2**) protein and polyphenol components' deconvoluted spectra.

Table 3. Optimal and control product properties.

Parameter	OGP	Control
Intermolecular associations (%)	1.71 ± 0.09 [a]	0.00 ± 0.00 [b]
Intramolecular associations (%)	4.75 ± 0.05 [a]	0.00 ± 0.00 [b]
β-sheets (%)	13.17 ± 0.44 [a]	13.23 ± 0.58 [a]
α-helix (%)	17.25 ± 0.71 [a]	26.67 ± 1.89 [b]
β-turn (%)	13.22 ± 0.39 [a]	2.23 ± 0.59 [b]
Antiparallel β-sheets (%)	19.74 ± 0.93 [a]	5.44 ± 0.85 [b]
Hydrated crystallin starch structure (%)	32.16 ± 0.37 [a]	34.66 ± 1.90 [a]
Short-ordered crystallin starch structure (%)	27.26 ± 0.20 [a]	27.25 ± 0.20 [a]
Amorphous starch structure (%)	34.26 ± 0.13 [a]	34.00 ± 0.31 [a]
Protein content (% dm)	14.29 ± 0.10 [a]	13.93 ± 0.09 [b]
Lipid content (% dm)	0.21 ± 0.02 [a]	0.13 ± 0.02 [b]
Ash (% dm)	0.80 ± 0.08 [a]	0.63 ± 0.01 [a]
Carbohydrates (% dm)	83.07 ± 0.22 [a]	85.28 ± 0.11 [b]
Radical scavenging activity (%)	38.74 ± 1.14 [a]	20.15 ± 0.26 [b]
RDS (% dm)	54.38 ± 0.24 [a]	69.78 ± 0.69 [b]
SDS (% dm)	19.61 ± 0.95 [a]	17.35 ± 0.20 [b]

OGP—optimal formulation of wheat flour with grape peels, RDS—rapid digestible starch, SDS—slowly digestible starch, dm—dry matter, a–b means in the same row followed by different letters are significantly different ($p < 0.05$).

2.4.2. Pasta Chemical Properties

The chemical compositions of the OGP and control sample are presented in Table 3.

GP addition to wheat flour caused a significant ($p < 0.01$) increase of the protein, lipid, ash and carbohydrate contents (Table 3) of pasta. The OGP sample presented higher radical scavenging activity (38.74%) compared to the control (20.15%). On the other hand, cooked pasta RDS significantly decreased ($p < 0.01$) when GP was added, while SDS increased compared to the control.

2.4.3. Microstructure Analysis

Dry pasta microstructure analysis revealed a well-developed matrix comprised of a gluten network, which encompassed starch grains and fiber fractions (Figure 2). In both the control and OGP samples round and lenticular starch shapes with smooth surfaces were observed. The addition of GP resulted in a compact dough structure in which the proteins embedded the fine particles of fibers and starch grains.

(**a**)   (**b**)

Figure 2. Dry pasta microstructure of (**a**) control and (**b**) optimal sample with grape peels (OGP): SG—starch grain; P—protein matrix; F—fiber.

Dough ingredients influenced the pasta extrusion process, which resulted in surface structure changes. The addition of GP caused a slight increase of pasta surface roughness, as can be seen in Figure 3. OGP pasta presented an uneven surface compared to the control. The alternation of high and low regions was given by the use of the *rigatoni* mold.

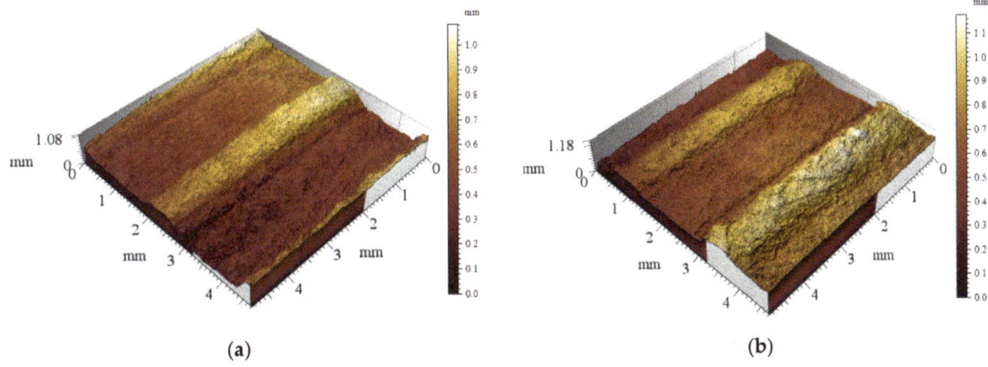

(**a**)   (**b**)

Figure 3. Three-dimensional dry pasta surface of (**a**) control and (**b**) optimal sample with grape peels (OGP).

3. Discussion

3.1. Effects of Grape Peels on Flour, Dough and Pasta Quality

GP caused an increase of flour slurry peak viscosity, a similar trend being reported by Mironeasa et al. [34] for wheat flour supplemented with GP and can be related to the affinity of GP for water, which may determine a viscosity increase. Probably, the pectin content of GP can be another factor responsible for the viscosity increase since it may assist starch swelling [35]. Masoodi et al. [36] also reported a reduction of wheat flour peak viscosity when more than 5% apple pomace was incorporated.

The dynamic complex modulus' proportional increase with GP level could be explained by the strengthening effect of GP on the gluten network due to the presence of non-starch polysaccharides, tannins and polyphenols, which can form complexes with proteins [37]. Additionally, the oxidation of sulfhydryl groups to disulfides caused by the raised content of oxidizing compounds can possibly contribute to the complex modulus increase [3]. In this study, GP caused more cohesive dough. Mironeasa et al. [34] reported increased dough cohesiveness when small particle sizes of GP were used probably due to the ability of fine particles to better absorb water. GP fibers present many hydroxyl groups in their structure, which will determine higher interactions with water trough hydrogen bonds [38]. The sugars present in GP can also influence dough cohesiveness due to their water solubility [39]. Saad et al. [28] showed that pasta soft wheat dough supplemented with cucumber pomace rheological properties changed compared to the control, an increase of the extensibility and water absorption and a decrease of elasticity being observed.

Dry pasta chroma decreased as the amount of GP was higher. In the literature, Smith and Yu [40] reported higher b* (yellow nuance) and lower a* (red nuance) values of bread supplemented with grape pomace. High amounts of sugars found in GP can promote Maillard reactions and, along with the polyphenol's presence, may determine C* decrease. Additionally, the natural pigments from GP are responsible for color intensity decrease [41]. Other studies reveled that the addition of by-products such as dragonhead seeds, coconut residue or *Moringa oleifera* L. leaf [19,20,24] produced significant changes in pasta color, depending on the pigments present in the ingredient added. The quality of the final product depends on the chemical composition of the ingredient added, soluble dietary fibers such as inulin and fructooligosaccharides playing an important role in pasta network strength [42]. Polyphenol interactions with proteins can determine stronger dough structure [18,43], leading to higher fracturability values.

The higher cooking loss obtained in this study for pasta enriched with GP can be attributed to the polymer interactions in the gluten matrix and/or to the competition of proteins for water, leading to starch loss [44]. A similar trend of cooking loss was reported for fettuccini pasta supplemented with grape marc [5] and can be due to the dietary fiber content of the added ingredient. Xu et al. [22] reported also a proportional increase of pasta cooking loss with the addition level of apple pomace increase. Pasta texture can be a good predictor for consumer preferences. The results obtained showed a decreasing trend of pasta chewiness with GP level increase. The interference of GP components with starch, which led to a reduction of starch gelatinization during cooking, can be associated with lower chewiness [45]. The use of a small particle size (<180 µm) for GP flour can be an advantage for pasta quality by diminishing the negative impact on the texture. A study regarding the influence of apple pomace on wheat pasta showed that this fortification caused a decrease of pasta chewiness as the level was higher [22], similar trend being reported also by Chen et al. [33] for pasta enriched with more than 1% grape seeds.

The nutritional value of wheat pasta was increased by the incorporation of GP. The raised resistant starch content of pasta enriched with GP can be attributed to the interactions between polyphenols and starch [46,47] through non-covalent linkages formation. Furthermore, starch digestion diminution produced by the formation of linkages between polyphenols and starch can occur from the starch molecular structure modifications. A similar RS increase was reported by Simonato et al. [25] for pasta fortified with olive pomace. For instance, some studies revealed that these interactions can contribute to the ordered structure or development of starch crystallin areas, inducing RS formation [48,49]. On the other hand, the polyphenolics content of pasta increased with GP level increase. Gaita et al. [17] also reported higher polyphenolic contents for pasta supplemented with grape peels, the increase being directly proportional to the amount added. A study performed by Michalak-Majewska et al. [23] showed that the addition of onion skin powder caused an increase in total phenolic compounds, flavonoids content and antioxidant activity. The phenolic compounds present in grape by-products are highly accessible and available for metabolization in the human intestine [50]. In addition, the bioavailability polyphenols in pasta are directly related to the hydrogen, ionic, covalent, and hydrophobic interactions with proteins [3]. The addition of fiber-rich ingredients can be a useful technique to increase the nutritional value of the final product. Wheat pasta fiber content was enhanced by GP addition. Acun and Gül [51] also reported raised fiber content of cookies enhanced with seedless grape pomace. Similar to our study, Saad et al. [28] showed that the addition of cucumber pomace in soft wheat pasta induced the increase of fibers and polyphenol contents. Fibers are known to have some biological functions, such as activity against cancer, amelioration of gastrointestinal systems functioning, improvement of cardiovascular system activity, and decrease of cholesterol and glycaemia levels in blood [52]. Balli et al. [53] found that durum wheat tagliatelle with grape and olive pomace resulted in improved quality in terms of organoleptic and nutritional properties, with high levels of phenolic compounds and increased fibers content.

3.2. Control and Optimal Product Properties

In this study, the optimal amount of GP was found to be 4.62%; at this level acceptable technological characteristics and superior nutritional value compared to the control were obtained. The optimal sample with GP (OGP) and the control made of wheat flour were characterized.

FTIR spectra showed the interactions between components of the composite flours. The peak at 3284 cm^{-1} was attributed to the -OH stretching vibrations, while the peak at 2926 cm^{-1} was assigned to the C-H stretching and showed higher absorbances for OGP probably due to the phenolic compound's presence [54]. The phenolic compounds found in GP have specific absorptions at 1712–1704 cm^{-1} corresponding to the carbonyl stretching and 1609–1608 and 1519–1516 cm^{-1} given by the stretching vibrations of C=C [55] (Figure 1b2). The galloyl group's presence could possibly be observed at about 1747 cm^{-1} (Figure 1b2), which corresponds to the stretching vibrations of carbonyl groups (C=O) [54]. In the case of OGP, an increase of peak intensities at 1149 and 1077 cm^{-1} compared to the control could be due to the intake of small hemicellulose, cellulose and pectin [30] found in GP. Starch structure characterized by the vibrations given at 1022 cm^{-1} for amorphous and at 1047 cm^{-1} for short-ordered regions [30] was not significantly affected by the addition of GP. The absorption at 1700–1600 cm^{-1} was assigned to the Amide I fraction, which could offer information about the secondary structure of proteins [30]. GP addition in wheat flour caused inter- and intramolecular associations compared to the control, which lacked absorbance for these structures. Lower α-helix conformations were observed for OGP compared to the control, while β-turn and antiparallel β-sheet structures were present in higher proportion. These changes could be possibly related to the protein–polyphenols interactions in the dough matrix, a similar opinion being reported by Ertürk and Meral [56]. Sivam et al. [57] also reported lower α-helices and lower intermolecular associations when polyphenols were added to bread. Chen et al. [33] obtained a reduction of β-turn conformational composition of gluten proteins with grape seeds' level increase, while β-sheet conformations increased. The α-helix secondary structures were dominant for both the OGP and control samples, similar results being obtained by Nawrocka et al. [58], who studied the influence of dietary fibers on gluten proteins. Dough rheological properties are directly influenced by protein secondary structure. Our results indicated a stronger and more cohesive dough structure of OGP, which can be related to the formation of protein–fiber complexes. The band at 1670 cm^{-1} is due to non-hydrogen linkages of carbonyl groups in the β-turn structures of proteins; when carbonyl groups are hydrogen bonded with other compounds, a decrease of their absorption band to smaller wavenumbers should be observed [58]. The results of the present study revealed that in the control sample the protein carbonyl groups were bonded trough non-hydrogen linkages, a fact evidenced by the presence of the 1670 cm^{-1} band, while in OGP the carbonyl groups would have formed hydrogen bonds with fibers or other GP components, a fact suggested by the shift to the left of the absorption band. This band shift could possibly be attributed also to the presence of polysaccharides such as pectin from GP [58]. On the other hand, the appearance of the absorption band at 1661 cm^{-1} in the OGP sample (Figure 1b2) could be related to the development of intramolecular and intermolecular hydrogen linkages between glutamine side chains and peptide groups, which was probably due to the influence of the GP fiber-rich ingredient [58]. The band at 1625 cm^{-1} observed for OGP can be associated with the hydrogen bonding of protein aggregates and/or of polypeptide chains complexed with phenolic compounds [59] from GP, as shown by their chemical composition.

GP raised the protein, lipid, ash and carbohydrate contents of pasta due to their intake of nutrients. Compared to our study, Saad et al. [28] showed that cucumber pomace addition in soft wheat noodles increased the mineral and polyphenols content, but decreased protein and carbohydrates. The addition of GP caused higher radical scavenging activity compared to the control, in agreement with the polyphenols content, which presented a raised value for pasta samples with GP. Gaita et al. [17] also reported higher antioxidant capacity of pasta enriched with grape peels. Cooked pasta RDS significantly decreased ($p < 0.01$) when GP were added, while SDS increased compared to the control. Similar trends of RDS and SDS were reported by Simonato et al. [25] when wheat pasta was fortified with olive pomace, which may be due to the starch content reduction caused by the addition of the fiber-rich ingredient. These results could be related to the starch–polyphenol interactions that may occur during pasta making. It has been demonstrated that polyphenols can reduce starch digestion rates due to their interactions trough hydrophobic forces with amylose and the linear fraction of amylopectin and/or to the inhibition effects on enzymes [49]. On the other hand, fibers from GP could compete with starch granule for water, reducing starch gelatinization, thereby causing starch digestibility limitation [25].

GP incorporation resulted in a compact dough structure in which the proteins embedded the fine particles of fibers and starch grains. Similar results were presented by Tolve et al. [1] for durum wheat pasta with grape pomace. The denser gluten network of pasta with GP is probably due to the intake of protein, cellulose and polysaccharides, which can act as fillers in the gluten matrix, while the polyphenols present in GP may interact with gluten proteins to support the formation of a gluten matrix, similar observations being made by Chen et al. [33] for pasta enriched with 1% grape seeds. Huang et al. [60] also reported a dense structure of wheat noodles formed of starch granules embedded in a developed fiber matrix. Pasta surface roughness was higher when GP was added. This increase could be possibly due to the difference in the water absorption capacity of the composite flour compared to the control, a difference caused by the presence of fibers from GP [34]. According to the study of Chen et al. [33], the incorporation of more than 3% grape seeds in wheat noodles induced the appearance of jagged edges and uneven mesh structure of pasta caused by the non-gluten components.

4. Materials and Methods

4.1. Materials

The research was conducted on 650 white wheat (*Triticum aestivum*) flour (WWF) type from the 2019 harvest, provided by Dizing S.R.L. (Brusturi, Neamt, Romania). Wheat flour chemical components reported to dry matter were: lipids content of 1.11 ± 0.03%, protein content of 14.41 ± 0.21%, carbohydrates 81.26 ± 0.16%, ash content of 0.60 ± 0.05%, moisture of 14.01 ± 0.16%, total polyphenols content of 105.59 ± 4.42 µg GAE/g. The falling number was 404 ± 1.73 s, wet gluten content was 29.75 ± 0.15%, dry gluten was

$10.04 \pm 0.13\%$, gluten deformation index was 6.17 ± 0.29 mm and the water absorption capacity was $59.54 \pm 0.31\%$. Grape pomace from the Fetească Regală variety was provided by Iași Research and Development Center for Viticulture and Vinification and was dried in a convection oven at 50 °C for 18 h. Grape peels were manually separated, ground and sieved to obtain the particle size of <180 μm. Grape peels contained $2.50 \pm 0.19\%$ lipids, $9.85 \pm 0.05\%$ proteins, $25.25 \pm 0.05\%$ fibers, $55.73 \pm 0.20\%$ carbohydrates, $4.47 \pm 0.03\%$ ash, $8.00 \pm 0.08\%$ moisture and 1448.69 ± 15.39 μg GAE/g polyphenols, $0.40 \pm 0.01\%$ pectin reported to dry matter (dm) and a water absorption capacity of $271.10 \pm 4.34\%$. Wheat flour was sieved before mixing in order to achieve a particle size of <300 μm. Composite flours were mixed for 15 min in a Yucebas Y21 machine (Izmir, Turkey).

4.2. Pasta Processing

Pasta dough was mixed in a Kitchen Aid mixer (Whirlpool Corporation, Benton Harbor, MI, USA) by adding the amount of water calculated in order to obtain 40% moisture. Pasta modeling was performed after 15 min of the dough resting at room temperature, by using a rigatoni mold of the Kitchen Aid machine. Pasta drying was performed in a convection oven, according to the method described by Bergman et al. (1994) as follows: 30 min drying in open air at room temperature, followed by a first step of drying for 60 min at 40 °C, a second for 120 min at 80 °C and a third for 120 min at 40 °C.

4.3. Evaluation of GP Effects on WWF and Pasta Quality

4.3.1. Flour Pasting Properties

Flour pasting behavior in terms of peak viscosity η_{max} (Pa·s) was simulated on a dynamic rheometer Thermo-HAAKE, MARS 40 (Karlsruhe, Germany) with a Peltier temperature controller, by using a cup-cylinder geometry. The method described by Ahmed et al. [61] was adapted as follows: the slurry was kept at 50 °C for 60 s, followed by a heating at 95 °C for 222 s, keeping at 95 °C for 210 s, cooling at 50 °C for 228 s and keeping at 50 °C for 120 s.

4.3.2. Fundamental Rheological Behavior

Laminated dough samples were rested for 30 min for internal strain removal and tested for linear viscoelastic region (LVR) by using a Thermo-HAAKE, MARS 40 (Karlsruhe, Germany). A frequency sweep test was performed for the determination of the complex modulus G* (Pa). For this purpose, the sample was placed between the parallel plates at a 3 mm gap and a vaseline layer was applied on the exposed edges for moisture loss prevention. The frequency was varied from 0.1 to 20 Hz, at a strain of 15 Pa that was in the LVR.

4.3.3. Dough Texture

For dough cohesiveness (Co) determination, a texture analysis was performed by using a Perten TVT-6700 texturometer (Perten Instruments, Hägersten, Sweden). A double compression was applied to dough balls of 50 g weight, at 50% height, a speed of 5.0 mm/s and a trigger force of 20 g [62].

4.3.4. Dry Pasta Color

Pasta chroma C* Equation (11) of the CIE Lab system was determined by reflectance by using a Konica Minolta CR-400 (Tokyo, Japonia) colorimeter.

$$C^* = \sqrt{a^{*2} + b^{*2}} \tag{11}$$

where C*—chroma, a*—red or green nuance, b*—yellow or blue nuance.

4.3.5. Pasta Fracturability

Pasta fracturability as the maximum force F (g) required to break a dry pasta piece was determined with a Perten TVT-6700 texturometer (Perten Instruments, Sweden). An aluminum break rig set adjusted to 13 mm width was used, the test speed being set to 3 mm/s and the trigger force to 50 g [62].

4.3.6. Total Polyphenolics Content

The extracts were prepared according to the method described by Melilli et al. [63]. An amount of 2 g of grinded uncooked dry pasta was extracted with 20 mL of methanol 80% (*v/v*) in a sonication bath at 37 °C and 45 Hz for 40 min, and then the mix was filtered. Total polyphenols content (TPC) (µg GAE/g dm) of dry pasta was evaluated by using the Folin–Ciocalteu method [64]. The extract was diluted in a ratio of 1:4 with distilled water and mixed with Folin–Ciocalteu reagent (1N) and sodium carbonate 20% (*v/v*) and left to rest in darkness for 40 min. The absorbance was read at 725 nm on an UV–VIS–NIR Shimadzu 3600 (Tokyo, Japan) spectrophotometer. TPC was calculated from a calibration curve ($R^2 = 0.99$) made with gallic acid (GAE).

4.3.7. Total Dietary Fiber Content

For dietary fiber content TDF (% dm) estimation, a FOSS 6500 NIR (FOSS NIRSystems, Silver Springs, FL, USA) infrared spectrometer was used. Uncooked dry pasta samples were ground before the analysis and the spectra were collected at room temperature. For calibration, off the shelf INGOT commercial calibrations (AUNIR, Towcester, UK) were used. Standard materials provided by AUNIR were used for bias corrections [65].

4.3.8. Pasta Cooking Behavior

The loss of solids CL (%) during pasta cooking was determined gravimetrically by evaporation of the water that resulted after boiling 10 g of pasta in 200 mL of water for the optimum cooking time previously established [66].

4.3.9. Boiled Pasta Texture

Pasta chewiness Ch (J) was determined by double cycle compression on one piece of pasta by using a Perten TVT-6700 device (Perten Instruments, Sweden) equipped with a 35 mm cylinder probe, at 50% of the sample height, a test speed of 5.0 mm/s and a trigger force of 20 g [62].

4.3.10. Rapid Digestible Starch (RDS), Slowly Digestible Starch (SDS) and Resistant Starch (RS) Contents

The international AOAC 2017.16 method was used for RDS, SDS and RS determination from boiled pasta, by using Megazyme kit. After 20 min (for RDS), 120 min (for SDS) or 240 min (for RS) of sample digestion with α-amylase and amyloglucosidase the reaction was stopped and the mix was digested again with amyloglucosidase. The resulting glucose was determined by using GOPOD reagent and reading of the absorbance at 510 nm. The results were reported as percent to dry matter.

4.4. Optimization of Grape Peels Level and Models Validation

GP level optimization was performed by using the trial version of Design Expert software (Stat-Ease, Inc., Minneapolis, USA). For this purpose, the multiple response optimization and the desirability function were used and the goals were selected as follows: G*, Co, C*, F, TPC, TDF and RS were maximized, CL and Ch were minimized, and η_{max}* was kept in range. The experimental design matrix containing mean values of three replications of the responses is presented in Table 4.

Table 4. The effects of GP on the responses used in the experimental design.

GP (%)	η_{max} (Pa·s)	G* (Pa)	Co (adim.)	C* (adim.)	F (g)	CL (%)	Ch (J)	RS (% dm)	TPC (µg GAE/g dm)	TDF (% dm)
1.00	0.56 ± 0.03	82,936.67 ± 4914.90	0.39 ± 0.01	21.46 ± 0.63	4181.00 ± 103.00	4.55 ± 0.30	4034.97 ± 9.10	3.20 ± 0.08	102.00 ± 7.11	0.02 ± 0.01
2.00	0.62 ± 0.05	90,176.67 ± 7961.22	0.39 ± 0.01	20.94 ± 0.47	4357.00 ± 199.02	5.43 ± 0.18	3975.14 ± 168.70	4.10 ± 0.05	118.72 ± 2.76	0.35 ± 0.05
3.00	0.74 ± 0.12	96,045.00 ± 1395.00	0.40 ± 0.00	20.58 ± 0.49	4519.00 ± 139.54	5.72 ± 0.17	3700.57 ± 140.00	4.31 ± 0.09	124.91 ± 3.36	0.60 ± 0.00
4.00	0.81 ± 0.04	99,103.33 ± 7081.74	0.41 ± 0.01	19.65 ± 0.26	4998.33 ± 15.50	6.09 ± 0.31	3621.45 ± 100.87	4.57 ± 0.01	129.12 ± 2.00	1.05 ± 0.05
5.00	0.83 ± 0.16	122,600.00 ± 6080.60	0.42 ± 0.01	18.78 ± 0.23	5685.67 ± 54.78	7.28 ± 0.36	3589.50 ± 119.78	4.67 ± 0.04	149.27 ± 2.02	1.30 ± 0.00
6.00	0.87 ± 0.07	131,893.33 ± 8144.26	0.43 ± 0.01	18.54 ± 0.25	5961.00 ± 17.00	7.99 ± 0.14	3478.03 ± 102.51	4.88 ± 0.16	157.02 ± 4.38	1.50 ± 0.00

GP—grape peels level, η_{max}—peak viscosity, G*—complex modulus, Co—cohesiveness, C*—Chroma, F—fracturability, Ch—chewiness, CL—cooking loss, RS—resistant starch, TPC—total polyphenols content, TDF—total dietary fiber.

For model validation, pasta was made using the optimal level of GP obtained and the response values were checked. The real values of the optimum sample characteristics were compared to the control made of untreated wheat flour. For the evaluation of differences among the predicted and the experimental results of the optimal solution, and among the optimal and control sample, *t* tests for two samples ($p < 0.05$) were performed.

4.5. Determination of Control and Optimal Product Properties

4.5.1. Chemical Composition and Antioxidant Activity

The chemical composition was determined according to the Romanian and International standard methods and the results are reported to dry matter: moisture (SR EN ISO 712/2010), ash (SR ISO 2171/2002), protein (SR EN ISO 20483/2007) and lipids (SR 91/2007). TDF was determined by NIR (as described in Section 4.3.6) and the carbohydrates were calculated by difference.

The radical scavenging activity (%) was evaluated by using 2, 2-di (4-tert-octylphenyl)-1-picrylhydrazyl (DPPH). The extract prepared as described in Section 4.3.5 (0.5 mL) was diluted with methanol 80% (0.5 mL) and mixed with 5 mL of DPPH. After resting 30 min in the darkness the absorbance was read at 517 nm on an UV–VIS–NIR Shimadzu 3600 (Tokyo, Japan) spectrophotometer [67].

4.5.2. ATR-FT-IR Analysis of Flour

FT-IR spectra of the flours were collected in triplicate in the range of 650 to 4000 cm^{-1} by using a Thermo Scientific Nicolet iS20 (Waltham, MA, USA) spectrometer, at a resolution of 8 cm^{-1} by 64 scans. The fractions of amide I (1835–1585 cm^{-1}), polyphenols (1516–1747 cm^{-1}) and starch (800–1300 cm^{-1}) were identified, the data being processed with OMNIC software. The starch, polyphenols and protein structures were assessed according to previous studies [31,68–70]. Fourier deconvolution was applied in order to characterize starch and protein structures.

4.5.3. Microstructure

Electronic scanning microscopy was employed for flour microstructure evaluation with a VEGA II LSH scanning electronic microscope (Tescan, Brno, Czech Republic). The acceleration tension was 30 kV and the magnification 1000×, the samples being fixed with adhesive carbon bands.

Pasta surface structure analysis was performed on a Mahr CWM100 microscope (Gottingen, Germany). The data collected were processed with Mountain Map trial version software (Digital Surf, Besançon, France).

4.6. Statistical Analysis

All of the analyses in the present study were performed in triplicate. XLSTAT for Excel 2021 version (Addinsoft, New York, USA) was used for statistical analysis of the data. In order to evaluate the significant differences ($p < 0.05$) among samples the *t*-test was used.

For the assessment of GP addition effects on WWF and pasta quality, mathematical modeling of the results was carried out by using the trial version of Design Expert software (Stat-Ease, Inc., Minneapolis, USA), by applying response surface methodology (RSM). The effect of GP level factor on the responses (η_{max}, G*, Co, C*, Ch, F, CL, TPC, TDF and RS) was observed by using a D-optimal design with one factor varied at six levels (1, 2, 3, 4, 5, 6%) and three replications. The model's fitting was evaluated through a sequential Fisher test, coefficients of determination (R^2) and adjusted coefficient of determination (*Adj.-R^2*), at a 95% confidence level. The most suitable model was selected according to the highest *Adj.-R^2* value.

5. Conclusions

The addition of grape peels increased the amounts of nutrients in wheat pasta, especially fibers, resistant starch and polyphenols, which are compounds with many health benefits. The optimal quantity of grape peels that can be added to wheat flour was found to be 4.62%. Dough cohesiveness, dry pasta fracturability and cooked pasta chewiness were enhanced compared to the control, while the cooking losses were within acceptable limits (<12%). Starch digestibility of pasta was positively influenced by grape peels, higher resistant starch, slowly digestible starch and lower rapid digestible starch being obtained. Interactions between wheat flour and grape peels were shown trough FT-IR analysis, significant changes of protein structures being observed. A compact microstructure of pasta was noticed at 4.62% addition of small particle sizes (<180 μm). Thus, this study can be helpful for both processors and consumers, underlying the opportunity to enhance the nutritional value of wheat pasta by incorporating grape peels, with benefits for human health and environmental waste management. This research was performed on a laboratory scale, and thus more investigations at the industrial level are needed.

Supplementary Materials: The following are available online at https://www.mdpi.com/article/10.3390/plants10050926/s1, Figure S1: Effect of GP level on peak viscosity (ηmax); Figure S2: Effect of GP level on dough: a. complex modulus (G*), b. cohesiveness (Co); Figure S3: Effect of GP level on dry pasta: a. chroma (C*), b. fracturability (F); Figure S4: Effect of GP level on pasta: a. cooking loss (CL), b. chewiness (Ch); Figure S5: Effect of GP level on pasta: a. resistant starch content (RS), b. total polyphenols content (TPC), c. total dietary fiber (TDF).

Author Contributions: The authors contributed equally to this research. All of the authors examined and approved the contents of the article. All authors have read and agreed to the published version of the manuscript.

Funding: This work was funded by Stefan cel Mare University of Suceava, Romania.

Institutional Review Board Statement: Not applicable.

Informed Consent Statement: Not applicable.

Data Availability Statement: Not applicable.

Conflicts of Interest: The authors declare no conflict of interest.

References

1. Tolve, R.; Pasini, G.; Vignale, F.; Favati, F.; Simonato, B. Effect of Grape Pomace Addition on the Technological, Sensory, and Nutritional Properties of Durum Wheat Pasta. *Foods* **2020**, *9*, 354. [CrossRef] [PubMed]
2. Hogervorst, J.C.; Miljić, U.; Puškaš, V. Extraction of Bioactive Compounds from Grape Processing By-Products. In *Handbook of Grape Processing By-Products*; Elsevier BV: Amsterdam, The Netherlands, 2017; pp. 105–135.
3. Iuga, M.; Mironeasa, S. Potential of grape byproducts as functional ingredients in baked goods and pasta. *Compr. Rev. Food Sci. Food Saf.* **2020**, *19*, 2473–2505. [CrossRef] [PubMed]
4. Spinelli, S.; Padalino, L.; Costa, C.; del Nobile, M.A.; Conte, A. Food by-products to fortified pasta: A new approach for optimization. *J. Clean. Prod.* **2019**, *215*, 985–991. [CrossRef]
5. Sant'Anna, V.; Christiano, F.D.P.; Marczak, L.D.F.; Tessaro, I.C.; Thys, R.C.S. The effect of the incorporation of grape marc powder in fettuccini pasta properties. *LWT* **2014**, *58*, 497–501. [CrossRef]
6. Minarovičová, L.; Karovičová, J.; Kohajdová, Z.; Kuchtová, V. The Chemical Composition of Grape Fibre. *Potravin. Slovak J. Food Sci.* **2015**, *9*, 53–57. [CrossRef]
7. Iuga, M.; Ropciuc, S.; Mironeasa, S. Antioxidant Activity and Total Phenolic Content of Grape Seeds and Peels from Romanian Varieties. Food Environ. *Saf. J.* **2017**, *16*, 276–282.
8. Troilo, M.; Difonzo, G.; Paradiso, V.; Summo, C.; Caponio, F. Bioactive Compounds from Vine Shoots, Grape Stalks, and Wine Lees: Their Potential Use in Agro-Food Chains. *Foods* **2021**, *10*, 342. [CrossRef]
9. Olejar, K.J.; Ricci, A.; Swift, S.; Zujovic, Z.; Gordon, K.C.; Fedrizzi, B.; Versari, A.; Kilmartin, P.A. Characterization of an Antioxidant and Antimicrobial Extract from Cool Climate, White Grape Marc. *Antioxidants* **2019**, *8*, 232. [CrossRef]
10. Deng, Q.; Penner, M.H.; Zhao, Y. Chemical composition of dietary fiber and polyphenols of five different varieties of wine grape pomace skins. *Food Res. Int.* **2011**, *44*, 2712–2720. [CrossRef]
11. Meini, M.-R.; Cabezudo, I.; Boschetti, C.E.; Romanini, D. Recovery of phenolic antioxidants from Syrah grape pomace through the optimization of an enzymatic extraction process. *Food Chem.* **2019**, *283*, 257–264. [CrossRef] [PubMed]

12. Oliveira, D.A.; Salvador, A.A.; Smânia, A.; Smânia, E.F.; Maraschin, M.; Ferreira, S.R. Antimicrobial activity and composition profile of grape (*Vitis vinifera*) pomace extracts obtained by supercritical fluids. *J. Biotechnol.* **2013**, *164*, 423–432. [CrossRef]

13. Zhu, F.; Du, B.; Zheng, L.; Li, J. Advance on the bioactivity and potential applications of dietary fibre from grape pomace. *Food Chem.* **2015**, *186*, 207–212. [CrossRef] [PubMed]

14. Mironeasa, S. *Valorisation of Secondary Products from Wine Making*; Publishing House Performantica: Iasi, Romania, 2017.

15. Bender, A.B.B.; Speroni, C.S.; Salvador, P.R.; Loureiro, B.B.; Lovatto, N.M.; Goulart, F.R.; Lovatto, M.T.; Miranda, M.Z.; Silva, L.P.; Penna, N.G. Grape Pomace Skins and the Effects of Its Inclusion in the Technological Properties of Muffins. *J. Culin. Sci. Technol.* **2017**, *15*, 143–157. [CrossRef]

16. Mironeasa, S.; Iuga, M.; Zaharia, D.; Mironeasa, C. Optimization of grape peels particle size and flour substitution in white wheat flour dough. *Sci. Study Res. Chem. Chem. Eng. Biotechnol. Food Ind.* **2019**, *20*, 29–42.

17. Gaita, C.; Alexa, E.; Moigradean, D.; Conforti, F.; Poiana, M.-A. Designing of high value-added pasta formulas by incorporation of grape pomace skins. *Rom. Biotechnol. Lett.* **2020**, *25*, 1607–1614. [CrossRef]

18. Iuga, M.; Mironeasa, C.; Mironeasa, S. Oscillatory Rheology and Creep-Recovery Behaviour of Grape Seed-Wheat Flour Dough: Effect of Grape Seed Particle Size, Variety and Addition Level. *Bull. Univ. Agric. Sci. Veter. Med. Cluj Napoca. Food Sci. Technol.* **2019**, *76*, 40–51. [CrossRef]

19. Simonato, B.; Tolve, R.; Rainero, G.; Rizzi, C.; Sega, D.; Rocchetti, G.; Lucini, L.; Giuberti, G. Technological, nutritional, and sensory properties of durum wheat fresh pasta fortified with Moringa oleifera L. leaf powder. *J. Sci. Food Agric.* **2021**, *101*, 1920–1925. [CrossRef]

20. Sykut-Domańska, E.; Zarzycki, P.; Sobota, A.; Teterycz, D.; Wirkijowska, A.; Blicharz-Kania, A.; Andrejko, D.; Mazurkiewicz, J. The potential use of by-products from coconut industry for production of pasta. *J. Food Process. Preserv.* **2020**, *44*, 1–9. [CrossRef]

21. Sobota, A.; Wirkijowska, A.; Zarzycki, P. Application of vegetable concentrates and powders in coloured pasta production. *Int. J. Food Sci. Technol.* **2020**, *55*, 2677–2687. [CrossRef]

22. Xu, J.; Bock, J.E.; Stone, D. Quality and textural analysis of noodles enriched with apple pomace. *J. Food Process. Preserv.* **2020**, *44*, 44. [CrossRef]

23. Michalak-Majewska, M.; Teterycz, D.; Muszyński, S.; Radzki, W.; Sykut-Domańska, E. Influence of onion skin powder on nutritional and quality attributes of wheat pasta. *PLoS ONE* **2020**, *15*, e0227942. [CrossRef] [PubMed]

24. Zarzycki, P.; Teterycz, D.; Wirkijowska, A.; Kozłowicz, K.; Stasiak, D.M. Use of moldavian dragonhead seeds residue for pasta production. *LWT* **2021**, *143*, 111099. [CrossRef]

25. Simonato, B.; Trevisan, S.; Tolve, R.; Favati, F.; Pasini, G. Pasta fortification with olive pomace: Effects on the technological characteristics and nutritional properties. *LWT* **2019**, *114*, 108368. [CrossRef]

26. Giuberti, G.; Rocchetti, G.; Lucini, L. Interactions between phenolic compounds, amylolytic enzymes and starch: An updated overview. *Curr. Opin. Food Sci.* **2020**, *31*, 102–113. [CrossRef]

27. Wang, Y.; Chao, C.; Huang, H.; Wang, S.; Wang, S.; Wang, S.; Copeland, L. Revisiting Mechanisms Underlying Digestion of Starches. *J. Agric. Food Chem.* **2019**, *67*, 8212–8226. [CrossRef]

28. Saad, A.M.; El-Saadony, M.T.; Mohamed, A.S.; Ahmed, A.I.; Sitohy, M.Z. Impact of cucumber pomace fortification on the nutritional, sensorial and technological quality of soft wheat flour-based noodles. *Int. J. Food Sci. Technol.* **2021**, *2021*, 1–14. [CrossRef]

29. Tuoc, T.; Glasgow, S. On the texture profile analysis test. In Proceedings of the Chemeca 2012: Quality of life through chemical engineering, Wellington, New Zealand, 23–26 September 2012; pp. 749–760.

30. González, M.; Vernon-Carter, E.; Alvarez-Ramirez, J.; Carrera-Tarela, Y. Effects of dry heat treatment temperature on the structure of wheat flour and starch in vitro digestibility of bread. *Int. J. Biol. Macromol.* **2021**, *166*, 1439–1447. [CrossRef]

31. Marti, A.; Bock, J.E.; Pagani, M.A.; Ismail, B.; Seetharaman, K. Structural characterization of proteins in wheat flour doughs enriched with intermediate wheatgrass (*Thinopyrum intermedium*) flour. *Food Chem.* **2016**, *194*, 994–1002. [CrossRef] [PubMed]

32. Gao, X.; Tong, J.; Guo, L.; Yu, L.; Li, S.; Yang, B.; Wang, L.; Liu, Y.; Li, F.; Guo, J.; et al. Influence of gluten and starch granules interactions on dough mixing properties in wheat (*Triticum aestivum* L.). *Food Hydrocoll.* **2020**, *106*, 105885. [CrossRef]

33. Chen, S.-X.; Ni, Z.-J.; Thakur, K.; Wang, S.; Zhang, J.-G.; Shang, Y.-F.; Wei, Z.-J. Effect of grape seed power on the structural and physicochemical properties of wheat gluten in noodle preparation system. *Food Chem.* **2021**, *355*, 129500. [CrossRef]

34. Mironeasa, S.; Iuga, M.; Zaharia, D.; Mironeasa, C. Rheological Analysis of Wheat Flour Dough as Influenced by Grape Peels of Different Particle Sizes and Addition Levels. *Food Bioprocess Technol.* **2018**, *12*, 228–245. [CrossRef]

35. Gull, A.; Prasad, K.; Kumar, P. Nutritional, antioxidant, microstructural and pasting properties of functional pasta. *J. Saudi Soc. Agric. Sci.* **2018**, *17*, 147–153. [CrossRef]

36. Masoodi, F.A.; Chauhan, G.S.; Tyagi, S.M.; Kumbhar, B.; Kaur, H. Effect of Apple Pomace Incorporation on Rheological Characteristics of Wheat Flour. *Int. J. Food Prop.* **2001**, *4*, 215–223. [CrossRef]

37. Samohvalova, O.; Grevtseva, N.; Brykova, T.; Grigorenko, A. The effect of grape seed powder on the quality of butter biscuits. *East. Eur. J. Enterp. Technol.* **2016**, *3*, 61. [CrossRef]

38. Kohajdová, Z.; Karovičová, J.; Magala, M.; Kuchtová, V. Effect of apple pomace powder addition on farinographic properties of wheat dough and biscuits quality. *Chem. Pap.* **2014**, *68*, 1059–1065. [CrossRef]

39. Chevallier, S.; Colonna, P.; Buléon, A.; della Valle, G. Physicochemical Behaviors of Sugars, Lipids, and Gluten in Short Dough and Biscuit. *J. Agric. Food Chem.* **2000**, *48*, 1322–1326. [CrossRef] [PubMed]

40. Smith, I.N.; Yu, J. Nutritional and Sensory Quality of Bread Containing Different Quantities of Grape Pomace from Different Grape Cultivars. *EC Nutr.* **2015**, *2*, 291–301.
41. Aksoylu, Z.; Çağindi, Ö.; Köse, E. Effects of Blueberry, Grape Seed Powder and Poppy Seed Incorporation on Physicochemical and Sensory Properties of Biscuit. *J. Food Qual.* **2015**, *38*, 164–174. [CrossRef]
42. Mehta, R.S. Addressing texture challenges in baked goods with fiber. *Food Texture Des. Optim.* **2014**, 245–280. [CrossRef]
43. Tkacz, K.; Wojdyło, A.; Nowicka, P.; Turkiewicz, I.; Golis, T. Characterization in vitro potency of biological active fractions of seeds, skins and flesh from selected *Vitis vinifera* L. cultivars and interspecific hybrids. *J. Funct. Foods* **2019**, *56*, 353–363. [CrossRef]
44. Sun, K.-N.; Liao, A.-M.; Zhang, F.; Thakur, K.; Zhang, J.-G.; Huang, J.-H.; Wei, Z.-J. Microstructural, Textural, Sensory Properties and Quality of Wheat–Yam Composite Flour Noodles. *Foods* **2019**, *8*, 519. [CrossRef] [PubMed]
45. Bustos, M.C.; Paesani, C.; Quiroga, F.; León, A.E. Technological and sensorial quality of berry-enriched pasta. *Cereal Chem. J.* **2019**, *96*, 967–976. [CrossRef]
46. Camelo-Méndez, G.A.; Agama-Acevedo, E.; Tovar, J.; Bello-Pérez, L.A. Functional study of raw and cooked blue maize flour: Starch digestibility, total phenolic content and antioxidant activity. *J. Cereal Sci.* **2017**, *76*, 179–185. [CrossRef]
47. Rocchetti, G.; Giuberti, G.; Busconi, M.; Marocco, A.; Trevisan, M.; Lucini, L. Pigmented sorghum polyphenols as potential inhibitors of starch digestibility: An in vitro study combining starch digestion and untargeted metabolomics. *Food Chem.* **2020**, *312*, 126077. [CrossRef] [PubMed]
48. Chi, C.; Li, X.; Zhang, Y.; Chen, L.; Li, L.; Wang, Z. Digestibility and supramolecular structural changes of maize starch by non-covalent interactions with gallic acid. *Food Funct.* **2017**, *8*, 720–730. [CrossRef]
49. Sun, L.; Miao, M. Dietary polyphenols modulate starch digestion and glycaemic level: A review. *Crit. Rev. Food Sci. Nutr.* **2020**, *60*, 541–555. [CrossRef]
50. Gomes, T.M.; Toaldo, I.M.; Haas, I.C.D.S.; Burin, V.M.; Caliari, V.; Luna, A.S.; de Gois, J.S.; Bordignon-Luiz, M.T. Differential contribution of grape peel, pulp, and seed to bioaccessibility of micronutrients and major polyphenolic compounds of red and white grapes through simulated human digestion. *J. Funct. Foods* **2019**, *52*, 699–708. [CrossRef]
51. Acun, S.; Gül, H. Effects of grape pomace and grape seed flours on cookie quality. *Qual. Assur. Saf. Crop. Foods* **2014**, *6*, 81–88. [CrossRef]
52. Wyrwisz, M.K.J. The Application of Dietary Fiber in Bread Products. *J. Food Process. Technol.* **2015**, *6*, 6. [CrossRef]
53. Balli, D.; Cecchi, L.; Innocenti, M.; Bellumori, M.; Mulinacci, N. Food by-products valorisation: Grape pomace and olive pomace (pâté) as sources of phenolic compounds and fiber for enrichment of tagliatelle pasta. *Food Chem.* **2021**, *355*, 129642. [CrossRef]
54. Zhao, X.; Zhu, H.; Zhang, G.; Tang, W. Effect of superfine grinding on the physicochemical properties and antioxidant activity of red grape pomace powders. *Powder Technol.* **2015**, *286*, 838–844. [CrossRef]
55. Silva, S.D.; Feliciano, R.P.; Boas, L.V.; Bronze, M.R. Application of FTIR-ATR to Moscatel dessert wines for prediction of total phenolic and flavonoid contents and antioxidant capacity. *Food Chem.* **2014**, *150*, 489–493. [CrossRef] [PubMed]
56. Ertürk, B.; Meral, R. The impact of stabilization on functional, molecular and thermal properties of rice bran. *J. Cereal Sci.* **2019**, *88*, 71–78. [CrossRef]
57. Sivam, A.; Sun-Waterhouse, D.; Perera, C.; Waterhouse, G. Exploring the interactions between blackcurrant polyphenols, pectin and wheat biopolymers in model breads; a FTIR and HPLC investigation. *Food Chem.* **2012**, *131*, 802–810. [CrossRef]
58. Nawrocka, A.; Szymanskachargot, M.; Miś, A.; Ptaszyńska, A.A.; Kowalski, R.; Waśko, P.; Gruszecki, W.I. Influence of dietary fibre on gluten proteins structure—A study on model flour with application of FT-Raman spectroscopy. *J. Raman Spectrosc.* **2015**, *46*, 309–316. [CrossRef]
59. Nawrocka, A.; Miś, A.; Niewiadomski, Z. Dehydration of gluten matrix as a result of dietary fibre addition—A study on model flour with application of FT-IR spectroscopy. *J. Cereal Sci.* **2017**, *74*, 86–94. [CrossRef]
60. Huang, D.-W.; Chan, Y.-J.; Huang, Y.-C.; Chang, Y.-J.; Tsai, J.-C.; Mulio, A.; Wu, Z.-R.; Hou, Y.-W.; Lu, W.-C.; Li, P.-H. Quality Evaluation, Storage Stability, and Sensory Characteristics of Wheat Noodles Incorporated with Isomaltodextrin. *Plants* **2021**, *10*, 578. [CrossRef]
61. Ahmed, J.; Ramaswamy, H.S.; Ayad, A.; Alli, I. Thermal and dynamic rheology of insoluble starch from basmati rice. *Food Hydrocoll.* **2008**, *22*, 278–287. [CrossRef]
62. Ungureanu-Iuga, M.; Dimian, M.; Mironeasa, S. Development and quality evaluation of gluten-free pasta with grape peels and whey powders. *LWT* **2020**, *130*, 109714. [CrossRef]
63. Melilli, M.G.; Pagliaro, A.; Scandurra, S.; Gentile, C.; di Stefano, V. Omega-3 rich foods: Durum wheat spaghetti fortified with Portulaca oleracea. *Food Biosci.* **2020**, *37*, 100730. [CrossRef]
64. FAO/IAEA. *Quantification of Tannins in Tree Foliage. A Laboratory Manual For the FAO/IAEA Coordinated Research Project on Use of Nuclear and Related Techniques to Develop Simple Tannin Assays for Predicting and Improving the Safety and Efficiency of Feeding Rumina*; IAEA: Vienna, Austria, 2000.
65. Haughey, S.A.; Chevallier, O.P.; McVey, C.; Elliott, C.T. Laboratory investigations into the cause of multiple serious and fatal food poisoning incidents in Uganda during 2019. *Food Control.* **2021**, *121*, 107648. [CrossRef]
66. Giménez, M.; González, R.; Wagner, J.; Torres, R.; Lobo, M.; Samman, N. Effect of extrusion conditions on physicochemical and sensorial properties of corn-broad beans (*Vicia faba*) spaghetti type pasta. *Food Chem.* **2013**, *136*, 538–545. [CrossRef] [PubMed]
67. Bajerska, J.; Mildner-Szkudlarz, S.; Górnaś, P.; Seglina, D. The effects of muffins enriched with sour cherry pomace on acceptability, glycemic response, satiety and energy intake: A randomized crossover trial. *J. Sci. Food Agric.* **2015**, *96*, 2486–2493. [CrossRef]

68. Chavez-Murillo, C.E.; Orona-Padilla, J.L.; Millan, J.D.L.R. Physicochemical, functional properties and ATR-FTIR digestion analysis of thermally treated starches isolated from black and bayo beans. *Starch Stärke* **2019**, *71*, 1–25. [CrossRef]
69. Arslan, F.N.; Akin, G.; Elmas, Ş.N.K.; Üner, B.; Yilmaz, I.; Janssen, H.-G.; Kenar, A. FT-IR spectroscopy with chemometrics for rapid detection of wheat flour adulteration with barley flour. *J. Consum. Prot. Food Saf.* **2020**, *15*, 245–261. [CrossRef]
70. Lucarini, M.; Durazzo, A.; Kiefer, J.; Santini, A.; Lombardi-Boccia, G.; Souto, E.; Romani, A.; Lampe, A.; Nicoli, S.F.; Gabrielli, P.; et al. Grape Seeds: Chromatographic Profile of Fatty Acids and Phenolic Compounds and Qualitative Analysis by FTIR-ATR Spectroscopy. *Foods* **2019**, *9*, 10. [CrossRef]

MDPI
St. Alban-Anlage 66
4052 Basel
Switzerland
Tel. +41 61 683 77 34
Fax +41 61 302 89 18
www.mdpi.com

Plants Editorial Office
E-mail: plants@mdpi.com
www.mdpi.com/journal/plants